Water and Environmental Engineering: Entropy Theory and its Application

Water and Environmental Engineering: Entropy Theory and its Application

Edited by Cohen Foster

SYRAWOOD
PUBLISHING HOUSE

New York

Published by Syrawood Publishing House,
750 Third Avenue, 9th Floor,
New York, NY 10017, USA
www.syrawoodpublishinghouse.com

Water and Environmental Engineering: Entropy Theory and its Application
Edited by Cohen Foster

International Standard Book Number: 978-1-64740-429-1 (Hardback)

Cataloging-in-publication Data

Water and environmental engineering : entropy theory and its application / edited by Cohen Foster.
   p. cm.
Includes bibliographical references and index.
ISBN 978-1-64740-429-1
1. Hydraulic engineering. 2. Environmental engineering. 3. Entropy.
4. Water--Thermal properties--Mathematical models. 5. Hydraulics.
I. Foster, Cohen.
TC145 .W38 2023
627--dc23

TABLE OF CONTENTS

Permissions

List of Contributors

Index

PREFACE

Water and environmental engineering is an area of engineering that aims to address environmental issues related to air, water and soil. It deals with the development and management of water resources; designing hydraulic structures such as dams and tunnels; water quality engineering; water resources engineering; outdoor and indoor air quality engineering; ocean engineering; and hazardous waste management. Entropy is referred to as a measure of the disorder or randomness of a system. It has been used to solve numerous issues in geographical, earth, and environmental sciences. In the field of water and environment engineering, it has numerous applications in areas such as water distribution networks, sediment transport, river flow forecasting, and water monitoring network design. This book aims to shed light on some of the unexplored aspects of entropy and its applications in water and environmental engineering. It will also provide interesting topics for research, which interested readers can take up. A number of latest studies have been included to keep the readers up-to-date with the global concepts in this area of study.

This book is a comprehensive compilation of works of different researchers from varied parts of the world. It includes valuable experiences of the researchers with the sole objective of providing the readers (learners) with a proper knowledge of the concerned field. This book will be beneficial in evoking inspiration and enhancing the knowledge of the interested readers.

In the end, I would like to extend my heartiest thanks to the authors who worked with great determination on their chapters. I also appreciate the publisher's support in the course of the book. I would also like to deeply acknowledge my family who stood by me as a source of inspiration during the project.

Editor

PREFACE

Intuitionistic Fuzzy Entropy for Group Decision Making of Water Engineering Project Delivery System Selection

Xun Liu [1],*[ID], Fei Qian [2], Lingna Lin [1], Kun Zhang [2] and Lianbo Zhu [1]

[1] School of Civil Engineering, Suzhou University of Science and Technology, Suzhou 215000, China; linlingna1010@126.com (L.L.); lbzhu@usts.edu.cn (L.Z.)

[2] Institute of Engineering Management, Hohai University, Nanjing 211100, China; qf2019@hhu.edu.cn (F.Q.); dreamerzk@126.com (K.Z.)

* Correspondence: liuxun8127@usts.edu.cn

Abstract: The project delivery mode is an extremely important link in the life cycle of water engineering. Many cases show that increases in the costs, construction period, and claims in the course of the implementation of water engineering are related to the decision of the project delivery mode in the early stages. Therefore, it is particularly important to choose a delivery mode that matches the water engineering. On the basis of identifying the key factors that affect the decision on the project delivery system and establishing a set of index systems, a comprehensive decision of engineering transaction is essentially considered to be a fuzzy multi-attribute group decision. In this study, intuitionistic fuzzy entropy was used to determine the weight of the influencing factors on the engineering transaction mode; then, intuitionistic fuzzy entropy was used to determine the weight of decision experts. Thus, a comprehensive scheme-ranking model based on an intuitionistic fuzzy hybrid average (IFHA) operator and intuitionistic fuzzy weighted average (IFWA) operator was established. Finally, a practical case analysis of a hydropower station further demonstrated the feasibility, objectivity, and scientific nature of the decision model.

Keywords: water engineering; project delivery system; intuitionistic fuzzy entropy; group decision making

1. Introduction

For a water engineering project, the construction process is essentially the process of exchanges between the owner and contractor to obtain a construction product, but the particularity of the project itself makes the project delivery different from general commodity transactions, as engineering projects must rely on a certain mode of delivery. The project delivery system (PDS) not only defines the roles and responsibilities of each participant in the project, but also determines the payment method of the owner and the risk allocation of each participant, which provides a framework for the organization and implementation of the project [1]. Several PDSs that can be selected for the owner including design–bid–build (DBB), design–build (DB), construction management at risk (CM-at risk), or construction management as general contractor (GC), engineering–procurement–construction (EPC), and integrated project delivery (IPD) [2–6]. Each model of PDS has its own characteristics and requirements. There is not a universal PDS suitable for all types of water engineering projects under different situations. The decision making of PDS is an important link in the whole life cycle of a water engineering project, and an appropriate PDS can effectively improve the project performance [7,8]. Therefore, choosing and tailoring the most appropriate needs of the PDS to customers is a crucial task in early stage of any water engineering project [9].

Previous researchers have carried out extensive studies on the choice and decision of PDS. Molenaar et al. [10] established five regression models through multiple regression to predict the cost, duration, expected consistency, management responsibility, and overall user satisfaction of the DB model, respectively. However, the model has strict hypothetical conditions and poor prediction accuracy, which was difficult to apply and popularize in practice. Khalil introduced his decision of PDS into an analytic hierarchy process (AHP) to establish the AHP decision model of PDS [11]; Mafakheri et al. combined the AHP method with a rough set, compared PDS by the AHP method, and sorted the scheme by the rough set method [12]. However, the AHP method relies too much on the subjective judgment of experts and has high uncertainty. In view of the difficulty of predicting PDS performance, historical case experience plays an important role in PDS decision making. Therefore, case-based reasoning (CBR) was also applied to the selection decision of PDS [13–15]. Luu et al. [13,15] established an index system of PDS decisions, and constructed the PDS decision support system based on case experiences. In this system, the owner can retrieve the case by calculating the similarity of the water engineering project. Nevertheless, the uniqueness of water engineering projects makes it difficult for two completely similar water engineering projects to exist, and the standard has to be simplified in CBR analysis, which greatly reduces the effectiveness of decision making. Lo et al. [16] used data envelopment analysis (DEA) to analyze the performance index efficiency of DBB, DB, CM, and design build maintain (DBM) in highway projects. This method has the advantage of objectivity, but it cannot directly solve the decision-making problem of PDS. Ling et al. [17] analyzed the index correlation based on data from an actual case, and used the artificial neural network (ANN) method to construct the project performance prediction model under the DB model, but it is difficult to obtain the index data of this method in practical application. Chen et al. [18] proposed a PDS selection decision model combining DEA and ANN based on the advantages of a DEA-BND (Bound Variable) model. Oyetunji et al. [8] put forward the fuzzy simple multi-attribute rating technique with swing weights (SMARTS) selection method for project delivery.

Overall, these methods are helpful to solve the problem of the PDS decision to a certain extent, but the choice of project delivery mode is often made intuitively according to the past experience and knowledge of decision makers, as well as the information and data of the water engineering project. The inherent law between the decision-making attribute and the transaction mode was not explored. In addition, most of the studies on the decision-making model were based on the subjective scoring and evaluation of experts due to the different experience of experts, different knowledge background, and the uniqueness and one-off characteristics of water engineering projects. Therefore, there is great subjectivity in the determination of attribute weight and expert weight, which would eventually lead to the deviation of decision results, affecting the implementation of water engineering projects.

Group decision-making systems such as peer-to-peer (P2P) systems can be easily modeled as Fuzzy logic [19–21]. This opened a new area of research in decision making utilizing fuzzy sets (FSs) starting with Type-1 FSs, Type-2 FSs, and finally with intuitionistic fuzzy sets [20]. The comprehensive decision for project delivery is essentially a fuzzy multi-attribute group decision [21]; many researchers have done a lot work on fuzzy decision making for PDS selection [5,21–24], but the characteristics and fuzziness of the expert group were not considered yet. Therefore, it is necessary to select the proper PDS selection method to avoid the existing deficiencies. Studies have shown that fuzzy entropy is one of the most effective decision weights methods [25,26]. In this paper, based on fuzzy entropy theory, a group decision-making model to support PDS selection is proposed. In order to reduce the information lost in the overall judgment and improve the objectivity and fairness of group decision-making, the intuitionistic fuzzy hybrid average (IFHA) operator and intuitionistic fuzzy weighted average (IFWA) operator on the PDS decision are addressed to calculate the final decision weights. The present study aims to develop a more accurate and reliable PDS selection method. The main parts of this paper are organized as follows: Section 2 lists 15 key factors that affect the decision of water engineering PDS through a literature review, and establishes a set of index systems; Section 3 dealt with preliminaries; Section 4 presents the fuzzy group decision model for the selection of PDS; Section 5 presents a practical water engineering project case study that illustrates the applicability of this method; Section 6 provides conclusions.

2. Influencing Factors Indexes of PDS

An analysis of influencing factors on a water engineering PDS is the basis of scientific decision-making, which is also a hot spot in the theoretical research on water engineering project delivery mode. Scholars have carried out extensive studies on the influencing factors of PDS through theoretical analysis and case analysis. However, water engineering projects have the characteristics of a large investment scale, long construction cycle, and many uncertain factors, so there should be many influencing factors in the choice of PDS. At present, a unified index system of influencing factors for PDS selection has not been formed yet; however, it is obvious that although there are differences in the emphasis and quantity of the existing PDS index system, the main influencing factors of PDS selection can be summarized into three categories of owner characteristics, project characteristics, and external environment, with a total 15 indicators, which include owner liability, owner participation, the owner's own ability, risk allocation, owner design control, project scale, project complexity, project type, project scope clarity, project flexibility, project disputes, market competition, accessibility of materials, availability of technology, and the impact of laws and regulations. The details are shown in Table 1.

Table 1. Index system of influencing factors for a water engineering project delivery system (PDS).

Level I Indexes	Level II Indexes	Meanings of Level II Indexes	Source
Owner's Characteristics	Owner liability	The employer's expectation of liability for as few of the participants as possible	[7,11,13,17,18,24,27–36]
	Owner participation	Willingness and degree of participation of owners during the whole life cycle of the project	[10,11,17,28,30,31,34]
	Owner ability	Owner's own ability, such as decision-making ability, project control and organization ability, and project management ability	[13,17,28,31,34,37–39]
	Risk allocation	Expected commitment of owners to risks and losses (that is, whether it is shared equally with the contractor, or whether the owner bears most of the risk, or the contractor bears the majority of the risk)	[12,13,21,28,34]
	Owner design control	The willingness and degree of the owner to participate in the design	[11,31,32,34,38,40,41]
Project characteristics	Project scale	Compared with the average scale of the engineering project in the industry	[7,12,13,17,18,28,30–32,41–49]
	Project complexity	Whether the project needs a breakthrough in construction methods, technology and management, the complexity of technology, the uncertainty of the project, the observability of the characteristic values of engineering products, and so on	[8,10–13,17,18,24,27,31–33,39–44,48–54]
	Project type	What types of projects (e.g., housing construction projects, infrastructure projects, industrial projects, etc.)	[7,10,15,17,18,30,31,44,47,48,55–57]

Table 1. *Cont.*

Level I Indexes	Level II Indexes	Meanings of Level II Indexes	Source
Project characteristics	Project scope clarity	Clarity of project scoping	[8,10,11,18,21,31,38,40–45,48,50,51,58]
	Project flexibility	Flexibility of expected design and construction changes in the implementation of the project	[8,10,11,13,17,18,21,24,27–29,31–33,39,44,52,54]
	Project disputes	The severity of potential disputes in the course of project construction (e.g., serious disputes, etc.)	[7,17,18,24,27,28,31,32,41,44,49,53,59]
External environment	Market competition	Competition level in the contractor market	[8,11,12,15,18,21,29–31,49,52]
	Accessibility of materials	The extent to which the necessary raw materials for the project are difficult to purchase in the market	[8,10,15,21,30,38]
	Availability of technology	The degree of difficulty in obtaining the necessary technology for project construction in the market	[8,10,15,17,18,21,30,31,49]
	Impact of laws and regulations	The limitation of the perfection of laws and regulations on the PDS	[8,11,12,15,17,18,21,30,32,48,49,53]

3. Preliminaries

In this section, the concept of intuitionistic fuzzy theory and a series of relevant algorithms are introduced.

Fuzzy set theory was first proposed by Zadeh (1965) [60] and has been widely used. The idea of a fuzzy set is to extend the eigenfunction with a value of only 1 or 0 to a membership function with arbitrary values in the unit closed interval [0, 1]. However, this kind of membership function value is only a single number; thus, the meaning of approval, disapproval, or hesitation cannot be expressed. Due to the fuzziness and uncertainty of objective things, Atanasov (1986) [61] extended the Zadeh fuzzy set to consider the membership, the non-membership degree, and the hesitation degree at the same time. Entropy, originally a thermodynamic unit, was later applied to information theory in 1940, and is a measure for uncertainty. Greater entropy represents more uncertain information in the single evaluation result of the decision expert; thus, a smaller weight should be given to such an expert. The cross entropy based on fuzzy set theory was proposed by Shang and Jiang [62] to describe the difference between two fuzzy sets. Vlachos and Sergiadis [26] put forward the cross-entropy of the intuitionistic fuzzy set (IFS) to describe the degree of difference between the intuitionistic fuzzy sets. When x and y are two discrete distributions, the relative entropy can be a measure of the degree of coincidence [63]. Some basic concepts and definitions of IFSs are presented as follows:

Definition 1 [61,64]. *Let x be a non-empty set. An IFS$_A$ is an object having the form:*

$$A = \{\langle x, \mu_A(x), v_A(x)\rangle | x \in X\} \tag{1}$$

where the mapping is presented as "$\mu_A : X \to [0,1]$" and "$v_A : X \to [0,1]$" under the condition "$0 \leq \mu_A(x) + v_A(x) \leq 1$" for each $x \in X$. $\mu_A(x)$ and $v_A(x)$ are defined as the degree of membership and the degree of non-membership, respectively, of element $x \in X$ to set A.

Obviously, if $v_A(x) = 1 - \mu_A(x)$, every IFS(A) on a non-empty set X becomes a fuzzy set.

In a more simple way, Xu and Yager [65] regarded $\alpha = (\mu_\alpha, v_\alpha)$ as an intuitionistic fuzzy number and used it to represent an intuitionistic fuzzy set, where, $\mu_\alpha \in [0,1]$, $v_\alpha \in [0,1]$, $\mu_\alpha + v_\alpha \leq 1$.

Definition 2 [61,64,66]. *Let* $\alpha = (\mu_\alpha, v_\alpha)$ *and* $\beta = (\mu_\beta, v_\beta)$ *as any two intuitionistic fuzzy numbers; then, the algorithm for intuitionistic fuzzy numbers is as follows:*

(1)　$\overline{\alpha} = (v_\alpha, \mu_\alpha)$

(2)　$\alpha \oplus \beta = (\mu_\alpha + \mu_\beta - \mu_\alpha\mu_\beta, v_\alpha v_\beta)$

(3)　$\lambda\alpha = (1 - (1 - \mu_\alpha)^\lambda, v_\alpha^\lambda), \lambda > 0$

(4)　$\alpha \otimes \beta = (\mu_\alpha\mu_\beta, v_\alpha + v_\beta - v_\alpha v_\beta)$

(5)　$\alpha^\lambda = (\mu_\alpha^\lambda, 1 - (1 - v_\alpha)^\lambda), \lambda > 0$

Definition 3 [66]. *As for the two intuitionistic fuzzy numbers:* $\alpha_1 = (\mu_{\alpha_1}, v_{\alpha_1})$ *and* $\alpha_2 = (\mu_{\alpha_2}, v_{\alpha_2})$, $s(\alpha_1) = \mu_{\alpha_1} - v_{\alpha_1}$ *and* $s(\alpha_2) = \mu_{\alpha_2} - v_{\alpha_2}$ *are the scoring functions of* α_1 *and* α_2 *respectively;* $h(\alpha_1) = \mu_{\alpha_1} + v_{\alpha_1}$ *and* $h(\alpha_2) = \mu_{\alpha_2} + v_{\alpha_2}$ *are the exact function of* α_1 *and* α_2 *respectively, so:*

(1)　*if* $s(\alpha_1) < s(\alpha_2)$, *then* $\alpha_1 < \alpha_2$;

(2)　*if* $s(\alpha_1) = s(\alpha_2)$, *there are three situations:*

　　(a)　$h(\alpha_1) = h(\alpha_2)$, *then* $\alpha_1 = \alpha_2$;

　　(b)　$h(\alpha_1) < h(\alpha_2)$, *then* $\alpha_1 < \alpha_2$;

　　(c)　$h(\alpha_1) > h(\alpha_2)$, *then* $\alpha_1 > \alpha_2$.

Definition 4 [67]. *Let* $x_i, y_i \geq 0, i = 1, 2, \ldots, n$, *and* $1 = \sum_{i=1}^{n} x_i \geq \sum_{i=1}^{n} y_i$, *then:*

$$h(X, Y) = \sum_{i}^{n} x_i \lg(x_i / y_i) \tag{2}$$

The above equation was described as the relative entropy of X relative to Y, where, $X = (x_1, x_2, \ldots, x_n)$, $Y = (y_1, y_2, \ldots, y_n)$. *Hence, relative entropy can be used to measure the degree of compliance between X and Y.*

Definition 5 [66]. *The intuitionistic fuzzy mixed mean (IFHA) operator is a mapping:* $\Theta^n \to \Theta$, *it makes* $IFHA_{\omega,w}(\alpha_1, \alpha_2, \cdots, \alpha_n) = w_1\dot{\alpha}_{\sigma(1)} \oplus w_2\dot{\alpha}_{\sigma(2)} \oplus \cdots \oplus w_n\dot{\alpha}_{\sigma(n)}$, *where* $w = (w_1, w_2, \ldots, w_n)^T$ *is the IFHA operator weight vector,* $w_j \in [0,1](j = 1, 2, \ldots, n)$, $\sum_{j=1}^{n} w_j = 1$. $\dot{\alpha}_j = n\omega_j\alpha_j(j = 1, 2, \ldots, n)$, $(\dot{\alpha}_{\sigma(1)}, \dot{\alpha}_{\sigma(2)}, \ldots, \dot{\alpha}_{\sigma(n)})$ *weighted intuitionistic fuzzy array* $(\dot{\alpha}_1, \dot{\alpha}_2, \ldots, \dot{\alpha}_n)$ *a replacement, it makes* $\dot{\alpha}_{\sigma(j)} \geq \dot{\alpha}_{\sigma(j+1)}(j = 1, 2, \ldots, n-1)$, $\omega = (\omega_1, \omega_2, \ldots, \omega_n)^T$ *as* $\alpha_j(j = 1, 2, \ldots, n)$ *weight vector,* $\omega_j \in [0,1](j = 1, 2, \ldots, n)$, $\sum_{j=1}^{n} \omega_j = 1$, *n is the balance factor. Let* $\dot{\alpha}_{\sigma(j)} = (\mu_{\dot{\alpha}_{\sigma(j)}}, v_{\dot{\alpha}_{\sigma(j)}}), (j = 1, 2, \cdots, n)$, *then:*

$$IFHA_{\omega,w}(\alpha_1, \alpha_2, \ldots, \alpha_n) = \left(1 - \prod_{j=1}^{n} (1 - \mu_{\dot{\alpha}_{\sigma(j)}})^{w_j}, \prod_{j=1}^{n} (v_{\dot{\alpha}_{\sigma(j)}})^{w_j}\right) \tag{3}$$

Definition 6 [66]. *Let* $\alpha_j = (\mu_{\alpha_j}, v_{\alpha_j}), (j = 1, 2, \cdots, n)$ *be a set of intuitionistic fuzzy numbers, and let* $IFWA : \Theta^n \to \Theta$, *if* $IFWA_\omega(\alpha_1, \alpha_2, \ldots \alpha_n) = \omega_1\alpha_1 \oplus \omega_2\alpha_2 \oplus \ldots \oplus \omega_n\alpha_n$; *then, IFWA is named the intuitionistic fuzzy weighted average operator, among them,* $\omega = (\omega_1, \omega_2, \ldots \omega_n)^T$ *is an exponential weight vector of* $\alpha_j(j = 1, 2, \cdots, n)$, $\omega_j \in [0,1]$, $\sum_{j-=1}^{n} \omega_j = 1$.

Definition 7 [68]. *The defined $H : A \to [0,1]$ was the entropy of IFS: $A = \{< x, \mu_A(x), v_A(x) > | x \in X\}$, therefore, the following can be calculated:*

$$H(A) = -\frac{1}{n \ln 2} \sum_{i=1}^{n} [\mu_A(x) \ln \mu_A(x) + v_A(x) \ln v_A(x) - (1 - \pi_A(A)) \times \ln(1 - \pi_A(A)) - \pi_A(A) \ln 2] \quad (4)$$

4. Establishment of Intuitionistic Fuzzy Group Decision Model

4.1. Group Decision Model Description

Set up S experts in the group: $D = \{D_1, D_2, \ldots, D_s\}$ is an expert set, $M = \{M_1, M_2, \ldots, M_n\}$ and $C = \{C_1, C_2, \ldots, C_m\}$ are a set of schemes and decision attributes, respectively. The weight vector given to the experts by the subjective weighting method is $\xi = (\xi_1, \xi_2, \ldots, \xi_s)^T, 0 \leq \xi_k \leq 1, k = 1, 2, \ldots, s$ and $\sum_{k=1}^{s} \xi_k = 1$. The weight vector of the attribute is expressed as $\omega = (\omega_1, \omega_2, \ldots, \omega_m)^T, 0 \leq \omega_j \leq 1$, $j = 1, 2, \ldots, m$, and $\sum_{j=1}^{m} \omega_j = 1$. A certain scheme M_i is rated by expert D_k according to attribute C_j, and gets an intuitionistic fuzzy decision matrix $R_k = (r_{ij}^k)_{n \times m}$ where $r_{ij}^k = (\mu_{r_{ij}^k}, v_{r_{ij}^k})$, $\mu_{r_{ij}^k}$, $v_{r_{ij}^k}$, $\pi_{r_{ij}^k} = 1 - \mu_{r_{ij}^k} - v_{r_{ij}^k}$ represents the satisfaction, dissatisfaction, and hesitation of expert D_k for attribute M_i under attribute C_j, respectively, $k = 1, 2, \ldots, s, i = 1, 2, \ldots n, j = 1, 2, \ldots m$.

4.2. Intuitionistic Fuzzy Entropy Model for Decision Attribute Weight Determination

At present, in the field of intuitionistic fuzzy group decision making, scholars generally use entropy theory to determine the weight of decision attributes. Relative entropy determines the weight of the attribute by the relative entropy of the two attributes [63]. The calculation steps are as follows:

Step 1: Establishment of an intuitionistic fuzzy decision matrix.

The scores of each expert's evaluation constitute an intuitionistic fuzzy decision matrix:

$$R_k = \begin{pmatrix} (\mu_{r_{11}^k}, v_{r_{11}^k}) & \cdots & (\mu_{r_{1m}^k}, v_{r_{1m}^k}) \\ \vdots & \ddots & \vdots \\ (\mu_{r_{n1}^k}, v_{r_{n1}^k}) & \cdots & (\mu_{r_{nm}^k}, v_{r_{nm}^k}) \end{pmatrix} \quad (5)$$

Step 2: Take the i line from the above $R_k(k = 1, 2, \ldots, s)$, forming a new matrix $B_i = \left((\mu_{u_{ij}^k}, v_{u_{ij}^k})\right)_{s \times m}$ $(i = 1, 2, \ldots, n, k = 1, 2, \ldots s)$ to indicate that the i scheme given by each expert conforms to the judgment matrix of m attributes.

$$B_i = \begin{pmatrix} (\mu_{u_{i1}^1}, v_{u_{i1}^1}) & \cdots & (\mu_{u_{im}^1}, v_{u_{im}^1}) \\ \vdots & \ddots & \vdots \\ (\mu_{u_{i1}^s}, v_{u_{i1}^s}) & \cdots & (\mu_{u_{im}^s}, v_{u_{im}^s}) \end{pmatrix} \quad (6)$$

Step 3: Use Formula (4) to find the entropy of each attribute based on Equation (7).

$$H_j^i = -\frac{1}{s \ln 2} \sum_{k=1}^{s} \mu_{r_{ij}^k}(x) \ln \mu_{r_{ij}^k}(x) + v_{r_{ij}^k}(x) \ln v_{r_{ij}^k}(x) - (1 - \pi_{r_{ij}^k}(x) \times \ln((1 - \pi_{r_{ij}^k}(x)) - \pi_{r_{ij}^k}(x) \ln 2) \quad (7)$$

Step 4: Obtain the entropy weight of each attribute.

According to the entropy theory, if the entropy value for each criterion is smaller across alternatives, it should provide decision-makers with the useful information. Therefore, the criterion should be assigned a bigger weight; otherwise, such a criterion will be judged unimportant by most

Intuitionistic Fuzzy Entropy for Group Decision Making of Water Engineering Project...

7

decision-makers. In other words, such a criterion should be evaluated as a very small weight. If the information about weight w_j^i of the criterion C_j^i is completely unknown, the entropy weights for determining the criteria weight can be calculated as follows [69]:

$$w_j^i = \frac{1 - H_j^i}{n - \sum\limits_{j=1}^{m} H_j^i} \tag{8}$$

Further, obtain the objective weight of each attribute according to the judgment information under the i scheme $w_j^i = (\omega_1^i, \omega_2^i, \ldots, \omega_m^i)$. Then, the objective attribute weights of all the schemes can constitute a weight matrix, which is marked as:

$$\hat{w} = \begin{pmatrix} \omega_1^1 & \cdots & \omega_m^1 \\ \vdots & \ddots & \vdots \\ \omega_1^n & \cdots & \omega_m^n \end{pmatrix} \tag{9}$$

Step 5: Find the optimal weight of each attribute.

Set the optimal weight of the attribute as $w = (\omega_1, \omega_2, \ldots \omega_m)$; each row in the weight matrix can be thought of as the attribute weight probability distribution given by all decision-makers under each scenario. From the concept of relative entropy, the difference between the probability distribution ω and the optimal attribute weight should be small. Therefore, the following optimization model is constructed:

$$minRE(\omega) = \sum\limits_{i=1}^{n} \sum\limits_{j=1}^{m} \omega_j \lg\left(\frac{\omega_j}{\omega_j^i}\right)$$
$$s.t \sum\limits_{j=1}^{m} \omega_j = 1, \ \omega_j > 0 \tag{10}$$
$$j = 1, 2, \ldots, m$$

Then, the optimal solution of the model is the most weight of the attribute $\omega^* = (\omega_1^*, \omega_2^*, \ldots, \omega_m^*)$, where:

$$\omega_j^* = \frac{\prod\limits_{i=1}^{n} \omega_j^i}{\sum\limits_{j=1}^{m} \prod\limits_{i=1}^{n} \omega_j^i}, j = 1, 2, \ldots m \tag{11}$$

4.3. Intuitionistic Fuzzy Comprehensive Entropy Model Based on Decision Expert Weight

In the process of group decision making, it is of great significance to objectively determine the weight of experts for more reliable decision making. The idea of cross entropy of intuitionistic fuzzy sets is that if the cross entropy between the two experts is smaller—namely, if the difference between their scores is smaller—then the individual evaluation results are relatively good, and a larger weight is given; while in contrast, a smaller weight is given. This paper proposed a method to obtain expert weight by combining cross entropy and entropy. The specific steps are as follows:

Step 6: According to the attribute weight value calculated in step 5, combined with the original expert weight value, the IHWA in Definition 6 is used to aggregate the scheme information, and the expert D_k evaluation result y_{ki} for scheme M_i, as well as individual and the expert group evaluation result x_i are obtained.

$$y_{ki} = \overset{m}{\underset{j=1}{\oplus}} r_{ij}^k \omega_j = (1 - \prod\limits_{j=1}^{m} (1 - \mu_{r_{ij}^k})^{\omega_j}, \prod\limits_{j=1}^{m} (v_{r_{ij}^k})^{\omega_j}), k = 1, 2, \ldots, s; i = 1, 2, \ldots, n$$

$$x_i = \overset{s}{\underset{k=1}{\oplus}} y_{ki} \xi_k = (1 - \prod\limits_{k=1}^{s} [1 - (1 - \prod\limits_{j=1}^{m} (1 - \mu_{r_{ij}^k})^{\omega_j})]^{\xi_k}, \prod\limits_{k=1}^{s} [\prod\limits_{j=1}^{m} (v_{r_{ij}^k})^{\omega_j}]^{\xi_k}), i = 1, 2, \ldots, n$$

Step 7: According to the cross-entropy formula in reference [26,62], the cross entropy between the individual and group scoring results can be obtained from the individual evaluation result vector $Y_k = (y_{k1}, y_{k2}, \ldots, y_{kn})^T$ and the group evaluation result vector $X = (x_1, x_2, \ldots, x_n)^T$.

$$D(Y_k, X) = \sum_{i=1}^{n} \left[\mu_{ki} \ln \frac{\mu_{ki}}{\frac{1}{2}(\mu_{ki} + \mu_i)} + v_{ki} \ln \frac{v_{ki}}{\frac{1}{2}(v_{ki} + v_i)} \right] + \sum_{i=1}^{n} \left[\mu_i \ln \frac{\mu_i}{\frac{1}{2}(\mu_{ki} + \mu_i)} + v_i \ln \frac{v_i}{\frac{1}{2}(v_{ki} + v_i)} \right] \quad (12)$$

where $\mu_{ki} = 1 - \prod_{j=1}^{m} (1 - \mu_{r_{ij}^k})^{\omega_j}$, $v_{ki} = \prod_{j=1}^{m} (v_{r_{ij}^k})^{\omega_j}$, $\mu_i = 1 - \prod_{k=1}^{s} (1 - \mu_{ki})^{\xi_k}$, $v_i = \prod_{k=1}^{s} (v_{ki})^{\xi_k}$, $i = 1, 2, \ldots, n$.

Furthermore, the weight of experts based on cross entropy is obtained as follows:

$$r_k = \frac{1}{D(Y_k, X)} \Big/ \sum_{k=1}^{s} \frac{1}{D(Y_k, X)} \quad (13)$$

Then, $0 \le r_k \le 1; k = 1, 2, \ldots, s; \sum_{k=1}^{s} r_k = 1$.

Step 8: Then, the entropy value of the evaluation value r_k be E_k. E_k can be computed by the following equation [70]:

$$E_k = \frac{1}{n} \sum_{i=1}^{n} \frac{\min\{\mu_{ki}, v_{ki}\} + \pi_{ki}}{\max\{\mu_{ki}, v_{ki}\} + \pi_{ki}} \quad (14)$$

Furthermore, the weight of the expert based on entropy is $e_k = (1 - E_k)/(s - \sum_{k=1}^{s} E_k)$, where: $0 \le e_k \le 1; k = 1, 2, \ldots, s; \sum_{k=1}^{s} e_k = 1$.

Step 9: From the weight r_k of cross entropy and the weight e_k based on entropy, the expert weight is obtained by combining the weighting method:

$$\gamma_k = \alpha r_k + \beta e_k, k = 1, 2, \ldots, s \quad (15)$$

where $0 \le \alpha \le 1, \alpha + \beta = 1$.

4.4. Overall Scheme Sorting Model Based on IFHA and IHWA Operators

Step 10: The intuitionistic fuzzy hybrid average operator (IFHA) is used to aggregate the fuzzy evaluation value of each expert under each scheme.

$$IFHA_{\gamma_k, w}(r_{ij}^1, r_{ij}^2, \cdots, r_{ij}^s) = \left(1 - \prod_{k=1}^{s} (1 - \mu_{\dot{r}_{ij}^{\sigma(k)}})^{w_k}, \prod_{k=1}^{s} (v_{\dot{r}_{ij}^{\sigma(k)}})^{w_k} \right) \quad (16)$$

The fuzzy decision matrix between groups is obtained: $\ddot{R} = (\ddot{r}_{ij})_{n \times m}$.

Step 11: The comprehensive attribute values of each scheme are obtained by using the intuitionistic fuzzy weighted average operator (IFWA) by Definition 6.

$$IFWA_{\omega}(\ddot{r}_{ij}, \ddot{r}_{ij}, \ldots, \ddot{r}_{ij}) = \left(1 - \prod_{j=1}^{m} (1 - \mu_{\ddot{r}_{ij}})^{\omega_j}, \prod_{j=1}^{m} (v_{\ddot{r}_{ij}})^{\omega_j} \right) \quad (17)$$

Step 12: The score of the comprehensive attribute value is $s(\ddot{r}_i), i = 1, 2, \ldots, n$, from Definition 3, and the final ranking of the scheme is obtained.

5. Case Study Analysis

5.1. Background Description

In a large hydropower station project in China, the hydropower station adopts hybrid development. The normal storage water level of the reservoir is 398 m, the storage capacity is 63.3 million m^3, the design reference flow is 2640.9 m^3/s, the installed capacity of the power station is 772 MW, and the average annual power generation is 3.303 billion KWH. The main engineering quantities include (excluding temporary and diversion projects): earthwork excavation 16.1577 million m^3, stone excavation 2.1824 million m^3, concrete pouring 2.1112 million m^3, earthwork filling 3.2906 million m^3, masonry project 455,500 m^3, steel bar 41,700 t, curtain grouting 13,300 m, consolidation grouting 2.63 m, concrete impervious wall 100,200 m^2, metal structure installation 10,700 t, install 19 hoist sets.

The scale of the hydropower station project is relatively large and the geological condition is not very good. The owner intended to adopt a project delivery mode; alternative ones include the DBB mode, DB mode, and EPC mode. The decision index was the 15 key influencing factors index of the index system established in Section 2. The owner unit engaged five senior experts in the relevant fields, namely, the owner representative personnel D_1, the construction technology expert D_2, the cost engineer D_3, economic experts D_4, and environmental experts D_5.

5.2. Determination of Attribute Weight

According to the above description, the specific decision system can be expressed as follows: the set of decision makers composed of five experts is $D = \{D_1, D_2, \ldots, D_5\}$; on the basis of scientific research, the weight vector given to the five experts in advance is $\xi = (0.2, 0.1, 0.22, 0.17, 0.31)^T$ (only a preliminary assumption that the weight will vary according to different items). The scheme set of the three alternative trading modes is $M = \{M_1, M_2, M_3\}$. The attribute set of 15 evaluation indexes established according to Table 1 is $C = \{C_1, C_2, \ldots, C_{15}\}$. Expert D_k evaluates scheme M_i according to attribute C_j, and obtains the fuzzy decision matrix $R^k = (r_{ij}^k)_{5 \times 13} (k = 1, 2 \ldots 5; i = 1, 2, 3; j = 1, 2, \ldots, 15)$. The specific evaluation process for the project is as follows:

(1) Establishment of intuitionistic fuzzy matrix. Based on the rating of the three alternative models by five experts, the intuitionistic fuzzy matrix is as follows:

$R_1 =$

	C_1	C_2	C_3	C_4	C_5	C_6	C_7	C_8	C_9	C_{10}	C_{11}	C_{12}	C_{13}	C_{14}	C_{15}
M_1	(0.6,0.2)	(0.5,0.1)	(0.6,0.3)	(0.7,0.1)	(0.8,0.2)	(0.6,0.3)	(0.3,0.1)	(0.7,0.1)	(0.6,0.1)	(0.6,0.2)	(0.5,0.5)	(0.6,0.1)	(0.7,0.1)	(0.6,0.2)	(0.4,0.2)
M_2	(0.8,0.1)	(0.7,0.2)	(0.6,0.2)	(0.5,0.4)	(0.7,0.3)	(0.6,0.3)	(0.8,0.1)	(0.8,0.2)	(0.7,0.1)	(0.6,0.4)	(0.4,0.6)	(0.6,0.3)	(0.7,0.2)	(0.6,0.3)	(0.5,0.4)
M_3	(0.2,0.7)	(0.3,0.5)	(0.5,0.2)	(0.4,0.3)	(0.2,0.5)	(0.6,0.2)	(0.7,0.2)	(0.3,0.5)	(0.5,0.2)	(0.6,0.1)	(0.6,0.1)	(0.3,0.2)	(0.4,0.4)	(0.3,0.5)	(0.5,0.2)

$R_2 =$

	C_1	C_2	C_3	C_4	C_5	C_6	C_7	C_8	C_9	C_{10}	C_{11}	C_{12}	C_{13}	C_{14}	C_{15}
M_1	(0.5,0.2)	(0.6,0.1)	(0.5,0.4)	(0.6,0.3)	(0.8,0.1)	(0.7,0.2)	(0.4,0.1)	(0.6,0.3)	(0.7,0.1)	(0.7,0.1)	(0.6,0.2)	(0.5,0.4)	(0.8,0.1)	(0.7,0.2)	(0.5,0.4)
M_2	(0.6,0.3)	(0.6,0.1)	(0.7,0.2)	(0.4,0.5)	(0.3,0.6)	(0.4,0.5)	(0.2,0.7)	(0.6,0.4)	(0.2,0.6)	(0.7,0.2)	(0.4,0.5)	(0.7,0.2)	(0.6,0.3)	(0.2,0.6)	(0.4,0.5)
M_3	(0.3,0.2)	(0.4,0.4)	(0.6,0.2)	(0.3,0.6)	(0.5,0.1)	(0.6,0.2)	(0.6,0.2)	(0.5,0.4)	(0.7,0.2)	(0.4,0.1)	(0.5,0.4)	(0.6,0.1)	(0.3,0.4)	(0.7,0.1)	(0.6,0.2)

$R_3 =$

	C_1	C_2	C_3	C_4	C_5	C_6	C_7	C_8	C_9	C_{10}	C_{11}	C_{12}	C_{13}	C_{14}	C_{15}
M_1	(0.4,0.3)	(0.5,0.3)	(0.6,0.4)	(0.2,0.7)	(0.5,0.4)	(0.6,0.2)	(0.6,0.1)	(0.2,0.5)	(0.4,0.1)	(0.5,0.3)	(0.2,0.7)	(0.8,0.1)	(0.7,0.2)	(0.5,0.2)	(0.7,0.2)
M_2	(0.7,0.1)	(0.6,0.2)	(0.3,0.2)	(0.5,0.5)	(0.6,0.1)	(0.7,0.3)	(0.4,0.1)	(0.5,0.4)	(0.3,0.4)	(0.3,0.5)	(0.1,0.7)	(0.6,0.2)	(0.7,0.1)	(0.3,0.5)	(0.4,0.1)
M_3	(0.2,0.5)	(0.3,0.2)	(0.5,0.4)	(0.2,0.7)	(0.5,0.3)	(0.6,0.1)	(0.7,0.2)	(0.4,0.5)	(0.6,0.3)	(0.5,0.2)	(0.6,0.2)	(0.7,0.1)	(0.2,0.3)	(0.5,0.3)	(0.6,0.1)

$R_4 =$

	C_1	C_2	C_3	C_4	C_5	C_6	C_7	C_8	C_9	C_{10}	C_{11}	C_{12}	C_{13}	C_{14}	C_{15}
M_1	(0.5,0.2)	(0.4,0.1)	(0.2,0.6)	(0.3,0.5)	(0.6,0.1)	(0.4,0.1)	(0.5,0.2)	(0.3,0.4)	(0.5,0.2)	(0.2,0.6)	(0.3,0.1)	(0.7,0.3)	(0.3,0.3)	(0.7,0.1)	(0.5,0.4)
M_2	(0.6,0.2)	(0.5,0.4)	(0.4,0.5)	(0.4,0.5)	(0.5,0.2)	(0.6,0.1)	(0.5,0.4)	(0.2,0.6)	(0.5,0.2)	(0.6,0.1)	(0.5,0.4)	(0.3,0.1)	(0.5,0.3)	(0.3,0.4)	(0.3,0.4)
M_3	(0.2,0.6)	(0.5,0.3)	(0.4,0.2)	(0.2,0.7)	(0.5,0.3)	(0.6,0.1)	(0.7,0.2)	(0.4,0.5)	(0.6,0.3)	(0.5,0.2)	(0.6,0.2)	(0.7,0.1)	(0.2,0.3)	(0.6,0.2)	(0.6,0.1)

$R_5 =$

	C_1	C_2	C_3	C_4	C_5	C_6	C_7	C_8	C_9	C_{10}	C_{11}	C_{12}	C_{13}	C_{14}	C_{15}
M_1	(0.5,0.1)	(0.4,0.2)	(0.5,0.4)	(0.3,0.5)	(0.6,0.1)	(0.7,0.3)	(0.5,0.1)	(0.3,0.1)	(0.5,0.3)	(0.3,0.6)	(0.2,0.7)	(0.7,0.1)	(0.8,0.1)	(0.4,0.5)	(0.7,0.1)
M_2	(0.4,0.1)	(0.7,0.1)	(0.4,0.2)	(0.6,0.1)	(0.2,0.1)	(0.5,0.4)	(0.5,0.2)	(0.3,0.2)	(0.3,0.5)	(0.4,0.4)	(0.2,0.5)	(0.7,0.2)	(0.6,0.3)	(0.5,0.2)	(0.5,0.3)
M_3	(0.3,0.4)	(0.2,0.5)	(0.4,0.5)	(0.8,0.1)	(0.6,0.3)	(0.4,0.1)	(0.5,0.2)	(0.5,0.1)	(0.7,0.1)	(0.3,0.4)	(0.4,0.2)	(0.6,0.2)	(0.3,0.1)	(0.6,0.1)	(0.6,0.2)

(2) The X line of the above five intuitionistic fuzzy matrices is taken out, and the new matrix $B_i = \left(\left(\mu_{u_{ij}^k}, v_{u_{ij}^k} \right) \right)_{5 \times 15}$ $(i = 1, 2, 3, k = 1, 2, \ldots 5)$ represents the judgment matrix of 15 attributes of the first scheme by five experts, respectively. For example, take out the first line and construct a new matrix $B_1 = \left(\left(\mu_{u_{1j}^k}, v_{u_{1j}^k} \right) \right)_{5 \times 15}$ that represents the judgment matrix of five experts on the 15 attributes of the first scenario, which are shown as follows:

$$B_1 = \begin{pmatrix} & C_1 & C_2 & C_3 & C_4 & C_5 & C_6 & C_7 & C_8 & C_9 & C_{10} & C_{11} & C_{12} & C_{13} & C_{14} & C_{15} \\ R_1 & (0.6,0.2) & (0.5,0.1) & (0.6,0.3) & (0.7,0.1) & (0.8,0.2) & (0.6,0.3) & (0.3,0.1) & (0.7,0.1) & (0.6,0.1) & (0.6,0.2) & (0.5,0.5) & (0.6,0.1) & (0.7,0.1) & (0.6,0.2) & (0.4,0.2) \\ R_2 & (0.5,0.2) & (0.6,0.1) & (0.5,0.4) & (0.6,0.3) & (0.8,0.1) & (0.7,0.2) & (0.4,0.1) & (0.6,0.3) & (0.7,0.1) & (0.7,0.1) & (0.6,0.2) & (0.5,0.4) & (0.8,0.1) & (0.7,0.2) & (0.5,0.4) \\ R_3 & (0.4,0.3) & (0.5,0.3) & (0.6,0.4) & (0.2,0.7) & (0.5,0.4) & (0.6,0.2) & (0.6,0.1) & (0.2,0.5) & (0.4,0.1) & (0.5,0.3) & (0.2,0.7) & (0.8,0.1) & (0.7,0.2) & (0.5,0.2) & (0.7,0.2) \\ R_4 & (0.5,0.2) & (0.4,0.1) & (0.2,0.6) & (0.3,0.5) & (0.6,0.1) & (0.4,0.1) & (0.5,0.2) & (0.3,0.4) & (0.5,0.2) & (0.2,0.6) & (0.3,0.1) & (0.7,0.3) & (0.3,0.3) & (0.7,0.1) & (0.5,0.4) \\ R_5 & (0.5,0.1) & (0.4,0.2) & (0.5,0.4) & (0.3,0.5) & (0.6,0.1) & (0.7,0.3) & (0.5,0.1) & (0.3,0.1) & (0.5,0.3) & (0.3,0.6) & (0.2,0.7) & (0.7,0.1) & (0.8,0.1) & (0.4,0.5) & (0.7,0.1) \end{pmatrix}$$

(3) The entropies of 15 attributes are calculated, and the results are as shown in Table 2.

Table 2. The entropies of the 15 attributes for the three alternatives.

Alternatives	Indicator Entropy Value
M_1	$H_1^1 = 0.8874, H_2^1 = 0.8560, H_3^1 = 0.9461, H_4^1 = 0.8552, H_5^1 = 0.7390, H_6^1 = 0.8611, H_7^1 = 0.8388,$ $H_8^1 = 0.8759, H_9^1 = 0.8155, H_{10}^1 = 0.8446, H_{11}^1 = 0.8698, H_{12}^1 = 0.7550, H_{13}^1 = 0.7057,$ $H_{14}^1 = 0.7957, H_{15}^1 = 0.8257$
M_2	$H_1^2 = 0.7648, H_2^2 = 0.7956, H_3^2 = 0.9130, H_4^2 = 0.930, H_5^2 = 0.8803, H_6^2 = 0.9012, H_7^2 = 0.8196,$ $H_8^2 = 0.9039, H_9^2 = 0.8683, H_{10}^2 = 0.8873, H_{11}^2 = 0.8988, H_{12}^2 = 0.8551, H_{13}^1 = 0.8478,$ $H_{14}^2 = 0.8357, H_{15}^2 = 0.8942$
M_3	$H_1^3 = 0.9032, H_2^3 = 0.9634, H_3^3 = 0.9376, H_4^3 = 0.8089, H_5^3 = 0.9096, H_6^3 = 0.7975, H_7^3 = 0.8233,$ $H_8^3 = 0.9459, H_9^3 = 0.8360, H_{10}^3 = 0.8966, H_{11}^3 = 0.7832, H_{12}^3 = 0.7784, H_{13}^3 = 0.9235,$ $H_{14}^3 = 0.8477, H_{15}^3 = 0.8934$

(4) According to Formula (8), the entropy weight of each attribute of the information given by the expert is calculated, and the results are as follows:

$\omega_1^1 = 0.052, \omega_2^1 = 0.063, \omega_3^1 = 0.023, \omega_4^1 = 0.061, \omega_5^1 = 0.114, \omega_6^1 = 0.065, \omega_7^1 = 0.077, \omega_8^1 = 0.058,$ $\omega_9^1 = 0.088, \omega_{10}^1 = 0.076, \omega_{11}^1 = 0.065, \omega_{12}^1 = 0.124, \omega_{13}^1 = 0.158, \omega_{14}^1 = 0.079, \omega_{15}^1 = 0.084;$

$\omega_1^2 = 0.136, \omega_2^2 = 0.129, \omega_3^2 = 0.061, \omega_4^2 = 0.044, \omega_5^2 = 0.078, \omega_6^2 = 0.066, \omega_7^2 = 0.113, \omega_8^2 = 0.060,$ $\omega_9^2 = 0.085, \omega_{10}^2 = 0.071, \omega_{11}^2 = 0.064, \omega_{12}^2 = 0.086, \omega_{13}^2 = 0.086, \omega_{14}^2 = 0.089, \omega_{15}^2 = 0.104;$

$\omega_1^3 = 0.057, \omega_2^3 = 0.019, \omega_3^3 = 0.030, \omega_4^3 = 0.082, \omega_5^3 = 0.040, \omega_6^3 = 0.080, \omega_7^3 = 0.067, \omega_8^3 = 0.021,$ $\omega_9^3 = 0.061, \omega_{10}^3 = 0.040, \omega_{11}^3 = 0.082, \omega_{12}^3 = 0.083, \omega_{13}^3 = 0.069, \omega_{14}^3 = 0.094, \omega_{15}^3 = 0.063.$

5.3. Determination of Expert Weight

(1) According to the formula, the individual evaluation result $Y_k = (y_{k1}, y_{k2}, y_{k3}, y_{k4}, y_{k5})$ of expert D_k can be calculated, and the results are shown as follows:

$$y_{ki} = \begin{pmatrix} & D_1 & D_2 & D_3 & D_4 & D_5 \\ M_1 & (0.6435, 0.1416) & (0.6375, 0.2118) & (0.6584, 0.1953) & (0.5568, 0.2245) & (0.6311, 0.14874) \\ M_2 & (0.6404, 0.2726) & (0.5705, 0.3169) & (0.5741, 0.1946) & (0.4367, 0.1704) & (0.5538, 0.2306) \\ M_3 & (0.3928, 0.2496) & (0.6527, 0.1901) & (0.5775, 0.1808) & (0.5773, 0.1797) & (0.5424, 0.1934) \end{pmatrix}$$

The results of the expert group score can be obtained as follows:

$$X = ((0.6290, 0.1236), (0.5616, 0.2167), (0.5434, 0.1977), (0.7091, 0.1655), (0.5803, 0.2049))$$

(2) According to Formula (9), the weight of experts based on cross entropy is obtained as follows:

$$r_k = (0.1368, 0.0512, 0.1391, 0.1459, 0.5270)$$

(3) According to Formula (10), the weight of experts based on entropy is calculated as follows:

$$e_k = (0.1994, 0.2017, 0.2024, 0.1986)$$

(4) Setting $\alpha = 0.6$ and $\beta = 0.4$, the final weight of experts $\gamma_k = (0.1618, 0.1114, 0.1644, 0.1667, 0.3957)$ is calculated by the combined weight method.

5.4. Ranking of Overall Schemes and Patterns Comparison

(1) Information aggregation of the fuzzy evaluation value of all the experts under each program is carried out according to Formula (13), and the results are as follows:

$$R = \begin{pmatrix} & C_1 & C_2 & C_3 & C_4 & C_5 & C_6 & C_7 & C_8 & C_9 & C_{10} & C_{11} & C_{12} & C_{13} & C_{14} & C_{15} \\ M_1 & (0.475, 0.211) & (0.427, 0.166) & (0.489, 0.437) & (0.406, 0.407) & (0.637, 0.167) & (0.558, 0.222) & (0.455, 0.144) & (0.425, 0.251) & (0.501, 0.186) & (0.458, 0.355) & (0.356, 0.447) & (0.659, 0.162) & (0.677, 0.172) & (0.460, 0.563) & (0.439, 0.543) \\ M_2 & (0.617, 0.147) & (0.597, 0.201) & (0.457, 0.271) & (0.473, 0.357) & (0.478, 0.148) & (0.570, 0.294) & (0.519, 0.206) & (0.502, 0.348) & (0.415, 0.315) & (0.505, 0.313) & (0.322, 0.560) & (0.557, 0.222) & (0.605, 0.242) & (0.396, 0.342) & (0.592, 0.477) \\ M_3 & (0.217, 0.488) & (0.311, 0.372) & (0.427, 0.311) & (0.436, 0.400) & (0.464, 0.312) & (0.530, 0.143) & (0.624, 0.228) & (0.396, 0.370) & (0.586, 0.235) & (0.460, 0.244) & (0.563, 0.193) & (0.607, 0.174) & (0.381, 0.277) & (0.443, 0.236) & (0.372, 0.246) \end{pmatrix}$$

(2) According to Formula (14), the comprehensive attribute values of each scheme are calculated:

$$\ddot{r}_1 = (0.5730, 0.2075), \quad \ddot{r}_2 = (0.5318, 0.2216), \quad \ddot{r}_3 = (0.4904, 0.2579).$$

(3) According to Definition 3, the score of the comprehensive attribute value is calculated and sorted.

$$s(\ddot{r}_1) = 0.2798, \quad s(\ddot{r}_2) = 0.3215, \quad s(\ddot{r}_3) = 0.2801.$$

Furthermore, the final order of the scheme is obtained: $A > B > C$. Therefore, the second transaction mode of DBB is the optimal choice.

6. Conclusions

The PDS determines the project performance and is critical to water engineering project success. For a given water engineering project, selecting the proper PDS is one of the decisive factors. PDS selection is a typical multi-attribute decision-making problem that can be effectively solved by group decision making. IFS is always used to solve complex decision-making problems, especially multi-attribute group decision-making problems, under uncertain circumstances. Based on the IFS group decision-making model and intuitionistic fuzzy entropy, a new decision-making support method for PDS selection was proposed. In order to reduce the loss of judgment information and improve the objectivity and fairness of group decision making, two operators—IFHA and IFWA on PDS decision—were addressed to calculate the final decision weights. The case study showed that the method proposed in this paper is an effective approach and has potential practical application that would help water engineering project owners in PDS selection. The model proposed in this paper can also be applied to solve similar decision-making problems. The method proposed in this paper is an effective group decision method and can help the water engineering project owners in PDS selection, but the decision making that is influenced by expert subjective evaluation has to be considered, and how to reduce the influence of the subjective evaluations from experts is the next research direction.

Author Contributions: Formal analysis, L.L.; Supervision, L.Z.; Validation, K.Z.; Writing-original draft, X.L.; Writing-review and editing, F.Q.

Acknowledgments: The authors would like to appreciate the reviewers for all helpful comments, and to thank the foundation of Philosophy and Social Science Research in Colleges and Universities in Jiangsu Province (No. 2017SJB1364), Fundamental Research Funds for the Central Universities (No. 331711105), Jiangsu Provincial Construction System Science and Technology Project of Housing and Urban and Rural Development Department (No.2017ZD074), Jiangsu Province Joint Education Program High-Standard Example Project, for their supports.

References

1. Raouf, A.M.; Al-Ghamdi, S.G. Effectiveness of Project Delivery Systems in Executing Green Buildings. *J. Constr. Eng. Manag.* **2019**, *145*, 03119005. [CrossRef]
2. Mesa, H.A.; Molenaar, K.R.; Alarcón, L.F. Comparative analysis between integrated project delivery and lean project delivery. *Int. J. Proj. Manag.* **2019**, *37*, 395–409. [CrossRef]
3. Laurent, J.; Leicht, R.M. Practices for Designing Cross-Functional Teams for Integrated Project Delivery. *J. Constr. Eng. Manag.* **2019**, *145*, 05019001. [CrossRef]
4. Wu, P.; Xu, Y.; Jin, R.; Lu, Q.; Madgwick, D.; Hancock, C.M. Perceptions towards risks involved in off-site construction in the integrated design & construction project delivery. *J. Clean. Prod.* **2019**, *213*, 899–914.
5. An, X.; Wang, Z.; Li, H.; Ding, J. Project Delivery System Selection with Interval-Valued Intuitionistic Fuzzy Set Group Decision-Making Method. *Group Decis. Negot.* **2018**, *27*, 689–707. [CrossRef]
6. Li, H.; Qin, K.; Li, P. Selection of project delivery approach with unascertained model. *Kybernetes* **2015**, *44*, 238–252. [CrossRef]
7. Ojiako, U.; Johansen, E.; Greenwood, D. A qualitative re-construction of project measurement criteria. *Ind. Manag. Data Syst.* **2008**, *108*, 405–417. [CrossRef]
8. Oyetunji, A.A.; Anderson, S.D. Relative Effectiveness of Project Delivery and Contract Strategies. *J. Constr. Eng. Manag.* **2006**, *132*, 3–13. [CrossRef]
9. Khwaja, N.; O'Brien, W.J.; Martinez, M.; Sankaran, B.; O'Connor, J.T.; Hale, W. "Bill" Innovations in Project Delivery Method Selection Approach in the Texas Department of Transportation. *J. Manag. Eng.* **2018**, *34*, 05018010. [CrossRef]
10. Molenaar, K.R.; Songer, A.D. Model for Public Sector Design-Build Project Selection. *J. Constr. Eng. Manag.* **1998**, *124*, 467–479. [CrossRef]
11. Al Khalil, M.I. Selecting the appropriate project delivery method using AHP. *Int. J. Proj. Manag.* **2002**, *20*, 469–474. [CrossRef]
12. Mafakheri, F.; Dai, L.; Ślęzak, D.; Nasiri, F. Project Delivery System Selection under Uncertainty: Multicriteria Multilevel Decision Aid Model. *J. Manag. Eng.* **2007**, *23*, 200–206. [CrossRef]
13. Luu, D.T.; Ng, S.T.; Chen, S.E. Formulating Procurement Selection Criteria through Case-Based Reasoning Approach. *J. Comput. Civ. Eng.* **2005**, *19*, 269–276. [CrossRef]
14. Ribeiro, F.L. Project delivery system selection: A case-based reasoning framework. *Logist. Inf. Manag.* **2001**, *14*, 367–376. [CrossRef]
15. Luu, D.T.; Ng, S.T.; Chen, S.E.; Jefferies, M. A strategy for evaluating a fuzzy case-based construction procurement selection system. *Adv. Eng. Softw.* **2006**, *37*, 159–171. [CrossRef]
16. Lo, S.C.; Chao, Y. Efficiency assessment of road project delivery models. *AIP Conf. Proc.* **2007**, *963*, 1016–1019.
17. Ling, F.Y.Y.; Liu, M. Using neural network to predict performance of design-build projects in Singapore. *Build. Environ.* **2004**, *39*, 1263–1274. [CrossRef]
18. Chen, Y.Q.; Liu, J.Y.; Li, B.; Lin, B. Project delivery system selection of construction projects in China. *Expert Syst. Appl.* **2011**, *38*, 5456–5462. [CrossRef]
19. Azzedin, F.; Ridha, A.; Rizvi, A. Fuzzy trust for peer-to-peer based systems. *Proc. World Acad. Sci. Eng. Technol.* **2007**, *21*, 123–127.
20. Castillo, O.; Atanassov, K. Comments on fuzzy sets, interval type-2 fuzzy sets, general type-2 fuzzy sets and intuitionistic fuzzy sets. In *Recent Advances in Intuitionistic Fuzzy Logic Systems*; Springer: Berlin/Heidelberger, Germany, 2019; pp. 35–43.
21. Mostafavi, A.; Karamouz, M. Selecting Appropriate Project Delivery System: Fuzzy Approach with Risk Analysis. *J. Constr. Eng. Manag.* **2010**, *136*, 923–930. [CrossRef]
22. Al Nahyan, M.T.; Hawas, Y.E.; Raza, M.; Aljassmi, H.; Maraqa, M.A.; Basheerudeen, B.; Mohammad, M.S. A fuzzy-based decision support system for ranking the delivery methods of mega projects. *Int. J. Manag. Proj. Bus.* **2018**, *11*, 122–143. [CrossRef]
23. Martin, H.; Lewis, T.M.; Petersen, A.; Peters, E. Cloudy with a Chance of Fuzzy: Building a Multicriteria Uncertainty Model for Construction Project Delivery Selection. *J. Comput. Civ. Eng.* **2017**, *31*, 04016046. [CrossRef]
24. Ng, S.T.; Luu, D.T.; Chen, S.E.; Lam, K.C. Fuzzy membership functions of procurement selection criteria. *Constr. Manag. Econ.* **2002**, *20*, 285–296. [CrossRef]

25. Yuan, J.; Skibniewski, M.J.; Li, Q.; Zheng, L. Performance Objectives Selection Model in Public-Private Partnership Projects Based on the Perspective of Stakeholders. *J. Manag. Eng.* **2010**, *26*, 89–104. [CrossRef]

26. Vlachos, I.K.; Sergiadis, G.D. Intuitionistic fuzzy information—Applications to pattern recognition. *Pattern Recognit. Lett.* **2007**, *28*, 197–206. [CrossRef]

27. Love, P.; Skitmore, M.; Earl, G.; Skitmore, R. Selecting a suitable procurement method for a building project. *Constr. Manag. Econ.* **1998**, *16*, 221–233. [CrossRef]

28. Chan, A.P.C.; Ho, D.C.K.; Tam, C.M. Design and Build Project Success Factors: Multivariate Analysis. *J. Constr. Eng. Manag.* **2001**, *127*, 93–100. [CrossRef]

29. Cheung, S.-O.; Lam, T.-I.; Wan, Y.-W.; Lam, K.-C. Improving Objectivity in Procurement Selection. *J. Manag. Eng.* **2001**, *17*, 132–139. [CrossRef]

30. Luu, D.T.; Ng, S.T.; Chen, S.E. A case-based procurement advisory system for construction. *Adv. Eng. Softw.* **2003**, *34*, 429–438. [CrossRef]

31. Ling, F.Y.Y.; Chan, S.L.; Chong, E.; Ee, L.P. Predicting performance of design-build and design-bid-build projects. *J. Constr. Eng. Manag.* **2004**, *130*, 75–83. [CrossRef]

32. Mahdi, I.M.; Alreshaid, K. Decision support system for selecting the proper project delivery method using analytical hierarchy process (AHP). *Int. J. Proj. Manag.* **2005**, *23*, 564–572. [CrossRef]

33. Chan, C.T.W.; Chan, T.W.C. Fuzzy procurement selection model for construction projects. *Constr. Manag. Econ.* **2007**, *25*, 611–618. [CrossRef]

34. Liu, B.; Huo, T.; Shen, Q.; Yang, Z.; Meng, J.; Xue, B. Which owner characteristics are key factors affecting project delivery system decision making? Empirical analysis based on the rough set theory. *J. Manag. Eng.* **2014**, *31*, 1–12. [CrossRef]

35. Lin, H.; Zeng, S.; Ma, H.; Zeng, R.; Tam, V.W. An indicator system for evaluating megaproject social responsibility. *Int. J. Proj. Manag.* **2017**, *35*, 1415–1426. [CrossRef]

36. Zhang, Q.; Oo, B.L.; Lim, B.T.H. Drivers, motivations, and barriers to the implementation of corporate social responsibility practices by construction enterprises: A review. *J. Clean. Prod.* **2019**, *210*, 563–584. [CrossRef]

37. Kim, D.Y.; Han, S.H.; Kim, H. Discriminant Analysis for Predicting Ranges of Cost Variance in International Construction Projects. *J. Constr. Eng. Manag.* **2008**, *134*, 398–410. [CrossRef]

38. Lam, E.W.M.; Chan, A.P.C.; Chan, D.W.M. Determinants of successful design-build projects. *J. Constr. Eng. Manag.* **2008**, *134*, 333–341. [CrossRef]

39. Luu, D.T.; Ng, S.T.; Chen, S.E. Parameters governing the selection of procurement system—An empirical survey. *Eng. Constr. Arch. Manag.* **2003**, *10*, 209–218.

40. Doloi, H. Cost Overruns and Failure in Project Management: Understanding the Roles of Key Stakeholders in Construction Projects. *J. Constr. Eng. Manag.* **2013**, *139*, 267–279. [CrossRef]

41. Lu, W.; Hua, Y.; Zhang, S. Logistic regression analysis for factors influencing cost performance of design-bid-build and design-build projects. *Eng. Constr. Arch. Manag.* **2017**, *24*, 118–132. [CrossRef]

42. Baloi, D.; Price, A.D. Modelling global risk factors affecting construction cost performance. *Int. J. Proj. Manag.* **2003**, *21*, 261–269. [CrossRef]

43. Akinci, B.; Fischer, M. Factors Affecting Contractors' Risk of Cost Overburden. *J. Manag. Eng.* **1998**, *14*, 67–76. [CrossRef]

44. Liu, B.; Huo, T.; Liang, Y.; Sun, Y.; Hu, X. Key Factors of Project Characteristics Affecting Project Delivery System Decision Making in the Chinese Construction Industry: Case Study Using Chinese Data Based on Rough Set Theory. *J. Prof. Issues Eng. Educ. Pr.* **2016**, *142*, 5016003. [CrossRef]

45. Ibbs, C.W.; Kwak, Y.H.; Ng, T.; Odabasi, A.M. Project Delivery Systems and Project Change: Quantitative Analysis. *J. Constr. Eng. Manag.* **2003**, *129*, 382–387. [CrossRef]

46. Ibbs, W.; Chih, Y. Alternative methods for choosing an appropriate project delivery system (PDS). *Facilities* **2011**, *29*, 527–541. [CrossRef]

47. Konchar, M.; Sanvido, V. Comparison of U.S. Project Delivery Systems. *J. Constr. Eng. Manag.* **1998**, *124*, 435–444. [CrossRef]

48. Azhar, N.; Kang, Y.; Ahmad, I.U. Factors Influencing Integrated Project Delivery in Publicly Owned Construction Projects: An Information Modelling Perspective. *Procedia Eng.* **2014**, *77*, 213–221. [CrossRef]

49. Touran, A.; Gransberg, D.D.; Molenaar, K.R.; Ghavamifar, K. Selection of Project Delivery Method in Transit: Drivers and Objectives. *J. Manag. Eng.* **2011**, *27*, 21–27. [CrossRef]

50. Shane, J.S.; Molenaar, K.R.; Anderson, S.; Schexnayder, C. Construction Project Cost Escalation Factors. *J. Manag. Eng.* **2009**, *25*, 221–229. [CrossRef]

51. Shane, J.S.; Bogus, S.M.; Molenaar, K.R. Municipal Water/Wastewater Project Delivery Performance Comparison. *J. Manag. Eng.* **2013**, *29*, 251–258. [CrossRef]

52. Chan, A.P.C.; Yung, E.H.K.; Lam, P.T.I.; Tam, C.M.; Cheung, S.O. Application of Delphi method in selection of procurement systems for construction projects. *Constr. Manag. Econ.* **2001**, *19*, 699–718. [CrossRef]

53. Kumaraswamy, M.M.; Dissanayaka, S.M. Developing a decision support system for building project procurement. *Build. Environ.* **2001**, *36*, 337–349. [CrossRef]

54. Lam, K.C.; So, A.; Hu, T.; Ng, T.; Yuen, R.K.K.; Lo, S.M.; Cheung, S.O.; Yang, H. An integration of the fuzzy reasoning technique and the fuzzy optimization method in construction project management decision-making. *Constr. Manag. Econ.* **2001**, *19*, 63–76. [CrossRef]

55. Shen, L.-Y.; Lü, W.-S.; Yam, M.C.H. Contractor Key Competitiveness Indicators: A China Study. *J. Constr. Eng. Manag.* **2006**, *132*, 416–424. [CrossRef]

56. Sun, M.; Meng, X. Taxonomy for change causes and effects in construction projects. *Int. J. Proj. Manag.* **2009**, *27*, 560–572. [CrossRef]

57. De Marco, A.; Karzouna, A. Assessing the Benefits of the Integrated Project Delivery Method: A Survey of Expert Opinions. *Procedia Comput. Sci.* **2018**, *138*, 823–828. [CrossRef]

58. Creedy, G.D.; Skitmore, M.; Wong, J.K.W. Evaluation of Risk Factors Leading to Cost Overrun in Delivery of Highway Construction Projects. *J. Constr. Eng. Manag.* **2010**, *136*, 528–537. [CrossRef]

59. Gransberg, D.D.; Dillon, W.D.; Reynolds, L.; Boyd, J. Quantitative Analysis of Partnered Project Performance. *J. Constr. Eng. Manag.* **1999**, *125*, 161–166. [CrossRef]

60. Zadeh, L.A. Fuzzy sets. *Inf. Control* **1965**, *8*, 338–353. [CrossRef]

61. Atanassov, K.T. Intuitionistic fuzzy sets. *Fuzzy Sets Syst.* **1986**, *20*, 87–96. [CrossRef]

62. Shang, X.-G.; Jiang, W.-S. A note on fuzzy information measures. *Pattern Recognit. Lett.* **1997**, *18*, 425–432. [CrossRef]

63. Zhou, Y.; Wu, F.J. Combination weighting approach in multiple attribute decision making based on relative entropy. *Oper. Res. Manag. Sci.* **2006**, *15*, 48–53.

64. De, S.K.; Biswas, R.; Roy, A.R. Some operations on intuitionistic fuzzy sets. *Fuzzy Sets Syst.* **2000**, *114*, 477–484. [CrossRef]

65. Xu, Z.; Yager, R.R. Some geometric aggregation operators based on intuitionistic fuzzy sets. *Int. J. Gen. Syst.* **2006**, *35*, 417–433. [CrossRef]

66. Xu, Z. Intuitionistic Fuzzy Aggregation Operators. *IEEE Trans. Fuzzy Syst.* **2007**, *15*, 1179–1187.

67. Szmidt, E.; Kacprzyk, J. Distances between intuitionistic fuzzy sets. *Fuzzy Sets Syst.* **2000**, *114*, 505–518. [CrossRef]

68. Gumus, S.; Kucukvar, M.; Tatari, O. Intuitionistic fuzzy multi-criteria decision-making framework based on life cycle environmental, economic and social impacts: The case of U.S. wind energy. *Sustain. Prod. Consum.* **2016**, *8*, 78–92. [CrossRef]

69. Ye, J. Fuzzy decision-making method based on the weighted correlation coefficient under intuitionistic fuzzy environment. *Eur. J. Oper. Res.* **2010**, *205*, 202–204. [CrossRef]

70. Das, S.; Dutta, B.; Guha, D. Weight computation of criteria in a decision-making problem by knowledge measure with intuitionistic fuzzy set and interval-valued intuitionistic fuzzy set. *Soft Comput.* **2016**, *20*, 3421–3442. [CrossRef]

2

Estimation of Soil Depth using Bayesian Maximum Entropy Method

Kuo-Wei Liao [1,*]⬤**, Jia-Jun Guo** [1]**, Jen-Chen Fan** [1]**, Chien Lin Huang** [2] **and Shao-Hua Chang** [1]

[1]   Department of Bioenvironmental Systems Engineering, National Taiwan University, No. 1, Section 4, Roosevelt Rd., Taipei 10617, Taiwan; Sco-phoenix@hotmail.com.tw (J.-J.G.); jcfan@ntu.edu.tw (J.-C.F.); r05622015@ntu.edu.tw (S.-H.C.)

[2]   Hetengtech Company Limited, New Taipei City 24250, Taiwan; cheneyhuang@hetengtech.com

*   Correspondence: kliao@ntu.edu.tw

Abstract: Soil depth plays an important role in landslide disaster prevention and is a key factor in slopeland development and management. Existing soil depth maps are outdated and incomplete in Taiwan. There is a need to improve the accuracy of the map. The Kriging method, one of the most frequently adopted estimation approaches for soil depth, has room for accuracy improvements. An appropriate soil depth estimation method is proposed, in which soil depth is estimated using Bayesian Maximum Entropy method (BME) considering space distribution of measured soil depth and impact of physiographic factors. BME divides analysis data into groups of deterministic and probabilistic data. The deterministic part are soil depth measurements in a given area and the probabilistic part contains soil depth estimated by a machine learning-based soil depth estimation model based on physiographic factors including slope, aspect, profile curvature, plan curvature, and topographic wetness index. Accuracy of estimates calculated by soil depth grading, very shallow (<20 cm), shallow (20–50 cm), deep (50–90 cm), and very deep (>90 cm), suggests that BME is superior to the Kriging method with estimation accuracy up to 82.94%. The soil depth distribution map of Hsinchu, Taiwan made by BME with a soil depth error of ±5.62 cm provides a promising outcome which is useful in future applications, especially for locations without soil depth data.

Keywords: slopeland; Bayesian Maximum Entropy; soil depth; physiographic factors

1. Introduction

Soil depth plays an important role in landslide disaster prevention and management. Soil depth is an important parameter in deterministic analysis for shallow landslide prediction and landslide mass estimation. In Taiwan, soil depth is also used as for slopeland control by local governments. The Taiwan Slopeland Conservation and Utilization Act, 2016 defines soil depth as the distance from land surface to a specified depth, which confines roots of plants from further growth and grades them in four levels, as shown in Table 1. The grade of a slopeland is a key factor in determining whether it is applicable for agriculture or forestation only. Because this grading system is widely used by engineers, it is adopted here as the soil category system.

Table 1. Soil depth grading standards by the Taiwan Slopeland Conservation and Utilization Act.

Soil Depth Grade	Ratings
Very deep	>90 cm
Deep	50–90 cm
Shallow	20–50 cm
Very shallow	<20 cm

There are two sources of soil depth information available now: the slopeland soil depth distribution maps by the Soil and Water Conservation Bureau, Council of Agriculture, Executive Yuan in 1995. These maps suffer not only from being outdated data with low accuracy and poor space estimation, but also from limited coverage. The other available soil depth source is the manually measured soil depth by the Taiwan Agricultural Research Institute, Council of Agriculture, Executive Yuan. In spite of its high accuracy, the measured soil depths also suffer from the same drawback of limited coverage. As soil depth is usually in close correlation with topography [1,2], many scholars have managed to estimate soil depth by the Digital Elevation Model (DEM), a 3D digital representation of ground surface topography or terrain, and geostatistical methods.

Soil depth estimation relied on the Kriging method in the past. For example, Bourennane et al. [3] estimated soil depth with the Kriging method aligned with linear regression and ended up with more accurate estimates of soil depth than the Kriging method alone. Kuriakose et al. [4] employed regression in the Kriging method to estimate soil depth in a smaller area (9.5 square kilometers in Western Ghats of Kerala, India). He divided the site into 20 m × 20 m cells and executed the estimation based on environment variables including elevation, slope, aspect, curvature, topographic humidity index, use of land, and distance between sample points and rivers. Compared with other methods, the Kriging method appears ideal in predicting the spatial distribution of soil depth.

Most traditional geostatistical methods not only have to rely on normal distribution and linear estimation, but are also limited to analysis and study on hard data by actual measurement. This results in some new time-space geostatistical methods, e.g., the Bayesian Maximum Entropy (BME method). BME is not limited by Gaussian distribution and linear estimation, and combines physics knowledge or other probabilistic (soft) data with the concept of Bayesian conditional probability to gradually enhance estimation information and estimate stochastic and spatial data [5–10] at the same time. Since its introduction in 1991, BME has been widely used in various fields of science and engineering and proven to be more accurate than traditional spatial statistical methods [7,11,12].

Focused on the Hsinchu district in Taiwan, this study estimates soil depth based on hard and soft soil depth data using BME. For the first time, the BME is used to estimate the soil depth. According to Taiwan's classification of soil depth, the estimated soil depth is then divided into four levels: shallow layer (<20 cm), shallow layer (20–50 cm), deep layer (50–90 cm), and deep layer (>90 cm). This study also compares the accuracy of the results of the traditional Kriging method against the soil depth maps of Taiwan with the goal of identifying a valid space distribution estimation method to provide soil depth data in Taiwan.

2. Research Material and Methodology

2.1. Research Material

2.1.1. Research District

This study takes the Hsinchu district in Taiwan for research. Located in the northwestern part of Taiwan island, Hsinchu city and county span 1532 square kilometers east of the Taiwan Strait. Most of the district is occupied by hills, terraces, and mountains and with population concentrated in the alluvial plains in the lower reaches of the river. Its annual average temperature is about 22.6 degrees, rainfall about 1718.1 mm in 115.5 days, and with rainy season from May to September. The elevation distribution and geographical location of Hsinchu area are shown in Figure 1.

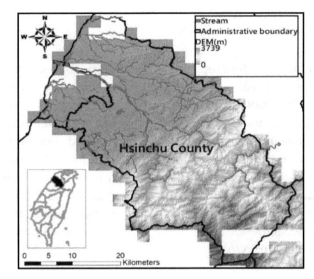

Figure 1. Elevation distribution and administrative division of Hsinchu district.

2.1.2. Soil Depth

This study uses 8217 soil depth records by the Agricultural Research Institute from drilling in this district in 2010, 2011, and 2016, as the basic data for the estimation model; the slopeland soil depth distribution maps by the Soil and Water Conservation Bureau in 1995 are taken as the control group. The soil drilling points and soil depth maps are shown in Figures 2 and 3, respectively.

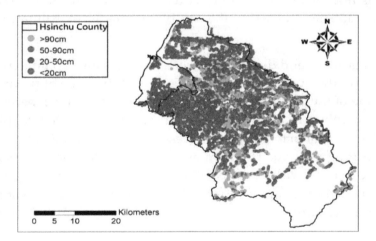

Figure 2. Locations of measured soil depth in Hsinchu County.

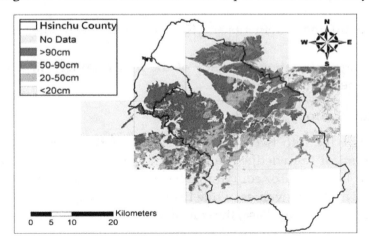

Figure 3. Current soil depth map for Hsinchu County.

2.2. Research Methodology

2.2.1. Kriging Estimation

The Kriging estimation method develops system equations based on characteristics possessed by regionalized variables to estimate position of any point in the space with known regionalized variable $Z(x_i)$ ($i = 1, 2, \ldots, n$) in a random field. For any position with a known relative distance, the Kriging method may be used to estimates its $Z^*(x_0)$. The method first presents the distance function of soil depth measurement (point data) in the semivariogram model and determines the weight coefficient of each point (λ_{0i}) with unbiased optimization from semivariogram. The semivariogram measures things nearby tend to be more similar than things that are farther apart, revealing the strength of statistical correlation as a function of distance. The soil depth then can be estimated using a linear combination of the weight coefficient and the measured data. The semivariance (γ) can be calculated by Equation (1), where h is distance between two separated points, $N(h)$ is the set of all pairwise Euclidean distances $i - j = h$, $|N(h)|$ is the number of pairs in $N(h)$.

$$\gamma(h) = \frac{\sum\limits_{N(h)} \left[z(x_i) - z(x_j)\right]^2}{2|N(h)|} \tag{1}$$

The difference between the simple Kriging (SK) and the ordinary Kriging (OK) is the assumption of stationarity. SK utilizes this assumption while OK does not. The focus of this research is to utilize BME for estimating the soil depth. Thus, whether or not the stationarity has an effect on the accuracy of the Kriging model is not further discussed. Thus, only OK with three different semivariogram models were investigated. The OK used in this study for soil depth estimation is executed in steps described below:

i.   Calculate the semivariance and distance between two measured points (ranging from meters to kilometers), calculate the average of semivariance and distance of points in the district as the representative value of the district, connect the latter to obtain the experimental semivariogram, overlay it with commonly used theoretical semivariogram models (index, spherical, and Gaussian).

ii.   Organize and calculate characteristics of the Best Linear Unbiased Estimator (BLUE) possessed by the ordinary Kriging method. The term of "best" indicates that compared to other unbiased and linear estimators, the lowest variance of the estimate is given

iii.   Acquire data Z_1 of one point out of n-each soil depth measurements, estimate Z_1 with Kriging method based on the remaining $n - 1$ points. Here, n (i.e., 8217) is the data points used in the Kriging model.

iv.   Replace Z_1 with another point Z_2 and repeat the same steps until all of the n-each points are estimated.

v.   Compare the soil depth measurements and Kriging method estimates according to the four soil depth gradings.

vi.   Calculate the average estimation accuracy of the index, spherical, and Gaussian theoretical semivariogram models in each watershed.

2.2.2. Bayesian Maximum Entropy Method (BME)

BME divides knowledge into two categories: General Knowledge Base (G-KB), containing time-space related statistical knowledge such as mean, covariance, and semivariation and the site-specific knowledge base (S-KB), containing hard and soft data types. The hard data type are actual measurements and exact values, while the soft ones are presented by a probability density function (PDF). The entire knowledge base, the comprehensive knowledge G-KB and the specific point knowledge S-KB combined, can be expressed as K = G ∪ S.

BME generally consists of three epistemological stages: a prior stage, a pre-posterior (or meta-prior) stage, and a posterior stage [7,13]. In the prior stage, the joint PDF $f_G(x_{map})$, given general knowledge G, is calculated via the maximum entropy theory. The variable x_{map} is a vector of points, x_{soft}, x_{hard}, and x_k, representing the information of the soft and hard data points and unknown values at the estimation point, respectively. The expected information is expressed as Equation (2):

$$\overline{\text{Info}_G[x_{map}]} = -\int dx_{map} f_G(x_{map}) \log f_G(x_{map}) \tag{2}$$

The general knowledge G in Equation (14) is expressed as $g_\alpha(x_{map})$, a set of functions of x_{map} such as the mean and covariance moments. To obtain the prior PDF of $f_G(x_{map})$, the expectation of Equation (2) is maximized with consideration of $g_\alpha(x_{map})$. Equation (3) is the object function if the Lagrange multipliers method (LMM) is adopted for the aforementioned maximization problem, in which μ_α is the Lagrange multiplier and the $E[g_\alpha(x_{map})]$ is the expected value of $g_\alpha(x_{map})$.

$$\begin{aligned}
\text{Max. } [f_G(x_{map})] &= -\int dx_{map} f_G(x_{map}) \log f_G(x_{map}) \\
&\quad -\sum_\alpha^N \mu_\alpha \left[\int g_\alpha f_G(x_{map}) f_G(x_{map}) dx_{map} - E[g_\alpha(x_{map})] \right]
\end{aligned} \tag{3}$$

At the pre-prior stage, new information, which can be hard data or soft data, is collected for the points to be estimated. The hard data could be actual measurements and the soft data could be in various forms but is not used at the prior stage [7]. At the posterior stage, the posterior PDF $f_K(x_k|x_{data})$ is derived using Bayesian theory, resulting in the following equations:

$$f_K(x_k|x_{data}) = A^{-1} \int dx_{soft} f_S(x_{soft}) f_G(x_{map}) \tag{4}$$

$$A = \int dx_{soft} f_S(x_{soft}) f_G(x_{data}) \tag{5}$$

in which x_{data} is a pointer for a context of knowledge, and $(x_k|x_{data})$ stands for the possible values x_k of the map in the context specified by x_{data}. In this study, the soil depth measurements and the physiographic factor out of 5 m DEM are source of estimates information which can be expressed in formula $S : X_{data} = (X_{hard}, X_{soft}) = (x_1, \ldots, x_n)$ according to information classification by S-KB; here $X_{hard} = (x_1, \ldots, x_{m_h})$ is the soil depth measurement of $P_i (i = 1, 2, \ldots, m_h)$; $X_{soft} = (x_{m_h+1}, \ldots, x_n)$ is the soil depth estimates on point $P_i (i = m_h + 1, \ldots, n)$. by the soil depth estimation model and soil depth relevant physiographic factor. The soil depth can be regarded as a space random field with the soil depth of any point in the field expressed by formula $X_P = X_s$ with $p = (s)$ where s is the space coordinate. This study takes distribution characteristics of soil depth in space into account and express soil depth of every point in space with f_{KB}, the PDF; where KB is the knowledge base (KB) used when constructing this PDF.

(1) Operation process

The input data of BME may contain hard data and soft data. The hard ones are soil depth measurements, while the soft ones are soil depth estimated from the built prediction model, such as Kriging with input of the soil depth measurement and their relevant physiographic factor on the DEM. After all estimates are included, set soft data to low frequency (data trend) and deduct it from the hard data to obtain high frequency (data residual); consider the covariance distribution of data residual as the input of the BME method to estimate soil depth residual difference of unknown points; combine the latter with the data trend established earlier to build the final distribution of soil depth in space.

(2) Soft data

This study employs the soil-drilling data of the Hsinchu district provided by the Agricultural Research Institute and relevant physiographic factors geologic factors as the input data for estimating effective soil depth and create low frequency data trend with AI model LSSVM (Least Squares Support Vector Machine) and the nonlinear model SVR (Support Vector Regression). The physiographic factors used include slope, aspect, profile curvature, plan curvature, and topographic wetness index. This study employs 80% of the soil drilling data to train the model before using it to estimate soil depth of the remaining 20% of the drilling points, and compares the estimates and actual soil depth measurements to assess both methods (LSSVM and SVR) to build up the estimation model for soil depth soft data.

i.   Pparation of Physiographic Factor

This study employs the DEM at 5 m resolution along with the physiographic factor adopted by Kuriakose et al. [4] to create physiographic factors by GIS software ArcGIS10.1, including slope, aspect, profile curvature, plan curvature, and topographic wetness index (TWI), in the Hsinchu district for soil depth estimation in future. Note that TWI is a function of both the slope and the upstream contributing area per unit width orthogonal to the flow direction. Table 2 displays the outcome of and correlation between physiographic factor preparation.

Table 2. Correlation between various factors and soil depth.

Physiographic Factor	Correlation with Soil Depth	Distribution Diagram
Slope	Steep slopes tend to have thinner soils which are hard to retain, while gradual slopes may have more pedogenesis and pileup	
Aspect	Different aspects may feature large differences in solar radiation, water evaporation, and soil moisture. For example, a sunny slope may have greater solar radiation and water evaporation and thinner soil due to more erosion by wind and rainfall than the sunless side	
Plan curvature	Perpendicular to the direction of the maximum slope which is related to the convergence and dispersion of water flowing through the surface	

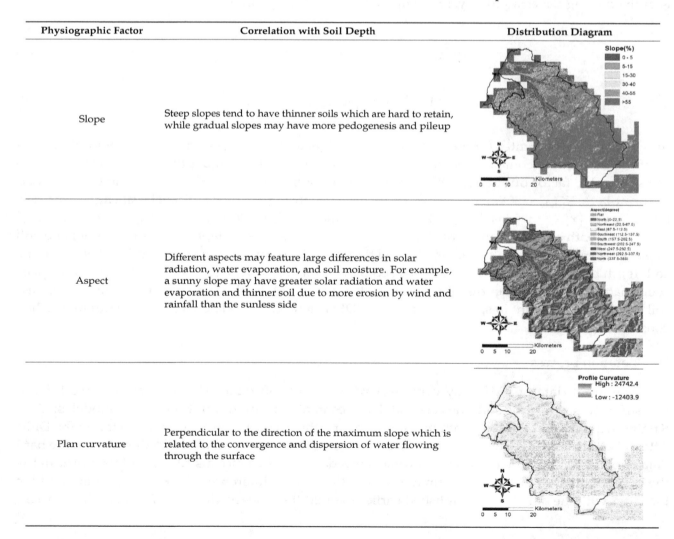

Table 2. *Cont.*

Physiographic Factor	Correlation with Soil Depth	Distribution Diagram
Profile curvature	Parallel to the direction of the maximum slope, that is related to acceleration and deceleration of the water flowing through ground surface and also erosion and accumulation of soil on the slope	
Topographic wetness index	A function of both the slope and the upstream contributing area per unit width orthogonal to the flow direction	

ii. LSSVM

Support Vector Machine (SVM) is a method for classification or regression. That is, one can train an SVM with a group of classified data and use it to estimate type of a piece of data in an unclassified group. Quadratic Programming (QP) usually adopted in solving the above optimization problem suffers from complex calculation as the restriction of SVM is an inequality. The LSSVM is an improved version of SVM with simpler and faster calculation. LSSVM was developed by Suykens, et al. [14] who introduced the concept of Least Squares Loss Function in SVM. A standard SVM, as described in Equation (6), solves a nonlinear classification problem by means of convex quadratic programs (QP).

$$\underset{w,b,\xi}{\text{minimize}}\ \tfrac{1}{2}w^T w + c \sum_{k=1}^{N} \xi_k$$
$$\text{Subject to} \begin{cases} y_k(w^T K(x_i) + b) \geq 1 - \xi_k \\ \xi_k \geq 0,\ i = 1, 2, \ldots, N \end{cases} \tag{6}$$

where w is a normal vector to the hyper-plane; c is a real positive constant; and ξ_k is the slack variable. If $\xi_k > 1$, the k-th inequality becomes violated compared to the inequality from the linearly separable case. y_k is the class; $[w^T K(x_i) + b]$ is the classifier; N is the number of data; and K is the kernel function. In the current study, the Gaussian radial basis function (RBF) kernel is used, as shown in Equation (7):

$$K(X, X_i) = e^{-\sigma(\|X - X_i\|)^2} \tag{7}$$

where X is the input vector, σ is the kernel function parameter; and X_i are the support vectors. LSSVM [14], instead of solving the QP problem, solves a set of linear equations by modifying the standard SVM, as described in Equation (8):

$$\min \tfrac{1}{2}w^T w + \tfrac{\gamma}{2} \sum_{k=1}^{N} e_k^2$$
$$\text{s.t. } y_k(w \cdot K(x_k) + b) = 1 - e_k,\ k = 1, \ldots, n \tag{8}$$

where γ is a constant number and e is the error variable. Compared to the standard SVM, there are two modifications leading to solving a set of linear equations. First, instead of inequality constraints, the LSSVM uses equality constraints. Second, the error variable is a squared loss function.

iii.   SVR

SVR differs from SVM in that SVM aims to find a plane to divide the data into two, while SVR is a plane accurately predicting data distribution. The main concept is to give a fault tolerance upper limit to execute regression analysis and leave a deterministic range of fault tolerance for errors of each data point. This prevents the regression model from over-fitting. With a similar mathematical model of SVM, the linear model of SVR may be converted into non-linear one with the kernel trick, which, in turn, can process more complex data with even better prediction results.

iv.   Model establishment and evaluation

The accuracy of these two soil depth estimation models (LSSVM and SVR) is determined by comparing actual soil depth data. Acquire 80% of physiographic factor and soil depth data of the Hsinchu district to train both models, the remaining 20% soil depth data is used to assess accuracy of models, in which 10-fold cross validation is adopted.

In addition to the LSSVM and SVR models, the K Nearest Neighbor (KNN) algorithm is also considered. KNN estimates the target soil depth by averaging 10 nearest soil depth values. This study employs three indices, R^2, MAPE, and hitting rate of the current soil depth grading, to assess accuracy of estimation by each model. An estimate is defined as "targeted" if both estimate and actual soil depth fall in the same soil depth grade. The formulae for the three indices are shown below:

$$R^2 1 - \frac{\sum (y_i - \hat{y})^2}{\sum (y_i - \overline{y})^2} \tag{9}$$

$$MAPE = \frac{\sum |y_i - \hat{y}|}{y_i} \times \frac{100}{N} \ (\%) \tag{10}$$

$$\text{Hitting rate}(\%) = \frac{m}{N} \tag{11}$$

where y_i is the actual soil depth, $\hat{y}$ is the soil depth estimates, $\overline{y}$ is the average actual soil depth, N is the number of data entries, and m is the number of actual and estimate soil depth pairs that fall in the same grade.

(3) Nested spatiotemporal covariance model

This study correlates residual data with the covariance model to estimate the high frequency data. The covariance model $C_{st}(h, \tau)$ is shown in Equation (12).

$$C_{st}(h, \tau) = \sum_{l=1}^{N} b_l C_{sl}(h; A_{sl}) C_{tl}(\tau; A_{tl}) \tag{12}$$

where s is space, t is time, h is spatial distance, τ is time distance, N is nest count, b_l denotes threshold value (sill), and $C_{sl}(h; A_{sl})$ is a space covariance model alone, $C_{tl}(\tau; A_{tl})$ is a time covariance model alone. Gaussian and Exponential are two commonly available patterns for covariance model, A_{sl} and A_{tl} are parameters required by C_{sl} and C_{tl}, i.e., the range of variances.

The nested covariance model is then used to determine model-data fitness based on the Akaike information criterion (AIC) as shown in Equation (13).

$$AIC = 2k + n ln\left(\frac{RSS}{n}\right) \tag{13}$$

where k is the number of parameters used, n is the number of measurements, and RSS is the residual sum of squares, and the smaller the AIC, the closer to the target.

2.2.3. Method of Accuracy Calculation

This study employs a confusion matrix to assess the accuracy of the Kriging method and BME. The estimates and actual measurement data are classified before inserting their counts in the confusion matrix as indicated in Table 3; soil depth grades are shown in rows of the confusion matrix, while actual soil depth measurement is shown in the columns. For example, in case soil depth estimation of a point is "very shallow" while the actual soil depth in grade "shallow" then insert it in cell (2) of Table 3 and counts this prediction to "invalid estimation". For another point with correct estimation for grade "deep", then add the value in cell (14) of Table 3, that is, this prediction is considered as a "valid estimation". Table 3 suggests the number of valid estimation points is in cell (1), (6), (11), and (16) while all the remaining cells are invalid estimations. If the total number of points is n, then the accuracy rate is expressed by Equation (14):

$$\text{Accuracy rate} = \frac{n_{(1)} + n_{(6)} + n_{(11)} + n_{(16)}}{n} \times 100\% \tag{14}$$

Table 3. Soil depth estimation confusion matrix (bold number: correct prediction).

		Soil Depth Measured (cm)			
		<20	20–50	50–90	>90
	<20	**(1)**	(2)	(3)	(4)
Soil Depth	20–50	(5)	**(6)**	(7)	(8)
Estimated (cm)	50–90	(9)	(10)	**(11)**	(12)
	>90	(13)	(14)	(15)	**(16)**

3. Results and Discussion

3.1. Existing Soil Map in Hsinchu District

The accuracy of the current slopeland soil map, provided by the Soil and Water Conservation Bureau, is estimated by comparing the measured results of soil drilling. The number of soil drilling measurements is 8217, while only 4768 points are available on slopeland soil map. Table 4 suggests only 1354 points have the same grades by measurement and mapping, i.e., the accuracy rate of the map is only 28.4%. Two causes may have contributed to this low rate. One is that the scale of slopeland soil map is 1/25,000 which is too coarse to reflect space distribution of soil depth; the other is that the map is not up-to-date and failed to consider the impact of relevant physiographic factors on soil depth. Note that although the total data number used here are different from those of Kriging and BME, the average accurate rate is used here and this should be a fair comparison since, for any case, the data number used is more than 4000 points. As shown in Figure 2, it is seen that points categorized into the same grade level often possess a closer distance. That is, although the comparison is performed within the same category, the distance effect is implicitly considered.

Table 4. Soil depth grading by measurements and slopeland soil map (bold number: correct prediction).

		Soil Depth Measured (cm)			
		<20	20–50	50–90	>90
	<20	**586**	199	111	11
Soil Depth Estimated by Soil	20–50	500	**233**	183	16
Depth Map (cm)	50–90	808	473	**461**	91
	>90	426	276	320	**74**
		Accuracy = 28.4%			

3.2. Kriging Method

Table 5 displays estimates of the three semivariogram models, Exponential model, Spherical, and Gaussian. All have similar accuracy rates, around 40%, with the exponential Kriging method being a little better at 40.4%. Figure 4 displays soil depth estimates by the exponential Kriging model. Detailed results of the exponential Kriging model are displayed in Figure 5, suggesting that overshoot or undershoot are more likely to happen in the "very shallow" and "very deep" grades than in grade "shallow" and "deep". The accuracy rate may be much higher, at 56.95% and 57.28%, respectively. The estimation of soil depth scores was better for grades "shallow" and "deep" and worse for "very shallow" and "very deep". The poorer estimates may have been caused by the estimation theory employed by the Kriging method. It is based on the characteristics of distribution of data in space without considering the impact of relevant physiographic factors. The semivariogram model-based estimates shown in Figure 6 do not indicate significant relation between soil depth data and space distribution, i.e., increasing distance does not lead to significant changes in semi-variance. Note that in Figure 6, the values of sill and nugget are 3.162×10^{-3} and 1.186×10^{-3}, respectively.

Table 5. Estimates by the Kriging method on three semivariogram models (bold number: correct prediction).

			Soil Depth Measured (cm)			
			<20	20–50	50–90	>90
		<20	**280**	54	1	0
	Exponential	20–50	1450	**1807**	469	174
	model	50–90	481	1151	**811**	789
		>90	42	143	143	**422**
			Accuracy = 40.4%			
		<20	**273**	49	1	
Soil Depth Estimated	Spherical	20–50	1449	**1804**	477	171
by the Kriging (cm)	model	50–90	490	1162	**804**	805
		>90	41	140	142	**409**
			Accuracy = 40.04%			
		<20	**270**	50	1	
	Gaussian	20–50	1451	**1792**	485	177
	model	50–90	493	1176	**796**	801
		>90	39	137	142	**407**
			Accuracy = 39.74%			

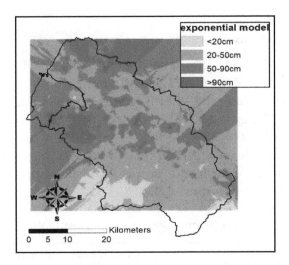

Figure 4. Soil depth estimates by the exponential Kriging model.

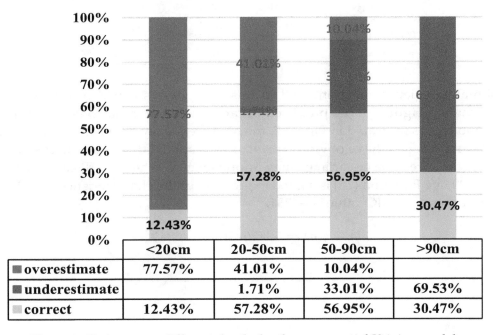

	<20cm	20-50cm	50-90cm	>90cm
■ overestimate	77.57%	41.01%	10.04%	
■ underestimate		1.71%	33.01%	69.53%
correct	12.43%	57.28%	56.95%	30.47%

Figure 5. Estimates at different depths by the exponential Kriging model.

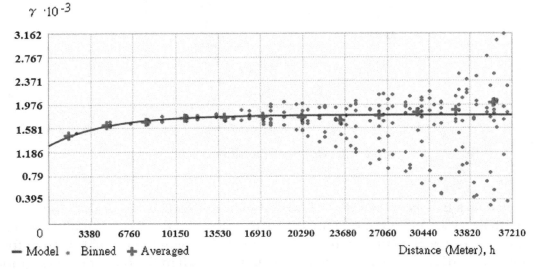

Figure 6. Analysis results of the exponential Kriging model semi-variogram.

3.3. BME

3.3.1. Soft Data and Covariance Model

Table 6 shows the results of R^2, mean absolute percentage error (MAPE), and estimated hitting rate by the K Nearest Neighbor (KNN) algorithm, LSSVM, nonlinear mode SVR, and their combinations. In spite of having a better hitting rate, the MAPE values of KNN suffer larger differences between measured and estimated data. Thus, the KNN results are considered as a factor in LSSVM and SVR, that is, the LSSVM+KNN and SVR+KNN models. As indicated in Table 6, the SVR+KNN model delivers the best hitting rate and least MAPE value. Thus, SVR+KNN is selected as the machine learning model for low frequency data in the proposed BME approach. That is, the outcomes of SVR+KNN is one of the inputs of the proposed BME.

Table 6. Accuracy of individual models.

Model	R^2	MAPE (%)	Hitting Rate
LSSVM	0.12	98.5	0.30
SVR	0.01	70.6	0.37
KNN	0.30	74.9	0.43
LSSVM+KNN	0.11	94.8	0.31
SVR+KNN	0.27	63.7	0.44

The residual covariance function used in this study is displayed in Equation (12), with its distribution shown in Figure 7. The proposed covariance function has one nest, exponential type, and is a space model. As indicated in Figure 7, the high-frequency data (soil depth residual) derived from hard data and soft data have obvious correlation in spatial distribution. The closer the points, the larger the covariance. The covariance may approach zero when the distance between points is extended to 12,000 m^2. Equation (15) is a good expression of the characteristics of soil depth residual in spatial distribution as its AIC value is −5731.

$$C_S(h) = 6.1\text{ExpC}(h, 9808.84) \tag{15}$$

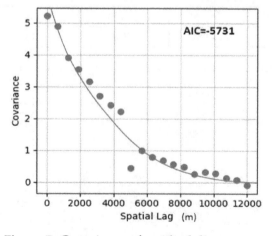

Figure 7. Covariance of residual discrepancy.

3.3.2. Estimates

Table 7 displays the accuracy rate of BME estimation. As shown, BME delivers a promising estimation with 82.94% accuracy rate. Differences between soil depth by actual measurement and estimation are two or fewer grades apart. There is no case where the estimates fall in the range of very shallow or very deep while their actual measurements fall in the other opposite grades. Cases of "two-grade" differences are few, 19 out of 8217. The BME method tends to overshoot as there are 1114 positions that suffer higher estimation, while only 288 ones end up with lower estimation. Most overshoot estimates are in the very shallow grade and the undershoot is in shallow grade. This is not the case in terms of percentage: rate of undershooting is highest in the deep grade. As shown in Figure 8, estimates for soil depth in very shallow grade has the most error, shallow and deep grade are about the same accuracy rate, and the very deep grade has makes the fewest error. That is, deeper soils give better estimates.

Table 7. Estimation outcomes of the Bayesian maximum entropy (BME) model (bold number: correct prediction).

		Soil Depth Measured (cm)			
		<20	20–50	50–90	>90
Soil Depth Estimated by BME (cm)	<20	**1632**	143	11	
	20–50	618	**2651**	80	3
	50–90	3	359	**1201**	51
	>90		2	132	**1331**
		Accuracy = 82.94%			

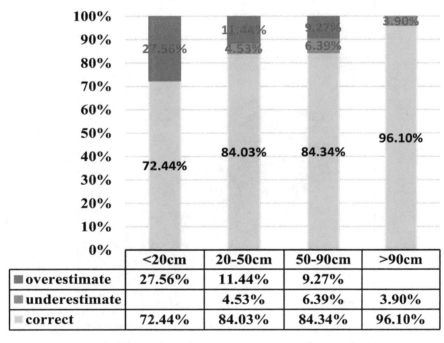

	<20cm	20-50cm	50-90cm	>90cm
overestimate	27.56%	11.44%	9.27%	
underestimate		4.53%	6.39%	3.90%
correct	72.44%	84.03%	84.34%	96.10%

Figure 8. Estimates at different depths by the BME model.

The BME method and physiographic factor data of the target points combined may be used to estimate the effective soil depth estimation and the variation and range of the error. At the 95% confidence level, the error range is displayed in Equation (16):

$$d\mu \pm 1.96 \sqrt{Var} \qquad (16)$$

where d is the error range of soil depth for a given grid point, μ is the mean value that is represented by the estimated soil depth for a given grid point, and Var is the variance for a given point. Distribution of the error range of grid points in the Hsinchu district is shown in Figure 9, which suggests that the greater the distances from the estimate point to the actual measurement, the greater the error range will be. As the maximum error range of soil depth estimated by BME in Hsinchu district is a mere 5.62 cm, it is acceptable for existing regulations in terms of soil depth grading. Figure 10 is the soil depth distribution of 5 m grid beads on the proposed BME. With a satisfactory accuracy rate, this map may serve as the reference for soil depth data acquisition in Hsinchu, Taiwan in the future.

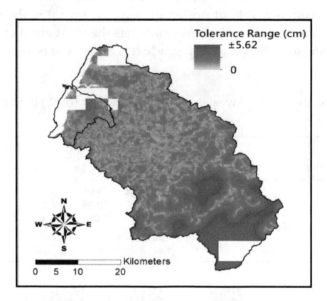

Figure 9. Distribution of soil depth estimation error by BME.

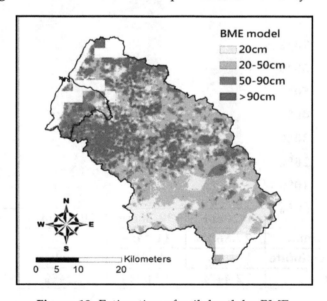

Figure 10. Estimation of soil depth by BME.

3.4. Comparison of Estimation Models

With an accuracy rate up to 82.94%, the BME method outperforms the Kriging method in soil depth estimation. In spite of taking data space distribution into account, the Kriging method failed to contain multiple physiographic factors into the analysis. In addition, the Kriging method assumes the actual data follow normal distribution. In the case of non-normal or non-linear space distribution pattern of actual data, it suffers estimation error. The BME method can contain normal and non-normal data and combine hard data and soft data (the soil depth estimation model based on physiographic factor) to obtain estimation. The proposed BME method was applied to the Hsinchu district. Based on the obtained promising results, BME is reported as a useful method that can show the depth of the soil in spatial distribution.

4. Conclusions

An appropriate soil depth estimation model is necessary in Taiwan for two reasons: existing soil depth distribution maps are out of date, while the supplemental measurement data available now fails to cover the entire country. This study establishes a soil depth estimation model based on the BME method with input of actual soil depth measurement, physiographic factors, and the selected machine learning model. The BME model was then compared with the traditional Kriging method. The results and conclusions of this study are as follows:

1. The Kriging method is commonly used for space estimation. It suffers in two aspects: first, the statistical assumption of normally distributed estimation data may be not the case of actual soil depth data; secondly, the Kriging method focuses on space distribution characteristics and does not take physiographic factors into account. In spite of being better than existing soil depth distribution diagrams available now in Taiwan, its accuracy rate is a mere 40%.

2. The BME method incorporates both hard data of soil depth measurement and physiographic factor (soft data)-based soil depth estimations to take both natural environmental factors and space distribution characteristics into account at the same time. The BME method not only performs without the normal distribution assumption, but also comes out with much better estimation (80%) than that of the Kriging method.

3. The soil depth distribution map of Hsinchu district in Taiwan produced by the BME method in this study gives soil depth estimation at grid points with an error range of a mere ±5.62 cm. This is acceptable for current soil depth grading standards and may be adopted for districts without soil depth data. This is of great help for disaster prevention and land management of slopeland in Taiwan.

Author Contributions: Conceptualization, K.-W.L. and J.-C.F.; methodology, K.-W.L. and J.-J.G.; software, J.-J.G. and S.-H.C.; validation, K.-W.L., J.-C.F. and C.L.H.; formal analysis, J.-J.G. and S.-H.C.; investigation, K.-W.L. and J.-J.G.; resources, C.L.H.; data curation, C.L.H.; writing—original draft preparation, K.-W.L. and J.-J.G.; writing—review and editing, K.-W.L.; visualization, J.-J.G.

Acknowledgments: Authors would like to thank the Soil and Water Conservation Bureau, Council of Agriculture, Executive Yuan of Taiwan for their support that enabled the smooth completion of this study.

References

1. Krezoner, W.R.; Olson, K.R.; Banwart, W.L.; Johnson, D.L. Soil, landscape, and erosion relationships in northwest Illinois watershed. *Soil Sci. Soc. Am. J.* **1989**, *53*, 1763–1771. [CrossRef]
2. Moore, I.D.; Burch, J.R. Sediment transport capacity of sheet and rill flow: Application of unit stream power theory. *Water Resour. Res.* **1986**, *22*, 1350–1360. [CrossRef]
3. Bourennane, H.; King, D.; Couturier, A. Comparison of kriging with external drift and simple linear regression for prediction soil horizon thickness with different sample densities. *Geoderma* **2000**, *97*, 255–271. [CrossRef]
4. Kuriakose, S.L.; Sankar, G.; Muraleedharan, C. History of landslide susceptibility and a chorology of landslide-prone areas in the Western Ghats of Kerala, India. *Environ. Geol.* **2009**, *57*, 1553–1568. [CrossRef]
5. Alizadeh, Z.; Mahjouri, N. A spatiotemporal Bayesian maximum entropy-based methodology for dealing with sparse data in revising groundwater quality monitoring networks: The Tehran region experience. *Environ. Earth Sci.* **2017**, *76*, 436. [CrossRef]
6. Christakos, G. Some Applications of the Bayesian, Maximum-Entropy Concept in Geostatistics. *Maximum Entropy Bayesian Methods* **1991**, *43*, 215–229.
7. Christakos, G.; Li, X.Y. Bayesian maximum entropy analysis and mapping: A farewell to kriging estimators? *Mathernatical Geol.* **1998**, *30*, 435–462. [CrossRef]

8.   Christakos, G.; Serre, M.L. BME analysis of spatiotemporal particulate matter distributions in North Carolina. *Atmos. Environ.* **2000**, *34*, 3393–3406. [CrossRef]

9.   Hajji, I.; Nadeau, D.F.; Music, B.; Anctil, F.; Wang, J. Application of the maximum entropy production model of evapotranspiration over partially vegetated water-limited land surfaces. *J. Hydrometeorol.* **2018**. [CrossRef]

10.   Yu, H.L.; Lee, C.H.; Chien, L.C. A spatiotemporal dengue fever early warning model accounting for nonlinear associations with hydrological factors: A Bayesian maximum entropy approach. *Stoch. Environ. Res. Risk Assess.* **2016**, *30*, 2127–2141. [CrossRef]

11.   Bogaert, P.; D'Or, D. Estimating soil properties from thematic soil maps: The Bayesian maximum entropy approach. *Soil Sci. Soc. Am. J.* **2002**, *66*, 1492–1500. [CrossRef]

12.   D'Or, D.; Bogaert, P.; Christakos, G. Application of the BME approach to soil texture mapping. *Stoch. Environ. Res. Risk Assess.* **2001**, *15*, 87–100. [CrossRef]

13.   Gao, S.; Zhu, Z.; Liu, S.; Jin, R.; Yang, G.; Tan, L. Estimating the spatial distribution of soil moisture based on Bayesian maximum entropy method with auxiliary data from remote sensing. *Int. J. Appl. Earth Obs. Geoinf.* **2014**, *32*, 54–66. [CrossRef]

14.   Suykens, J.A.K.; Vandewalle, J. Least squares support vector machine classifiers. *Neural Process. Lett.* **1999**, *9*, 293–300. [CrossRef]

3

Generalized Beta Distribution of the Second Kind for Flood Frequency Analysis

Lu Chen [1,*] **and Vijay P. Singh** [2]

[1]   College of Hydropower & Information Engineering, Huazhong University of Science & Technology, Wuhan 430074, China

[2]   Department of Biological and Agricultural Engineering & Zachry Department of Civil Engineering, Texas A&M University, College Station, TX 77843, USA; vsingh@tamu.edu

*   Correspondence: chen_lu@hust.edu.cn

Academic Editor: Kevin H. Knuth

Abstract: Estimation of flood magnitude for a given recurrence interval T (T-year flood) at a specific location is needed for design of hydraulic and civil infrastructure facilities. A key step in the estimation or flood frequency analysis (FFA) is the selection of a suitable distribution. More than one distribution is often found to be adequate for FFA on a given watershed and choosing the best one is often less than objective. In this study, the generalized beta distribution of the second kind (GB2) was introduced for FFA. The principle of maximum entropy (POME) method was proposed to estimate the GB2 parameters. The performance of GB2 distribution was evaluated using flood data from gauging stations on the Colorado River, USA. Frequency estimates from the GB2 distribution were also compared with those of commonly used distributions. Also, the evolution of frequency distribution along the stream from upstream to downstream was investigated. It concludes that the GB2 is appealing for FFA, since it has four parameters and includes some well-known distributions. Results of case study demonstrate that the parameters estimated by POME method are found reasonable. According to the RMSD and AIC values, the performance of the GB2 distribution is better than that of the widely used distributions in hydrology. When using different distributions for FFA, significant different design flood values are obtained. For a given return period, the design flood value of the downstream gauging stations is larger than that of the upstream gauging station. In addition, there is an evolution of distribution. Along the Yampa River, the distribution for FFA changes from the four-parameter GB2 distribution to the three-parameter Burr XII distribution.

Keywords: entropy theory; principle of maximum entropy (POME); GB2 distribution; flood frequency analysis

1. Introduction

Estimation of flood magnitude for a given recurrence interval T (T-year flood) at a given location is essential for the design of hydraulic and civil infrastructure facilities, such as dams, spillways, levees, urban drainage, culverts, road embankments, and parking lots. A key step in flood frequency estimation or analysis (FFA) is the selection of a suitable frequency distribution [1]. Commonly used distributions for flood frequency analysis include Gumbel, gamma, generalized extreme value (GEV), Pearson type III (P-III), log-Pearson type III (LP-III), Weibull, and log-normal (LN). Some of these distributions have been adopted in different countries. For example, the P-III distribution has been adopted in China and Australia as a standard method for hydrologic frequency analysis [2–4]. The LP-III distribution has been adopted in the United States and the GEV distribution in Europe.

Mielke and Johnson investigated the use of two special cases of the generalized beta distribution of the second kind, namely gamma and log normal distributions, for flood frequency analysis [5].

Wilks investigated the performance of eight three-parameter probability distributions for precipitation extremes using annual and partial duration data from stations in the northeastern and southeastern United States [6]. He found that the beta-κ distribution best described the extreme right tail of annual extreme series, and the beta-P distribution was best for the partial duration data.

Recently, some generalized frequency distributions have been used for hydrologic frequency analysis. For example, Perreault et al. presented a family of distributions, named Halphen distributions, for frequency analysis of hydrometeorological extremes [7]. Papalexiou and Koutsoyiannis used the generalized gamma distribution and generalized beta distribution of the second kind (GB2) for rainfall frequency analysis across the world and showed that these distributions were appropriate for worldwide rainfall data [8]. The greatest advantage of these generalized distributions is that they provide sufficient flexibility to fit a large variety of data sets, which facilitates the selection and comparison of different distributions. For instance, the GB2 distribution includes the exponential, Weibull, and gamma distributions as special cases. Since the GB2 distribution has four parameters, logically it should perform better than 3-parameter distributions, such as GEV, P-III, LP-III or LN-III. Papalexiou and Koutsoyiannis concluded that the GB2 distribution was a suitable model for rainfall frequency analysis because of its ability to describe both J-shaped and bell-shaped data [8]. The other advantages of the GB2 distribution can be summarized as: (1) the GB2 distribution can model positive or negative skewness which is an advantage over distributions, such as lognormal, with only positive skew; (2) it can jointly estimate both location and shape parameters, while many other distributions, such as exponential, logistic, normal, etc., usually focus on location only; and (3) it can better capture the long right or left tail. Because of these advantages, the GB2 distribution was employed in this study.

The second step in flood frequency analysis is to estimate parameters of the selected distribution. There are several standard parameter estimation methods, such as moments, maximum likelihood, L-moments, probability weighted moments, and least square. Among these methods, the maximum likelihood (ML) and L-moment methods are widely used in hydrology. In addition, the principle of maximum entropy (POME) has been applied to parameter estimation [9,10]. Singh and Guo indicated that POME method was comparable to ML and L-moment methods, and for certain situations, POME method was superior to these two methods [11]. Therefore, the POME method was considered in this study for parameter estimation.

Another aspect of FFA that is of interest is how the flood frequency distribution evolves from upstream to downstream along a river. The drainage area along the river increases from upstream to downstream. It is interesting to investigate if the same frequency distribution applies at all gauging stations along the stream.

The objective of this study therefore is to employ the GB2 distribution for flood frequency analysis (FFA). The specific objectives are to: (1) estimate the GB2 distribution parameters using the principle of maximum entropy; (2) evaluate the performance of the GB2 distribution and compare it with commonly used distributions in hydrology; (3) select the best distribution; and (4) discuss the evolution of frequency distribution and its parameters along the river.

2. GB2 Distribution

The generalized beta distribution of the second kind, denoted as GB2, is a four-parameter distribution and can be expressed as:

$$f(x) = \frac{r_3}{\beta B(r_1, r_2)} \left(\frac{x}{\beta}\right)^{r_1 r_3 - 1} \left(1 + \left(\frac{x}{\beta}\right)^{r_3}\right)^{-(r_2 + r_1)} \tag{1}$$

where $B(\cdot)$ is the beta function; β is the scale parameter, $\beta > 0$; and $r_1 > 0$, $r_2 > 0$, and $r_3 > 0$ are the shape parameters. Parameter r_3 represents the overall shape; parameter r_1 governs the left tail; parameter r_2 controls the right tail; and β is a scale parameter and depends on the unit of measurement. These parameters allow the distribution to be able to fit data having very different histogram shapes. It can simulate both the J-shaped and bell-shaped distributions. Parameters r_1 and

r_2 together determine the skewness of the distribution. The general shapes of GB2 probability density distribution were shown in Figure 1.

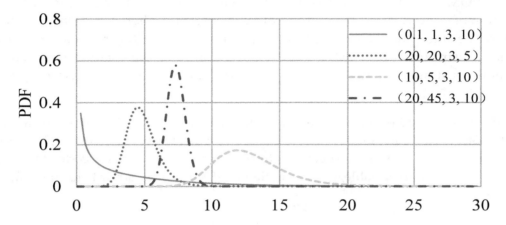

Figure 1. Shapes of PDF of GB2 distribution.

When analyzing extreme rainfall, Papalexiou and Koutsoyiannis showed that the GB2 distribution is a very flexible four-parameter distribution [8]. By fixing certain parameters, the GB2 distribution can yield some well-known distributions, such as the beta distribution of the second kind (B2), the Burr type XII, generalized gamma (GG), and so on. These distributions can be treated as special or limiting cases of the GB2 distribution, as shown in Figure 2. Some of these special cases have been applied in hydrological frequency analysis. For example, Shao et al. employed the Burr type XII distribution for flood frequency analysis [2].

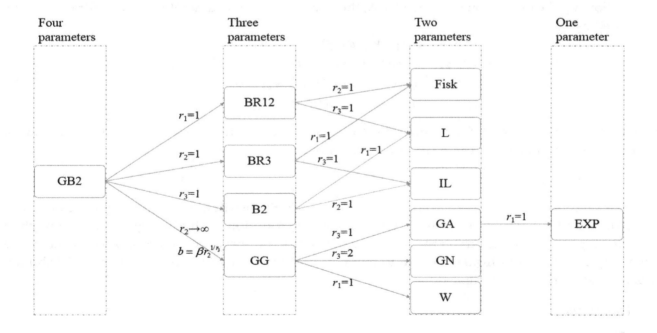

Figure 2. The GB2 distribution and its special cases (where BR12 means the Burr XII distribution; BR3 means the Burr III distribution; B2 means the beta distribution of second kind; Fisk means log-logistic distribution; L means the Lomax distribution; IL means inverse Lomax distribution; GA distribution means the gamma distribution; GN means the generalized normal distribution; W means the Weibull distribution and EXP means the exponential distribution).

3. Estimation of Parameters of GB2 Distribution by POME Method

The GB2 distribution parameters were determined using the principle of maximum entropy (POME). The POME method involves the following steps: (1) specification of constraints; (2) maximization of entropy using the method of Lagrange multipliers; (3) derivation of the relation between Lagrange multipliers and constraints; (4) derivation of the relation between Lagrange multipliers and distribution parameters; and (5) derivation of the relation between distribution parameters and constraints. These steps are discussed in Appendix A. Here only steps (1) and (5) are outlined.

Flood discharge is considered as a random variable X, which ranges from 0 to infinite. Its probability distribution function (PDF) and cumulative distribution function (CDF) are denoted as $f(x)$ and $F(x)$ respectively, where x is a specific value of X. Since constraints encode the information that can be given for the random variable, following Singh (1998), the constraints for the GB2 distribution can be expressed as:

$$\int_0^\infty f(x)\mathrm{d}x = 1 \tag{2a}$$

$$\int_0^\infty f(x)\ln x \mathrm{d}x = E(\ln x) \tag{2b}$$

$$\int_0^\infty f(x)\ln(1 + (\frac{x}{\beta})^{r_3})\mathrm{d}x = E(\ln(1 + (\frac{x}{\beta})^{r_3})) \tag{2c}$$

The first constraint is the total probability law, the second constraint is the mean of log values or the geometric mean, and the third constraint is the mean of log of scaled values raised to a power and then shifted by unity.

Following the derivation in Appendix A, the relation between parameters and constraints can be expressed as:

$$
\begin{aligned}
-\ln\beta - \frac{1}{r_3}\varphi(r_1) + \frac{1}{r_3}\varphi(r_2) &= -E(\ln x)\\
\beta^{r_3}\varphi(r_2) - \beta^{r_3}\varphi(r_1 + r_2) &= -E(\beta^{r_3}\ln(1 + (\frac{x}{\beta})^{r_3}))\\
-\ln\beta + \frac{1}{r_3^2}\varphi'(r_1) + \frac{1}{r_3^2}\varphi'(r_2) &= \mathrm{var}(\ln x)\\
\varphi'(r_2) - \varphi'(r_1 + r_2) &= \mathrm{var}(\ln(1 + (\frac{x}{\beta})^{r_3}))
\end{aligned}
\tag{3}
$$

where $\phi(.)$ is the digamma function; and $\phi'(.)$ is the trigamma function. Detailed information for deriving these relationships can be found in Appendix A.

4. Flood Frequency Analysis

For FFA, three problems were addressed. First, the GB2 distribution was tested using observed flood data, and was compared with commonly used distributions in hydrology. Second, a method for selecting the best distribution was discussed. Third, flood frequency analysis was carried out at several gauging stations from upstream to downstream, and the evolution of frequency distribution along the stream was investigated.

4.1. Flood Data

Flood data from eight gauging stations on the Colorado River and its tributaries, as shown in Figure 3, were considered to test the performance of the GB2 distribution and discuss the evolution of frequency distribution along the river. The Colorado River is the principal river of the Southwestern United States and northwest Mexico. It rises in the central Rocky Mountains, flows generally southwest across the Colorado Plateau and through the Grand Canyon. The basin boundary consists of mountains that are 13,000 to 14,000 feet (3962.4 m to 4267.2 m) high in Wyoming, Colorado, and Utah; and the boundary drops to elevations of less than 1000 feet (304.8 m) at Hoover Dam. The northern part of the river basin in Colorado and Wyoming is a mountainous plateau that ranges from 5000 to 8000 feet (1524 m to 2438 m) in elevation, which encompasses deep canyons, rolling valleys, and intersecting mountain ranges. The central and southern portions of the basin in eastern Utah, northwestern New Mexico, and northern Arizona consist of rugged mountain ranges interspersed with rolling plateaus and broad valleys. In general, the mountains in the southern part of the basin are much lower than those in the northern part. Of the eight gauging stations considered in this study, gauging stations or sites 1, 2 and 3 are on the Yampa River which is a secondary tributary of the Colorado River. Sites 4, 5, 6, 7 and 8 are on the mainstream of the Colorado River. Site 8 is near the location of the Hoover Dam. The data of these gauging stations is directly downloaded from USGS (United States Geological Survey) website. The characteristics of flow data of these gauging stations, including length of the data, mean, standard deviation, skewness, and kurtosis, were calculated, as shown in Table 1. Since there is a dam, named Glenn Canyon, regulating the river flow past Lees Ferry (shown in Figure 3), the characteristics of the flow at the Hoover dam (site 8) are quite different from those at sites 4, 5, 6 and 7 upstream. It can be seen from Table 1 that for sites 1 to 7 the mean values increase from upstream to downstream, as more rainfall or water flows into the river. Since the standard deviation is related to the flood magnitude, it also increases with the mean value. For site 8, considering the impact of reservoir operation, some streamflow was stored in the reservoir, which leads that the streamflow at site 8 is reduced. The skewness is positive for all gauging stations, indicating that the right tail is longer or fatter than the left side and the mass of distribution is concentrated on the left side. Kurtosis is a measure of the peakedness of the probability distribution. The skewness and kurtosis values in the mainstream are generally lower than those in the tributaries.

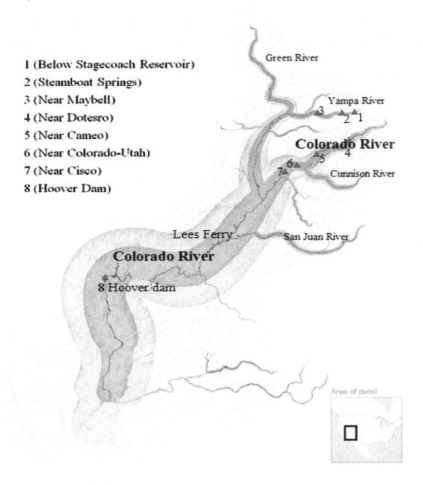

Figure 3. Locations of gauging stations on the Colorado River.

Table 1. Characteristics of the gauging stations used in the study.

River	No.	Gaging Station	Drainage Area (Square Miles)	Length of Data	Mean Value (ft³/s)	Standard Deviation	Skewness	Kurtosis
Yampa River	1	Below Stagecoach Reservoir	228	1957–2014	315	189	0.91	3.49
	2	Steamboat Springs	567	1904–2013	3630	1115	0.26	2.94
	3	Near Maybell	3383	1904–2013	10,419	3657	0.90	4.88
Colorado River	4	Near Dotsero	4390	1941–2013	9870	4450	0.39	2.59
	5	Near Cameo	7986	1934–2013	19,049	7687	0.26	2.68
	6	Near Colorado-Utah	17,847	1951–2013	26,714	13,936	0.84	3.53
	7	Near Cisco	24,100	1884–2013	34,329	16,520	0.36	2.31
	8	Hoover Dam	171,700	1934–2013	26,131	6831	1.37	5.83

4.2. Performance Measures

For evaluating the performance of the GB2 distribution, two measures were employed: (1) the root mean square deviation (RMSD); and (2) the Akaike information criterion (AIC). These methods assess the fitted distribution at a site by summarizing the deviations between observed discharges and computed discharges.

A frequently used method for assessing the goodness-of-fit of a function is the RMSD [12]. This method was used by NERC (1975) for ranking candidate distributions [13]. RMSD can be expressed as:

$$RMSD = \sqrt{\frac{1}{n}\sum_{i=1}^{n}\left(\frac{Q_{the}(i) - Q_{emp}(i)}{Q_{emp}(i)}\right)^2} \tag{4}$$

where n is the sample size; Q_{the} is the computed discharge at the ith plotting position. Q_{emp} denotes the observed i^{th} smallest discharge. The value of RMSD is from 0 to 1. The samller is, the better the distribution fits.

AIC is a measure of the relative quality of statistical models for a given set of data. It also includes a penalty that is an increasing function of the number of estimated parameters. The AIC value was calculated as [14]:

$$AIC = n(\ln{(MSE)}) + 2\,K \tag{5}$$

where K is the number of parameters of the distribution, and MSE was calculated by

$$MSE = \frac{1}{n}\sum_{i=1}^{n}(Q_{the}(i) - Q_{emp}(i))^2 \tag{6}$$

Given a set of candidate models for the data, the preferred model is the one with the minimum AIC value.

4.3. Evaluation of GB2 Distribution

Annual maximum flood peak data from four gauging stations, namely sites 2, 6, 7 and 8 in Figure 3, were selected. The empirical frequencies were calculated first. The purpose of defining the empirical distribution is to compare it with selected theoretical distributions in order to verify whether they fit sample data.

Many plotting positions are proposed, most of which can be expressed in general form:

$$P_i = \frac{i - a}{n + 1 - 2a} \tag{7}$$

where a is a constant having values from 0 to 0.5 in different formula, 0.5 for Hazen's formula, 0.3 for Chegadayev's formula, zero for Weibull's formula, 3/8 for Blom's formula, 1/3 for Tukey's formula, and 0.44 for Gringorten's formula.

Among these formulars, Gringorten's formular is recoganized by lots of researchers, especially for GEV, gumbel, exponential, Generalized pareto distributions which have been widely used for flood frequency analysis [15–20]. The Gringorten formula is also used for GB2 distribution. For normal, generalized normal and Gamma distributions, the Blom's formula is recommended [21,22]. For Pearson type 3 and log Pearson type 3 distributions, Weibull's formula is recommended [18,21]. The GB2 distribution was employed to fit the annual maximum (AM) series of the four sites. The distribution parameters were estimated using Equation (3) and given in Table 2. The fitted GB2 distributions and empirical frequency of each AM series are shown in Figure 4. In the left of Figure 4, the line represents the fitted distribution and circle the empirical frequencies of observations. Results show that the marginal distributions fit the empirical data well. Histograms of AM flood peak series fitted by the GB2 distribution for the gauging stations on the Colorado River are shown in the right section of Figure 4. It also indicates that the GB2 distribution can successfully be fitted to empirical histograms.

Several distributions, including normal, exponential, gamma, Gumbel, generalized normal, pearson type III, log Pearson type III, generalized Pareto, and generalized extreme-value that are commonly used in hydrology, were fitted to the AM series at this site. The L-moment method was used to estimate the parameters of these distributions.

Table 2. Parameters of the GB2 distribution for the gauging stations along the Colorado River.

Number	Location	r_1	r_2	r_3	β
4	Near Dotsero	1.58	60.30	1.75	85.11
5	Near Cameo	1.12	77.57	2.53	112.93
6	Near Colorado-Utah	3.94	83.08	0.94	69.05
7	Near Cisco	2.73	76.82	1.07	80.90
8	Hoover Dam	10.59	434.72	1.31	43.62

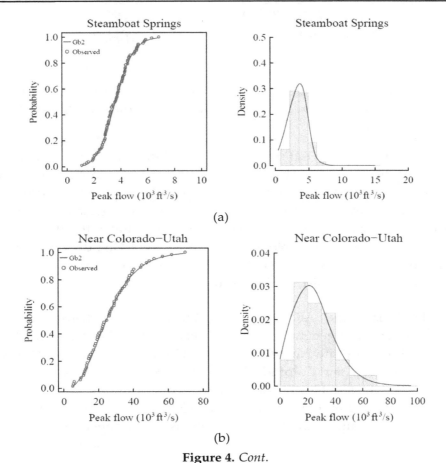

(a)

(b)

Figure 4. *Cont.*

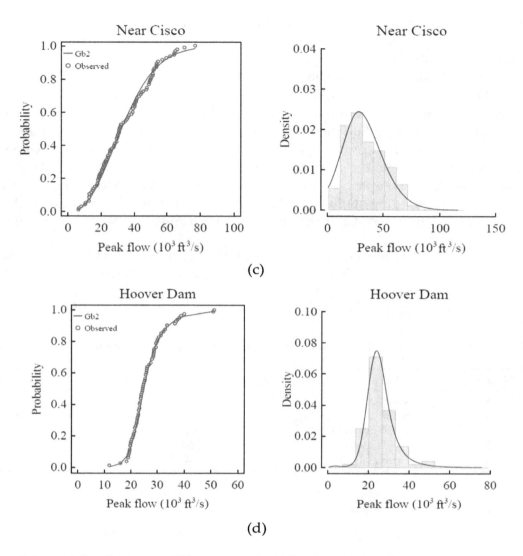

Figure 4. Marginal distributions and histograms of AM flood peak series fitted by the GB2 distribution for the gauging stations on the Colorado River. (**a**) Steamboat springs; (**b**) Near Colorado-Utah; (**c**) Near Cisco; (**d**) Hoover Dam.

Singh and Guo compared the POME method with the L-moment method, and indicated that the two methods are comparable [11,23,24]. Therefore no matter what method is used, it has little influence on the value of the T-year design discharge. The Kolmogorov-Smirnov test was used here to compare a sample with a reference probability distribution. The p-value was calculated and given in Table 3 as well. The higher or more close to 1 the p-value is the more similar the theoretical and empirical distributions are. It is indicated from Table 3 that the p-value of GB2 distribution is 1 or close to 1, which demonstrates that the GB2 distribution fit the data better. Table 3 also listed the RMSD and AIC values computed for the fitted GB2 distribution using Equations (4)–(7).

The smaller the RMSD and AIC values are, the better the distribution fits. For the site streamboat springs, the GB2 and generalized normal distributions have the smallest RMSD values, which is equal to 0.025. For the site Near Cisco, the GB2 has the smallest RMSE values, which is equal to 0.061. For the site Near Colorado-Utah, the GB2 and gamma distributions have the smallest RMSE value. For the site Hoover dam, the GB2 distribution has the smallest RMSE value. Since the GB2 distribution have more parameters, the AIC values of GB2 distribution are larger than those of generalized normal, Gamma and GEV distributions. Thus, generally GB2 distribution gives a getter fit.

Table 3. RMSE and AIC values of different distributions.

Number	Distribution	Steamboat Springs			Near Cisco			Near Colorada-Utah			Hoover Dam		
		p-Value	RMSE	AIC	p-Value	RMSE	AIC	p-Value	RMSE	AIC	p-Value	RMSE	AIC
1	GB2	0.976	**0.025**	**924.1**	1	**0.061**	**1384.1**	1	**0.047**	**852.7**	1	**0.036**	**1098.8**
2	Normal	0.926	0.043	992.9	0.839	0.194	1502	0.436	0.221	1031	0.436	0.081	1236.2
3	Exponential	0.409	0.122	1306.8	0.839	0.171	1669.6	0.919	0.145	1036.4	0.919	0.055	1152.6
4	Gamma	1	0.045	1005.1	1	0.064	1512.8	0.692	**0.047**	**842.5**	0.692	0.057	1192.9
5	Gumbel	0.976	0.066	1143.5	1	0.088	1546.2	0.978	0.107	869.1	0.978	0.039	1122.3
6	Generalized normal	0.844	**0.025**	**922.9**	1	0.137	1455.5	0.978	0.083	852.8	0.978	0.039	1106.7
7	Pearson type III	0.976	0.035	953.2	1	0.1	1425	1	0.054	895.6		0.058	1146.8
8	Log Pearson type III	0.976	0.034	951.3	1	0.106	1431.5	1	0.054	893.1	1	0.052	1133
9	Generalized Pareto	0.976	0.078	1158	1	0.062	1386.1	1	0.09	960.2	1	0.054	1169
10	GEV	0.976	0.027	929.6	1	0.138	1450.1	1	0.128	865.5	1	0.036	1096.8

In order to compare the POME with the current used method, the maximum likelihood (ML) method was also employed for the parameter estimation of GB2 distribution. Taking the site Near Colorada-Utah for an example, the estimated parameters by POME and ML method are given in Table 4. The p-value, RMSE and AIC values are also given in Table 4. It is indicated that the parameters obtained by the two method are more or less the same. And the RMSE and AIC values based on the POME method are smaller.

Table 4. Parameters estimated by POME and ML methods for site Near Colorada-Utah.

Methods	r_1	r_2	r_3	β	p-Value	RMSE	AIC
POME	2.14	24.78	1.40	157.18	1	0.0169	−357.76
ML	2.26	30.85	1.35	158.55	1	0.0170	−357.80

4.4. Flood Frequency Analysis

The Hoover dam is a multi-purpose dam, serving the needs of flood control, irrigation, water supply, and hydropower generation. Therefore, it was desired to determine the most appropriate distribution for FFA at the dam site. The T-year design flood at Hoover dam was calculated using each distribution, as given in Table 5, and it can be seen that different distributions yielded significantly different values. For example, the 1000-year design flood values calculated by the GB2 and gamma distributions were 76,702 and 50,485 ft^3/s, respectively. The RMSD and AIC values for GB2 distribution (Gamma distribution) were 0.036 (0.057) and 1098.8 (1192.9), respectively, which indicates that the performance of GB2 distribution is much better than that of the gamma distribution. It concludes that if the gamma distribution were used, the design flood would be underestimated and potential flood risk would be higher.

Table 5. Comparison of T-year design flood discharges (10^3 ft^3/s) calculated by different distributions for the Hoover dam site.

Number	Return Period	1000	500	100	50	10
1	GB2	76.702	67.914	51.198	34.138	30.125
2	Normal	45.800	44.451	40.938	34.288	31.488
3	Exponential	68.561	63.583	52.024	35.486	30.508
4	Gamma	50.485	48.424	43.314	34.613	31.320
5	Gumbel	58.926	55.332	46.973	34.799	30.912
6	Generalized normal	50.513	49.271	45.325	35.732	31.485
7	Pearson type III	60.025	56.451	47.985	35.145	30.926
8	Log Pearson type III	69.568	64.494	52.713	35.639	31.858
9	Generalized Pareto	64.809	59.870	49.084	34.893	30.695
10	GEV	57.809	54.766	47.324	35.270	31.072

4.5. Change in Flood Frequency Distribution with Change in Drainage Area

The GB2 distribution was applied for FFA along the main stem of the Colorado River. Four gauging stations (sites 4, 5, 6 and 7) from upstream to downstream were used, as shown in Figure 3 and Table 6. These gauging stations were selected, because all these stations are on the mainstream and no dam has been built on this reach. The drainage area and statistical characteristics (including mean, skewness and kurtosis of the annual maximum data) of these stations were calculated, as given in Table 1. The T-year design flood of these gauging stations was calculated, as shown in Figure 5, in which the x-axis represents the return periods and the y-axis represents the design flood values. Figure 5 shows that for a given return period, the design flood value of the downstream gauging stations is larger than that of the upstream gauging stations. The increasing rates of drainage area and T-year design flood values between the adjacent gauging stations were computed, as given in Table 6, which indicates that the percentage increase of the drainage area was nearly the same as that of the design flood values. For instance, with the increase of drainage area up to 45% from the gauging station near Dotsero to that near Cameo, the flood value increased by 43% on average. It is also seen that from upstream to downstream, when the drainage area increased by 45%, 55% and 26%, the flood value increased by 43%, 42%, and 16%, respectively. It seems that in a mountainous watershed, the upstream the reach is, the greater the impact the drainage area has on flood. This may be because that the runoff coefficient is generally larger in the steep area.

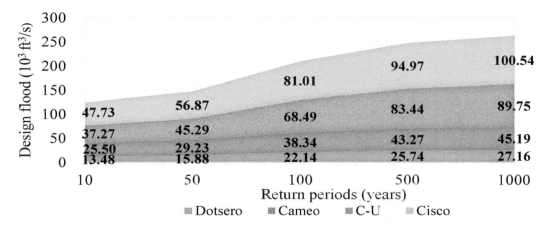

Figure 5. Flood values along the mainstream of the upper Colorado River.

Table 6. Statistical characteristics of the four gauging stations, the increasing rate of drainage area and flood discharge between adjacent gauging stations.

Number	Locations	Drainage Area (Square Miles)	Increase in Drainage Area (%)	Increase in in Flood Value (%)					
				1000	500	100	50	10	Mean
4	Near Dotsero	11370	45	40	41	42	47	46	43
5	Near Cameo	20683	55	50	48	44	32	35	42
6	Near Colorado-Utah	46228	26	11	12	15	22	20	16
7	Near Cisco	62419							

4.6. Evolution of Frequency Distribution along Stream

In order to determine the evolution of frequency distribution and its parameters along the river, data from the Yampa River were applied, because this river is taken as one of the west's last wild rivers and has only a few small dams and diversions. The Yampa River with a length of 402 km, located in northwestern Colorado, is a tributary of Green River and a secondary tributary of the Colorado River. Data from three gauging stations along this river, designated as sites 1, 2 and 3 in Figure 6, were used. The GB2 distribution was used to fit the AM series of each of the three gauging stations, as shown in Table 7. It can be seen that shape parameters r_1 and r_2 decreased along the river. The value of r_1 became close to be 1. When r_1 equals 1, the GB2 distribution becomes the Burr XII distribution [25]. This distribution has been shown to reasonably fit the income distribution data [20,26,27] and has recently been used in hydrology [2,28]. The PDF of Burr XII distribution can be written as:

$$f(x) = \frac{r_3}{bB(1,r_2)}\left(\frac{x}{b}\right)^{1\times r_3-1}\left(1+\left(\frac{x}{b}\right)^{r_3}\right)^{-(r_2+1)} = \frac{r_3 r_2}{b}\left(\frac{x}{b}\right)^{r_3-1}\left(1+\left(\frac{x}{b}\right)^{r_3}\right)^{-(r_2+1)} \quad (8)$$

where b is the scale parameter. The Burr XII distribution was also used to fit the data at the gauging station near Maybell of Yampa River. The estimated parameters of Burr XII distribution were: $r_2 = 1.94$, $r_3 = 4.19$, and $b = 12.33$. The fitting results of the GB2 and Burr distributions for the gauging station near Maybell are shown in Figure 7. For the gauging station near Maybell, parameters of the GB2 distribution estimated by POME method are nearly as the same as the parameters of the Burr XII distribution estimated by MLE method. Thus, Burr XII distribution instead of GB2 distribution can be used for FFA at that station. In other words, the distribution for FFA changes from the four-parameter GB2 distribution to the three-parameter Burr XII distribution along the Yampa River. There is an evolution of distribution along this river. From Equation (1), the value of scale parameter β increases with the mean value, because more water flows into the stream. Parameters r_1 and r_2 govern the left and right tails, respectively. The smaller the value of r_1, the fatter the left tail is; and the smaller the value of r_2, the fatter the right tail is. It can be seen from Table 7 that both r_1 and r_2 decrease along the stream, which demonstrates that both the left and right tails become fatter, and the PDF values become larger in these areas and lower in the central area.

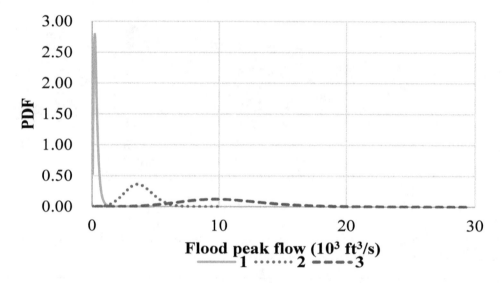

Figure 6. Evaluations of PDF of sites along the Yampa River.

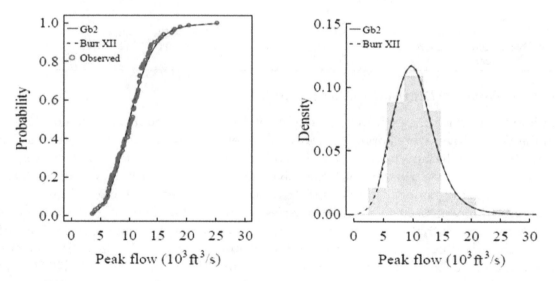

Figure 7. Marginal distribution and histograms of AM flood peak series fitted by the GB2 and Burr XII distributions for the gauging station near Maybell on the Yampa River.

Table 7. Parameters of the GB2 distribution for four gauging stations along the Yampa River.

Number	Location	r_1	r_2	r_3	β
1	Below stagecoach Reservoir	17.44	15.25	0.55	2.10
2	Steamboat springs	1.20	5.49	3.59	5.81
3	Near Maybell	1.14	2.07	3.92	12.11

5. Conclusions

The GB2 provides sufficient flexibility to fit a large variety of data sets. Papalexiou and Koutsoyiannis introduced this distribution in hydrology and used it for rainfall frequency analysis [8]. In this study, the generalized beta distribution of the second kind (GB2) is introduced for FFA for the first time. The POME method was proposed to estimate the parameters of GB2 distribution. Equations of POME method was deduced by ourselves and given in Appendix A. The Colorado River basin was selected as a case study to test the performance of GB2 distribution. Frequency estimates from the GB2 distribution were also compared with those of commonly used distributions in hydrology. In addition, some characteristics of FFA in mountainous areas are discussed. The conclusions can be summarized as follows:

(1)　Results demonstrate that the GB2 is appealing for FFA, since it has four parameters which allows the distribution to be able to fit data having very different histogram shapes, such as the J-shaped and bell-shaped distributions. And by fixing certain parameters, the GB2 distribution can yield some well-known distributions, such as the beta distribution of the second kind (B2), the Burr type XII, generalized gamma (GG), and so on.

(2)　The parameters estimated by POME method are found reasonable. Both the marginal distributions and histograms indicates that the GB2 distribution can successfully be fitted to empirical values using the POME method.

(3)　The performance of the GB2 distribution is better than that of the widely used distributions in hydrology. For the site streamboat springs, the GB2 and generalized normal distributions have the smallest RMSD values. For the site Near Cisco, the GB2 has the smallest RMSE values. For the site Near Colorado-Utah, the GB2 and gamma distributions have the smallest RMSE value. For the site Hoover dam, the GB2 distribution has the smallest RMSE value. Since the GB2 distribution have more parameters, the AIC values of GB2 distribution are larger than those of generalized normal, Gamma and GEV distributions. Thus, generally GB2 distribution gives a getter fit.

(4)　When using different distributions for FFA, significant different design flood values are obtained. It concludes that if the wrong distribution were used, the design flood would be underestimated and potential flood risk would be higher.

(5)　The design flood value increase with the drainage area. For a given return period, the design flood value of the downstream gauging stations is larger than that of the upstream gauging stations. In this study, the percentage increase of the drainage area was nearly the same as that of the design flood values. It seems that in a mountainous watershed, the upstream the reach is, the greater the impact the drainage area has on flood. This may be because that the runoff coefficient is generally larger in the steep area.

(6)　There is an evolution of distribution along this river. Along the Yampa River, the distribution for FFA changes from the four-parameter GB2 distribution to the three-parameter Burr XII distribution. And both r_1 and r_2 decrease along the stream, which demonstrates that both the left and right tails become fatter, and the PDF values become larger in these areas and lower in the central area, which means that when the drainage area become larger, the flood magnitudes has a more significant variation.

Acknowledgments: The project was financially supported by the National Natural Science Foundation of China (51679094, 51509273, 91547208 and 41401018), Fundamental Research Funds for the Central Universities (2017KFYXJJ194, 2016YXZD048).

Author Contributions: Vijay P. Singh conceived and designed the experiments; Lu Chen performed the experiments and analyzed the data; Lu Chen wrote the draft of the paper and Vijay P. Singh revised it. All authors have read and approved the final manuscript.

Appendix A. Estimation of Parameters of GB2 Distribution

The GB2 distribution parameters can be estimated by maximizing the Shannon entropy $H(X)$ which, for a random variable X, can be expressed as:

$$H(X) = -\int_0^\infty f(x) \log f(x) \mathrm{d}x \tag{A1}$$

where $f(x)$ is the probability density function (PDF). The principle of maximum entropy (POME) indicates that the most appropriate PDF is the one that maximizes the value of entropy, given available data and a set of known constraints [29].

Specification of Constraints: Following Singh, the constraints for the GB2 distribution can be expressed as

$$\int_0^\infty f(x) \mathrm{d}x = 1 \tag{A2a}$$

$$\int_0^\infty f(x) \ln x \mathrm{d}x = E(\ln x) \tag{A2b}$$

$$\int_0^\infty f(x) \ln(1 + (\tfrac{x}{\beta})^{r_3}) \mathrm{d}x = E(\ln(1 + (\tfrac{x}{\beta})^{r_3})) \tag{A2c}$$

Method of Lagrange Multipliers for Maximizing Entropy: In the search for an appropriate probability distribution for a given random variable, entropy should be maximized. In other words, the best fitted distribution is the one with the highest entropy. The method of Lagrange multipliers was used to obtain the appropriate probability distribution with the maximum entropy. Finally, the form of this distribution is given as:

$$f(x) = \exp(-\lambda_0 - \lambda_1 \ln(x) - \lambda_2' \ln(1 + (\tfrac{x}{\beta})^{r_3})) \tag{A3a}$$

in which λ_0, λ_1, and λ_2' are the Lagrange multipliers. Let $p = \beta^{-r_3}$. Then, Equation (A3a) can be written as

$$f(x) = \exp(-\lambda_0 - \lambda_1 \ln(x) - \lambda_2' \ln(1 + px^{r_3})) \tag{A3b}$$

Let $\lambda_2' = \frac{\lambda_2}{p}$ and $q = r_3$. Papalexiou and Koutsoyiannis defined the entropy-based PDF as:

$$f(x) = \exp(-\lambda_0 - \lambda_1 \ln(x) - \frac{\lambda_2}{p} \ln(1 + px^q)) \tag{A4}$$

Substitution of Equation (A4) in Equation (A2a) yields:

$$\int_0^\infty f(x) \mathrm{d}x = \int_0^\infty \exp(-\lambda_0 - \lambda_1 \ln(x) - \frac{\lambda_2}{p} \ln(1 + px^q)) \mathrm{d}x = 1 \tag{A5}$$

From Equation (A5):

$$\begin{aligned}
\exp(\lambda_0) &= \int_0^\infty \exp(-\lambda_1 \ln x - \lambda_2 \ln(1 + px^q)/p) \mathrm{d}x \\
&= \int_0^\infty \exp(-\lambda_1 \ln x) \exp(-\tfrac{\lambda_2}{p} \ln(1 + px^q)) \mathrm{d}x \\
&= \int_0^\infty x^{(-\lambda_1)} (1 + px^q)^{(-\frac{\lambda_2}{p})} \mathrm{d}x
\end{aligned} \tag{A6}$$

Let $t = px^q$. Then $x = (\frac{t}{p})^{\frac{1}{q}}$, and $dx = \frac{1}{pq}(\frac{t}{p})^{\frac{1}{q}-1}dt$. Thus, Equation (A6) can be expressed as:

$$
\begin{aligned}
\exp(\lambda_0) &= \int_0^\infty x^{(-\lambda_1)}(1 + px^q)^{(-\frac{\lambda_2}{p})}dx \\
&= \int_0^\infty (\tfrac{t}{p})^{\frac{-\lambda_1}{q}}(1 + t)^{\frac{-\lambda_2}{p}}\frac{1}{pq}(\tfrac{t}{p})^{\frac{1}{q}-1}dt \\
&= \int_0^\infty \tfrac{1}{q}p^{\frac{\lambda_1-1}{q}}t^{\frac{-\lambda_1}{q}}(1 + t)^{\frac{-\lambda_2}{p}}t^{\frac{1}{q}-1}dt
\end{aligned}
\tag{A7}
$$

Let $y = \frac{t}{1+t}$. Then $t = \frac{y}{1-y}$, and $dt = \frac{1}{(1-y)^2}dy$.
Since $y(0) = 0$ and $y(\infty) = 1$, $y \in [0, 1]$.

$$
\begin{aligned}
\exp(\lambda_0) &= \int_0^1 \tfrac{1}{q}p^{\frac{\lambda_1-1}{q}}(\tfrac{y}{1-y})^{\frac{-\lambda_1}{q}}(1 + \tfrac{y}{1-y})^{\frac{-\lambda_2}{p}}(\tfrac{y}{1-y})^{\frac{1}{q}-1}\frac{1}{(1-y)^2}dy \\
&= \int_0^1 \tfrac{1}{q}p^{\frac{\lambda_1-1}{q}}(\tfrac{y}{1-y})^{\frac{-\lambda_1+1}{q}-1}(1 + \tfrac{y}{1-y})^{\frac{-\lambda_2}{p}}\frac{1}{(1-y)^2}dy \\
&= \int_0^1 \tfrac{1}{q}p^{\frac{\lambda_1-1}{q}}(\tfrac{y}{1-y})^{\frac{-\lambda_1+1}{q}-1}(\tfrac{1}{1-y})^{\frac{-\lambda_2}{p}+2}dy \\
&= \int_0^1 \tfrac{1}{q}p^{\frac{\lambda_1-1}{q}}(y)^{\frac{1-\lambda_1}{q}-1}(1 - y)^{-\frac{1-\lambda_1}{q}+\frac{\lambda_2}{p}-1}dy \\
&= \tfrac{1}{q}p^{\frac{\lambda_1-1}{q}}B(\tfrac{1-\lambda_1}{q}, -\tfrac{1-\lambda_1}{q} + \tfrac{\lambda_2}{p})
\end{aligned}
\tag{A8}
$$

The Lagrange multiplier λ_0 can be calculated from Equation (A8) as:

$$
\lambda_0 = -\ln q + \frac{\lambda_1 - 1}{q}\ln(p) + \ln\Gamma(\frac{1-\lambda_1}{q}) + \ln\Gamma(-\frac{1-\lambda_1}{q} + \frac{\lambda_2}{p}) - \ln\Gamma(\frac{\lambda_2}{p})
\tag{A9}
$$

From Equation (A4), the other equation for calculating λ_0 can be defined as:

$$
\lambda_0 = \ln(\int_0^\infty \exp(-\lambda_1\ln x - \frac{\lambda_2}{p}\ln(1 + px^{r_3}))dx)
\tag{A10}
$$

Relation between Lagrange multipliers and constraints: Defining $a' = \frac{1-\lambda_1}{q}$ and $b' = -\frac{1-\lambda_1}{q} + \frac{\lambda_2}{p}$, differentiate Equation (A9) with respect to λ_1 and λ_2:

$$
\begin{aligned}
\frac{\partial\lambda_0}{\partial\lambda_1} &= \frac{\ln p}{q} + \frac{\partial\ln\Gamma(a')}{\partial a'}\frac{\partial a'}{\partial\lambda_1} + \frac{\partial\ln\Gamma(b')}{\partial(b')}\frac{\partial b'}{\partial\lambda_1} - \frac{\partial\ln\Gamma(a+b')}{\partial(a+b')}\frac{\partial(a+b')}{\partial\lambda_1} \\
&= \frac{\ln p}{q} - \tfrac{1}{q}\varphi(a') + \tfrac{1}{q}\varphi(b')
\end{aligned}
\tag{A11a}
$$

$$
\begin{aligned}
\frac{\partial\lambda_0}{\partial\lambda_2} &= \frac{\partial\ln\Gamma(b')}{\partial(b')}\frac{\partial b'}{\partial\lambda_2} - \frac{\partial\ln\Gamma(a'+b')}{\partial(a'+b')}\frac{\partial(a'+b')}{\partial\lambda_2} \\
&= \tfrac{1}{p}\varphi(b') - \tfrac{1}{p}\varphi(a' + b')
\end{aligned}
\tag{A11b}
$$

where $\varphi(.)$ is a digamma function. Differentiate Equation (A10) with respect to λ_1 and λ_2:

$$
\frac{\partial\lambda_0}{\partial\lambda_1} = \frac{\int_0^\infty \ln x\exp(-\lambda_1\ln x - \frac{\lambda_2}{p}\ln(1 + px^q))dx}{\int_0^\infty \exp(-\lambda_1\ln x - \frac{\lambda_2}{p}\ln(1 + px^q))dx} = -E(\ln x)
\tag{A12a}
$$

$$\frac{\partial \lambda_0}{\partial \lambda_2} = \frac{\int_0^\infty x^q \exp(-\lambda_1 \ln x - \frac{\lambda_2}{p} \ln(1+px^q))dx}{\int_0^\infty \exp(-\lambda_1 \ln x - \frac{\lambda_2}{p} \ln(1+px^q))dx} = -E\left(\frac{\ln(1+px^q)}{p}\right) \qquad (A12b)$$

Based on Equations (A11) and (A12), the relation between Lagrange multipliers and constraints can be expressed as:

$$\frac{\ln p}{q} - \frac{1}{q}\varphi(a) + \frac{1}{q}\varphi(b) = -E(\ln x) \qquad (A13a)$$

$$\frac{1}{p}\varphi(b) - \frac{1}{p}\varphi(a+b) = -E\left(\frac{\ln(1+px^q)}{p}\right) \qquad (A13b)$$

Since there are four parameters, Equations (A13a) and (A13b) are not sufficient for calculating parameters, and two additional equations are needed that are given as:

$$\frac{\partial^2 \lambda_0}{\partial^2 \lambda_1} = \frac{1}{q^2}\varphi'(a') + \frac{1}{q^2}\varphi'(b') = \mathrm{var}(\ln x) \qquad (A14a)$$

$$\frac{\partial^2 \lambda_0}{\partial^2 \lambda_2} = \varphi'(r_2) - \varphi'(r_1+r_2) = \mathrm{var}\left(\ln\left(1+\left(\frac{x}{\beta}\right)^q\right)\right) \qquad (A14b)$$

Relation between Lagrange multipliers and parameters: Substituting Equation (A8) in Equation (A4), it is known that:

$$f(x) = \frac{1}{\frac{1}{q}p^{\frac{\lambda_1-1}{q}} B\left(\frac{1-\lambda_1}{q}, -\frac{1-\lambda_1}{q} + \frac{\lambda_2}{p}\right)} x^{-\lambda_1}(1+px^q)^{-\frac{\lambda_2}{p}} \qquad (A15)$$

Equation (A15) is the GB2 distribution. Comparing Equation (1) with Equation (A15), the following equations can be obtained:

$$\begin{aligned}\lambda_1 &= 1 - r_1 q \\ \lambda_2 &= p\left(r_2 + \frac{1-\lambda_1}{q}\right) \\ p &= \left(\frac{1}{\beta}\right)^{r_3} \\ q &= r_3 \end{aligned} \qquad (A16)$$

Relation between parameters and constraints: Based on the relation between parameters and constraints, and parameters and Lagrange multipliers, the relation between parameters and constraints can be expressed as:

$$\begin{aligned} -\ln\beta - \frac{1}{r_3}\varphi(r_1) + \frac{1}{r_3}\varphi(r_2) &= -E(\ln x) \\ \beta^{r_3}\varphi(r_2) - \beta^{r_3}\varphi(r_1+r_2) &= -E\left(\beta^{r_3}\ln\left(1+\left(\frac{x}{\beta}\right)^{r_3}\right)\right) \\ -\ln\beta + \frac{1}{r_3^2}\varphi'(r_1) + \frac{1}{\gamma_3^2}\varphi'(r_2) &= \mathrm{var}(\ln x) \\ \varphi'(r_2) - \varphi'(r_1+r_2) &= \mathrm{var}\left(\ln\left(1+\left(\frac{x}{\beta}\right)^{r_3}\right)\right) \end{aligned} \qquad (A17)$$

References

1. Beven, K.J.; Hornberger, G.M. Assessing the effect of spatial pattern of precipitation in modeling streamflow hydrographs. *J. Am. Water Resour. Assoc.* **1982**, *18*, 823–829. [CrossRef]
2. Shao, Q.; Wong, H.; Xia, J.; Ip, W. Models for extremes using the extended three-parameter Burr XII system with application to flood frequency analysis. *Hydrol. Sci. J.* **2004**, *49*, 685–702. [CrossRef]
3. Chen, L.; Guo, S.L.; Yan, B.W.; Liu, P.; Fang, B. A new seasonal design flood method based on bivariate joint distribution of flood magnitude and date of occurrence. *Hydrol. Sci. J.* **2010**, *55*, 1264–1280. [CrossRef]
4. Chen, L.; Singh, V.P.; Guo, S.; Hao, Z.; Li, T. Flood coincidence risk analysis using multivariate copula functions. *J. Hydrol. Eng.* **2012**, *17*, 742–755. [CrossRef]

5. Mielke, P.W., Jr.; Johnson, E.S. Some generalized beta distributions of the second kind having desirable application features in hydrology and meteorology. *Water Resour. Res.* **1974**, *10*, 223–226. [CrossRef]

6. Wilks, D.S. Comparison of three-parameter probability distributions for representing annual extreme and partial duration precipitation series. *Water Resour. Res.* **1993**, *29*, 3543–3549. [CrossRef]

7. Perreault, L.; Bobée, B.; Rasmussen, P. Halphen distribution system. I: Mathematical and statistical properties. *J. Hydrol. Eng.* **1999**, *4*, 189–199. [CrossRef]

8. Papalexiou, S.M.; Koutsoyiannis, D. Entropy based derivation of probability distributions: A case study to daily rainfall. *Adv. Water Resour.* **2012**, *45*, 51–57. [CrossRef]

9. Singh, V.P. *Entropy Based Parameter Estimation in Hydrology*; Kluwer Academic Publishers: Dordrecht, The Netherlands, 1998.

10. Singh, V.P. *Entropy-Based Parameter Estimation Hydrology*; Springer: Dordrecht, The Netherlands, 1998.

11. Singh, V.P.; Guo, H. Parameter estimation for 3-parameter generalized Pareto distribution by the principle of maximum entropy (POME). *Hydrol. Sci. J.* **1995**, *40*, 165–181. [CrossRef]

12. Karim, A.; Chowdhury, J.U. A comparison of four distributions used in flood frequency analysis in Bangladesh. *Hydrol. Sci. J.* **1995**, *40*, 55–66. [CrossRef]

13. Natural Environment Research Council. *Flood Studies Report*; Natural Environment Research Council: London, UK, 1975; Volumes 1–5.

14. Zhang, L.; Singh, V.P. Bivariate flood frequency analysis using the copula method. *J. Hydrol. Eng.* **2006**, *11*, 150–164. [CrossRef]

15. Ross, R. Graphical method for plotting and evaluating weibull distribution data. In Proceedings of the 4th International Conference on Properties and Application of Dielectric Materials, Brisbane, Austrialia, 3–8 July 1994; pp. 250–253.

16. Cunnane, C. Unbiased plotting positions—A review. *J. Hydrol.* **1978**, *37*, 205–222. [CrossRef]

17. Makkonen, L. Notes and correspondence plotting positions in extreme value analysis. *J. Appl. Meteorol. Clim.* **2006**, *45*, 334–340. [CrossRef]

18. Shabri, A. A Comparison of plotting formulas for the pearson type III distribution. *J. Technol.* **2002**, *36*, 61–74. [CrossRef]

19. Gringorten, I.I. A plotting rule for extreme probability paper. *J. Geophys. Res.* **1963**, *68*, 813–814. [CrossRef]

20. Dagum, C. A New Model of Personal Income Distribution: Specification and Estimation. In *Modeling Income Distributions and Lorenz Curves*; Springer: New York, NY, USA, 2008; pp. 3–25.

21. Mehdi, F.; Mehdi, J. Determination of plotting position formula for the normal, log-normal, pearson(III), log-pearson(III) and gumble distribution hypotheses using the probability plot correlation coefficient test. *World Appl. Sci. J.* **2011**, *15*, 1181–1185.

22. Kim, S.; Shin, H.; Kim, T.; Taesoon, K.; Heo, J. Derivation of the probability plot correlation coefficient test statistics for the generalized logistic distribution. In Proceedings of the International Workshop Advances in Statistical Hydrology, Taormina, Italy, 23–25 May 2010; pp. 1–8.

23. Singh, V.P.; Guo, H. Parameter estimation for 2-parameter log-logistic distribtuion distribution (LLD2) by maximum entropy. *Civ. Eng. Syst.* **1995**, *12*, 343–357. [CrossRef]

24. Singh, V.P.; Guo, H. Parameter estimations for 2-parameter Pareto distribution by pome. *Stoch. Hydrol. Hydraul.* **1980**, *9*, 81–93.

25. Burr, I.W. Cumulative Frequency Functions. *Ann. Math. Stat.* **1942**, *13*, 215–232. [CrossRef]

26. Kleiber, C.; Kotz, S. *Statistical Size Distributions in Economics and Actuarial Sciences*; John Wiley & Sons: Hoboken, NJ, USA, 2003.

27. Singh, S.K.; Maddala, G.S. A function for size distribution of incomes. *Econometrica* **1976**, *44*, 963–970. [CrossRef]

28. Hao, Z.; Singh, V.P. Entropy-based parameter estimation for extended Burr XII distribution. *Stoch. Environ. Res. Risk Assess.* **2008**, *23*, 1113–1122. [CrossRef]

29. Singh, V.P. Hydrologic synthesis using entropy theory: Review. *J. Hydrol. Eng.* **2011**, *16*, 421–433. [CrossRef]

4

Investigation into Multi-Temporal Scale Complexity of Streamflows and Water Levels in the Poyang Lake Basin, China

Feng Huang [1,2,*], **Xunzhou Chunyu** [1,2], **Yuankun Wang** [3], **Yao Wu** [2,4], **Bao Qian** [5], **Lidan Guo** [6], **Dayong Zhao** [1,2] and **Ziqiang Xia** [1,2]

[1] State Key Laboratory of Hydrology-Water Resources and Hydraulic Engineering, Hohai University, Nanjing 210098, China; zodiacnix@163.com (X.C.); dyzhao@hhu.edu.cn (D.Z.); zqxia@hhu.edu.cn (Z.X.)

[2] College of Hydrology and Water Resources, Hohai University, Nanjing 210098, China; wuyao@jxsl.gov.cn

[3] School of Earth Sciences and Engineering, Nanjing University, Nanjing 210023, China; yuankunw@nju.edu.cn

[4] Poyang Lake Hydro Project Construction Office of Jiangxi Province, Nanchang 330046, China

[5] Bureau of Hydrology, Changjiang River Water Resources Commission, Wuhan 430012, China; jacber@163.com

[6] International River Research Centre, Hohai University, Nanjing 210098, China; ldguohhu@163.com

* Correspondence: hfeng0216@163.com

Academic Editors: Huijuan Cui, Bellie Sivakumar and Vijay P. Singh

Abstract: The streamflow and water level complexity of the Poyang Lake basin has been investigated over multiple time-scales using daily observations of the water level and streamflow spanning from 1954 through 2013. The composite multiscale sample entropy was applied to measure the complexity and the Mann-Kendall algorithm was applied to detect the temporal changes in the complexity. The results show that the streamflow and water level complexity increases as the time-scale increases. The sample entropy of the streamflow increases when the time-scale increases from a daily to a seasonal scale, also the sample entropy of the water level increases when the time-scale increases from a daily to a monthly scale. The water outflows of Poyang Lake, which is impacted mainly by the inflow processes, lake regulation, and the streamflow processes of the Yangtze River, is more complex than the water inflows. The streamflow and water level complexity over most of the time-scales, between the daily and monthly scales, is dominated by the increasing trend. This indicates the enhanced randomness, disorderliness, and irregularity of the streamflows and water levels. This investigation can help provide a better understanding to the hydrological features of large freshwater lakes. Ongoing research will be made to analyze and understand the mechanisms of the streamflow and water level complexity changes within the context of climate change and anthropogenic activities.

Keywords: complexity; streamflow; water level; composite multiscale sample entropy; trend; Poyang Lake basin

1. Introduction

The complexity of a hydrological time series, e.g., precipitation, streamflow, and water level means that there is a degree of uncertainty, randomness, or irregularity of the time series. Many researchers have studied the hydrological complexity [1–5]. Chou analyzed the complexity of the correlation between rainfall and runoff, using the multiscale entropy approach, and found that: (1) the entropy measures of rainfall are higher than those of runoff at all scale factors; (2) the entropy measures of the runoff coefficient series lie between the entropy measures of the rainfall and runoff at various scale factors; and (3) the entropy values of rainfall, runoff and runoff coefficient series increase as scale factors increase. Liu et al. studied the complexity features of regional groundwater depth and found that human activities are the main driving force, causing the complexity of regional groundwater depth [6]. However, there seems to have been few attempts made to investigate the hydrological complexities of large lakes. Therefore, this study looks to the Poyang Lake basin as the study area and investigates its streamflow and water level complexity.

Poyang Lake, which is located in the Yangtze River basin, is the largest freshwater lake in China. The lake is the major body of water in an important global ecoregion, and plays an important role in maintaining the water resources of the Yangtze River and a healthy aquatic ecosystem in the region [7]. The hydrological processes of Poyang Lake are essential to the lacustrine and wetland ecosystems; and the disturbance of hydrological processes may break the longstanding ecological balance and influence the distinctive biodiversity have in this region [8]. The hydrological changes of Poyang Lake have attracted worldwide attention [9–11]. The Poyang Lake basin has experienced six extreme droughts during the past 60 years, thus resulting in the reduction of water resources from the five tributaries flowing into Poyang Lake [12]. The Three Gorges Reservoir has changed the Yangtze River streamflow and has further impacted the interrelationship between the Yangtze River and Poyang Lake [13]. Since the start of operations of the Three Gorges Reservoir, the seasonal water level of Poyang Lake has had a great magnitude of fluctuation [14].

Based on these previous studies, the present study investigates the hydrological changes of Poyang Lake from another perspective: hydrological complexity. Using the streamflow and water level data in the Poyang Lake basin from 1954 to 2013, this study aims to accomplish the following objectives: to investigate the streamflow and water level complexity over various time-scales; and secondly to detect the temporal changes in the streamflow and water level complexity. The first objective will reveal the streamflow and water level complexity changes versus the time-scale changes. Will the complexity increase, keep stable or decrease when the time-scale increases from a daily scale to a seasonal scale? The second objective will reveal the streamflow and water level complexity changes in the recent 60 years. The 1954–2013 streamflows and water levels are divided into 51 subseries. The complexity of each subseries is analyzed for a specific time-scale, and a complexity series can be obtained. The trends in the complexity series are further investigated to reveal what the temporal changes are. The results will add an understanding to the hydrological features and the changes of Poyang Lake and will also provide scientific references for other similar large freshwater lakes.

2. Study Area and Data

Poyang Lake is located in the Jiangxi Province in the middle to lower reaches of the Yangtze River. It receives inflows from five main tributaries: the Ganjiang, Fuhe, Xinjiang, Raohe, and Xiushui rivers. The lake exchanges water with the Yangtze River through a narrow channel in the north (Figure 1). Six hydrometric stations (Table 1) monitor the inflow processes from the tributaries, and one hydrometric station (Hukou) monitors the water exchanges of the Poyang Lake and the Yangtze River. The streamflows of the Hukou station have both positive and negative values. Recorded positive values are when the lake discharges into the river. When the river water discharges into the lake, the streamflow data of the Hukou station are recorded as negative values. Because the streamflow data is dominated by the positive values, the streamflow processes at the Hukou station are described as the outflow processes of Poyang Lake. Four hydrometric stations monitor the water levels of Poyang Lake. The hydrological data are daily recorded and the period of all hydrological data spans from 1954 through 2013.

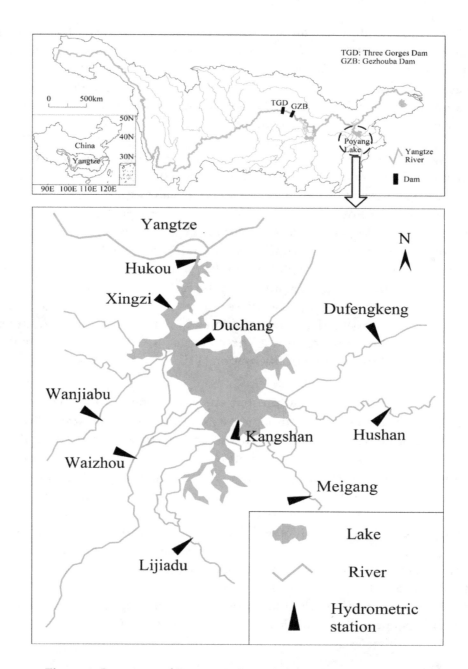

Figure 1. Locations of Poyang Lake and the hydrometric stations.

Table 1. Hydrometric stations in the Poyang Lake basin.

Hydrometric Station	Data Type	Location	Longitude and Latitude	Drainage Area (km²)
Waizhou	Streamflow	Ganjiang River	[115°50′E, 28°38′N]	80,948
Lijiadu	Streamflow	Fuhe River	[116°10′E, 28°13′N]	15,811
Meigang	Streamflow	Xinjiang River	[116°49′E, 28°26′N]	15,535
Hushan	Streamflow	Raohe River	[117°16′E, 28°55′N]	6374
Dufengkeng	Streamflow	Raohe River	[117°12′E, 29°16′N]	5013
Wanjiabu	Streamflow	Xiushui River	[115°39′E, 28°51′N]	3548
Kangshan	Water level	Poyang Lake	[116°25′E, 28°53′N]	/
Duchang	Water level	Poyang Lake	[116°11′E, 29°15′N]	/
Xingzi	Water level	Poyang Lake	[116°2′E, 29°27′N]	/
Hukou	Streamflow, Water level	Poyang Lake	[116°13′E, 29°45′N]	162,225

3. Methodology

3.1. Analysis Procedure

The streamflow and water level complexity are measured by sample entropy. Composite multiscale sample entropy is applied to reveal the streamflow and water level complexity over various time-scales. Both static and dynamic sample entropies over various time-scales are calculated to investigate the multiscale complexity [15]. The static sample entropy reveals different complexities over various time-scales, and the dynamic sample entropy reveals temporal complexity trends over various time-scales. First, the static sample entropy is analyzed, which ignores the complexity changes over time. The composite multiscale sample entropy is calculated using the entire hydrological data records. Second, the dynamic sample entropy is analyzed, which reflects the complexity changes over time. To obtain how the streamflow and water level complexity changed over time, a ten year sliding window was selected for the hydrological data, which divides the original entire hydrological data records. Thus, the 1954–2013 series is divided into 51 subseries: 1954–1963, 1955–1964, 1956–1965, ..., 2002–2011, 2003–2012, and 2004–2013. The composite multiscale sample entropy is calculated using each subseries, respectively. A sample entropy series can be obtained for a time-scale, which reflects the complexity changes over time. Temporal trends in the sample entropy series are further analyzed applying the Mann-Kendall algorithm.

3.2. Sample Entropy

The hydrological complexity means the degree of uncertainty or the rate of information production of the hydrological series, e.g., the streamflow and the water level. Techniques for measuring the hydrological complexity typically involve the calculations of: the Lyapunov exponent, correlation dimension, fractal dimension, Kolmogorov-Sinai entropy, spectral entropy, approximate entropy, and sample entropy [5]. The approximate entropy calculation solves the problem of insufficient number of data points, and is applicable to noisy, medium-sized data sets. However, it lacks relative consistency and its results depend on data length. Improved on the basis of the approximate entropy calculation, the sample entropy calculation is an unbiased estimation of the conditional probability that two similar sequences of m consecutive data points (m is the embedded dimension) will remain similar when one more consecutive point is included [16]. It is largely independent of data length and keeps relative consistency without counting self-matches [17]. For a time series of N points $\{x(i), i = 1, 2, \ldots, N\}$, the sample entropy ($SampEn$) is calculated by the following steps:

(1) Constitute vectors of m dimensions:

$$X(i) = \{x(i), x(i+1), \ldots, x(i+m-1)\} \quad (i = 1, 2, \ldots, N-m+1) \tag{1}$$

(2) Define the Euclidean distance between $X(i)$ and $X(j)$, $d[X(i), X(j)]$ as the maximum absolute difference of their corresponding scalar components:

$$d[X(i), X(j)] = \underset{k=0 \rightarrow m-1}{\text{Max}} \{|X(i+k) - X(j+k)|\} \tag{2}$$

(3) Take n_i^m as the number of sequences in the time series that match (without self-matching) the template with the length m within the tolerance criterion r. Then, define $C_i^m(r)$ and $C^m(r)$ separately as the following equations:

$$C_i^m(r) = \frac{n_i^m}{N-m} \tag{3}$$

$$C^m(r) = \frac{1}{N-m+1} \sum_{i=1}^{N-m+1} C_i^m(r) \tag{4}$$

(4) Change the dimension of the vector $X(i)$ to $m+1$ and calculate $C^{m+1}(r)$ similarly.

(5) Finally, *SampEn* is defined as:

$$SampEn = -\ln \frac{C^{m+1}(r)}{C^m(r)} \tag{5}$$

When calculating the sample entropy, the template length m is set to be 2, and the tolerance criterion r is set to be 0.15σ, where σ denotes the standard deviation of the original time series.

3.3. Composite Multiscale Sample Entropy

To completely measure the underlying dynamics of complex systems under consideration, displaying their disorderliness over multiple time-scales, Costa et al. proposed the multiscale entropy algorithm. This algorithm calculates the sample entropy of a time series over various time-scales [18]. Although the multiscale entropy algorithm has been successfully applied in a number of different fields (including the analysis of the human gait dynamics, heart rate variability, rainfall, and river streamflow), it encounters a problem in that the statistical reliability of the sample entropy of the coarse-grained series is reduced as the time-scale is increased. To overcome this limitation, Wu et al. proposed the concept of composite multiscale sample entropy, which better presents data in both simulation and real world data analysis [19].

The principles and calculation procedures of the composite multiscale sample entropy can be found in Wu et al. [19]. When the sample entropy is calculated with the template length $m = 2$, there are two and three coarse-grained time series divided from the original time series for scale factors of 2 and 3, respectively. The kth coarse-grained time series for a scale factor of τ, $y_k^{(\tau)}$ is defined as:

$$y_{k,j}^{(\tau)} = \frac{1}{\tau} \sum_{i=(j-1)\tau+k}^{j\tau+k-1} x_i \quad (1 \leq j \leq \tfrac{N}{\tau}, 1 \leq k \leq \tau) \tag{6}$$

At a scale factor of τ, the sample entropy of all coarse-grained time series are calculated and the composite multiscale sample entropy (*CMSE*) value is defined as the means of τ entropy values:

$$CMSE(x, \tau, m, r) = \frac{1}{\tau} \sum_{k=1}^{\tau} SampEn(y_k^{(\tau)}, m, r) \tag{7}$$

3.4. Mann-Kendall Algorithm

The trend detection methods include Spearman's rho test, Mann-Kendall test, seasonal Kendall test, linear regression test, and so on [20]. One of the most favored trend detection methods is the Mann-Kendall algorithm, which has been employed to detect the trends in hydrological series, including precipitation [21], runoff [22], sample entropy [5], coefficient of variation, and concentration degree [23]. Detailed information on the Mann-Kendall algorithm can be found in the published papers [24]. The Mann-Kendall test is based on the test statistic S:

$$S = \sum_{i=1}^{n-1} \sum_{j=i+1}^{n} \mathrm{sgn}(x_j - x_i) \tag{8}$$

where the x_j and x_i are the sequential data values, n is the length of the data set, and:

$$\mathrm{sgn}(\theta) = \begin{cases} 1 & if \quad \theta > 0 \\ 0 & if \quad \theta = 0 \\ -1 & if \quad \theta < 0 \end{cases} \tag{9}$$

When $n \geq 8$, the statistic S is approximately normally distributed with the mean and the variance as follows:

$$E(S) = 0 \tag{10}$$

$$Var(S) = \frac{n(n-1)(2n+5) - \sum_{i=1}^{n} t_i i(i-1)(2i+5)}{18} \tag{11}$$

where t_i is the number of ties of extent i. The standardized test statistic Z is computed by

$$Z = \begin{cases} \frac{S-1}{\sqrt{Var(S)}} & S > 0 \\ 0 & S = 0 \\ \frac{S+1}{\sqrt{Var(S)}} & S < 0 \end{cases} \tag{12}$$

Z follows the standard normal distribution with a mean of zero and variance of one. A positive or negative value of Z represents an increasing or decreasing trend, respectively. In a two-tailed test, the null hypothesis is no trend, which can be rejected at significance level α if $|Z| > Z_{\alpha/2}$. The $Z_{\alpha/2}$ is the value of the standard normal distribution with an exceedance probability of $\alpha/2$. The significance level α is set to be 0.05 in this study.

4. Results and Discussion

4.1. Multiscale Complexity of Streamflows

Figure 2 displays the composite multiscale sample entropy of the inflow and outflow data of Poyang Lake, respectively. The variations of sample entropy versus time-scale are similar. The sample entropy is increasing when the time-scale is increasing, revealing that the time series have obvious self-similarity and great complexity [25]. Similar results, i.e., the complexity of streamflows is increasing as the time-scale is increasing, have also been found in the multiscale entropy analysis of the streamflows of the Mississippi River in the United States [4], the East River in China [3], and the Yangtze River in China [2]. These previous studies attribute this phenomenon to the existence of the long-range correlation of the streamflow records. Larger sample entropy reflects more randomness and complicated systems, and vice versa. When the time-scale is increasing, the increasing sample entropy indicates an increase in the uncertainty and disorderliness, and a lessened regularity and predictability.

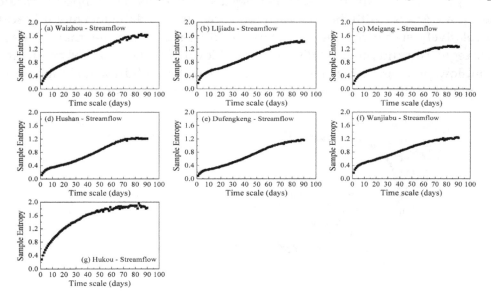

Figure 2. Composite multiscale sample entropy of the streamflows of Poyang Lake.

When comparing Figures 2a–f and 2g, it can be observed that the sample entropy values at different time-scales of the outflows are distributed above those of the inflows. When the sample entropy values of one time series are larger than those of another time series at most time-scales, it reveals that the former is more complex than the latter [26]. Figure 3 compares the average sample entropy of the inflows and outflows of Poyang Lake. The average sample entropy is the mean value of sample entropies at various time-scales. The average sample entropy of the streamflow data at the Hukou station is the largest in Figure 3, indicating that the outflows are more complex than the inflows of Poyang Lake. Due to the spatial and temporal distribution of precipitation, the Yangtze River and Poyang Lake have complex river-lake interactions. The streamflow processes at the Hukou station are determined by the water level in Poyang Lake (relative to the water level in the Yangtze River), which in turn is affected by both the inflow processes and the outflow processes [27]. Poyang Lake acts as a buffer at varying degrees for the Yangtze River streamflow. When the water level of the Yangtze River rises during the flood season, the Poyang Lake helps to absorb some flood and mitigates the peak streamflow of the Yangtze River [13]; hence, the streamflow processes at the Hukou station are more complex than the streamflow processes at the hydrometric stations of the tributaries.

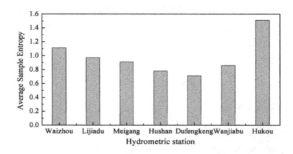

Figure 3. Average sample entropy of the streamflows of Poyang Lake.

4.2. Multiscale Complexity of Water Levels

Figure 4 displays the composite multiscale sample entropy of the water levels of Poyang Lake. The water levels, monitored at the Kangshan, Duchang, Xingzi, and Hukou stations, have similar change patterns in term of sample entropy versus time-scale. The sample entropy increases rapidly until the time-scale reaches about 30 days, after that, the increase becomes small. The sample entropy becomes relatively stable when the time-scale reaches about 30 days. The randomness and irregularity of the water levels are greater with the increased time-scale from a daily to a monthly scale. The water levels have a relatively stable complexity when the time-scale is larger than 30 days, indicating relatively stable regularity and predictability. The water level of Poyang Lake is an integrated response to the inflow, outflow, precipitation, evaporation, leakage, water withdraw, and topography of the lake. Hence, the complexity of the water levels may not be identical to that of the inflows and outflows.

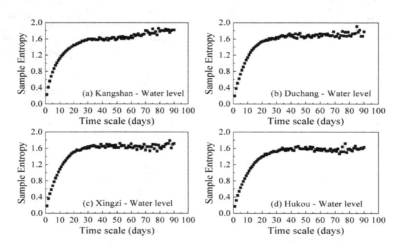

Figure 4. Composite multiscale sample entropy of the water levels of Poyang Lake.

4.3. Temporal Changes in Complexity of Streamflows and Water Levels

Figure 5 shows the statistical Z values of the Mann-Kendall algorithm, which reveals temporal trends in the complexity of the inflows and outflows of Poyang Lake. Figure 6 displays temporal variations in the complexity of the daily streamflows as examples. The sample entropy series of various time-scales at the Waizhou, Lijiadu, Meigang, Hushan, and Wanjiabu stations are dominated by increasing trends, indicating the increased inflow complexity of Poyang Lake. The sample entropy series of various time-scales at the Dufengkeng station are relatively stable. Only a significant decreasing trend is detected at the daily scale. The absolute Z values of the other time-scales are below the 0.05 significance level, indicating no significant trend. The sample entropy series over different time-scales at the Hukou station are also featured by increasing trends, indicating the increased outflow complexity of Poyang Lake.

Figure 7 shows temporal trends in the multiscale complexity of the water levels of Poyang Lake. Figure 8 displays temporal variations in the complexity of the daily water levels as examples. The sample entropy series over different time-scales are characterized by increasing tends, indicating the greater irregularity and randomness of the water levels.

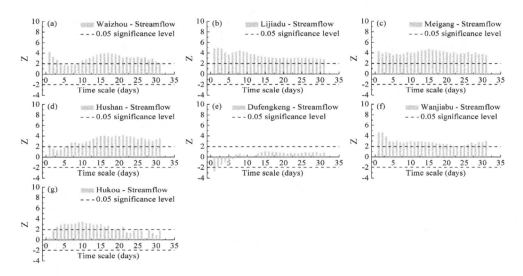

Figure 5. Trends in multiscale complexity of the streamflows of Poyang Lake.

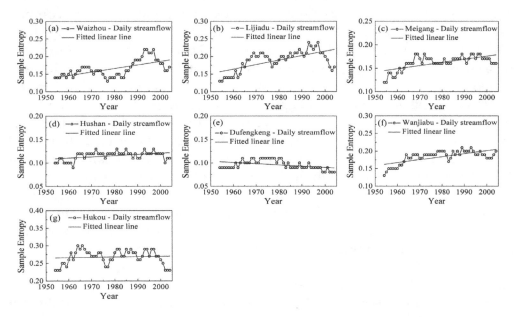

Figure 6. Temporal changes in complexity of the daily streamflows of Poyang Lake.

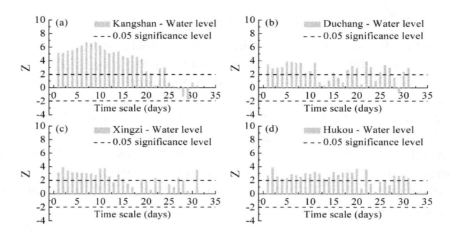

Figure 7. Trends in multiscale complexity of the water levels of Poyang Lake.

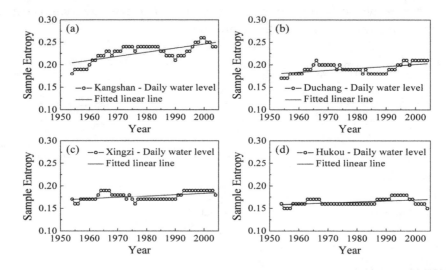

Figure 8. Temporal changes in complexity of the daily water levels of Poyang Lake.

4.4. Discussion on Temporal Changes in Streamflow and Water Level Complexity

Hydrological systems are open, dynamic, complex, giant, and nonlinear compound systems. For the Poyang Lake basin, the outputs of this hydrological system, which are described by the streamflows and the water levels, are highly nonlinear and complex, and are significantly impacted by both climate change and anthropogenic activities. Precipitation is a dominant impact factor of the Poyang Lake hydrological system. In the Jiangxi province, Huang et al. found remarkable differences among the meteorological stations with negative and positive precipitation trends at the annual, seasonal, and monthly scales. Significant increasing trends are mainly found during winter and summer, while significant decreasing trends are mostly observed during autumn [21]. Xiao et al. analyzed the spatial and temporal characteristics of rainfall across the Ganjiang River basin, the largest sub-basin of the Poyang Lake basin, and found significant increasing trends in the annual total rainfall amount [28]. Temperature is another relevant factor which directly affects evaporation. According to a study by Tao et al. of the Poyang Lake basin, the annual mean of the daily minimum temperature has increased significantly, while no significant trend has been detected in the annual mean of daily maximum temperature; thus, resulting in a significant decrease in the diurnal temperature range [29]. Ye et al. detected a significant decreasing trend in the annual reference evapotranspiration in the Poyang Lake basin [30]. Based on prior studies, more attention will be paid in future research to elucidate the complexity features of precipitation, temperature, evaporation, and the mechanisms on how the complexity of those impact factors affect the complexity of the outputs of the Poyang Lake hydrological system.

Including the climate factors, anthropogenic activities are also very important factors affecting the hydrological processes. The anthropogenic activities mainly include dam and reservoir construction, water withdraw, land use, and land cover changes, along with sand extraction from river and lake beds. Mei et al. argued that the average contributions of precipitation variation, human activities in the Poyang Lake catchment, and the Three Gorges Reservoir regulation to the Poyang Lake recession can be quantified as 39.1%, 4.6% and 56.3%, respectively [31]. Zhang et al. found that human-induced and climate-induced influences on streamflows are different in the five Poyang Lake sub-basins. Climate change is the major driving factor for the streamflow increases within the Ganjiang, Xinjiang, and Raohe River basins; however, anthropogenic activities are the principal driving factors for the streamflow increase of the Xiushui River basin and for the streamflow decrease of the Fuhe River basin [32].

Dam and reservoir construction is one of the most important anthropogenic activities, and has attracted worldwide attention due to its significant influence on hydrological processes [33–37]. In the Poyang Lake basin, thousands of reservoirs have been constructed, including 27 large reservoirs ($>10^8$ m^3 in storage capacity) [11]. The reservoirs in the Poyang Lake basin may affect the Poyang Lake hydrological complexity. The reservoirs in the upper Yangtze River basin, especially the Three Gorges Reservoir, may also affect the Poyang Lake hydrological complexity through disturbing the natural river-lake interrelationships. Some interesting conclusions have been obtained in terms of the impact of dam and reservoir construction on hydrological complexity. Huang et al. attributed the loss of streamflow complexity of the upper reaches of the Yangtze River to the underlying surface condition change, which has been influenced by human activities, especially reservoir construction. Huang et al. also argued that the reservoir operation makes the streamflow more regular and self-similar, leading to the streamflow complexity loss [5]. Similar conclusions were obtained in the hydrological complexity analysis of the Colorado River in the United States [38] and the Sao Francisco River in Brazil [39]. Those studies elucidated that dam and reservoir construction induced significant changes in streamflow dynamics, including an increase in regularity and a loss of complexity. However, the study in the East River (China) obtained the opposite conclusion. Zhou et al. argued that reservoir construction greatly increases the complexity of hydrological processes because of reservoir-induced noise of the streamflow [3]. More research is needed to systematically analyze the impact of reservoirs on hydrological complexity.

The enhancement of the streamflow and water level complexity in the Poyang Lake basin is the combined result of climate change and anthropogenic activities. Explaining how climate change and anthropogenic activities affect the hydrological complexity while at the same time distinguishing their individual contributions are our ongoing research objectives.

5. Conclusions

Based on the long term observed daily streamflow and water level data of Poyang Lake, the streamflow and water level complexity over various time-scales and its temporal changes are investigated using the composite multiscale sample entropy and the Mann-Kendall algorithm. The following conclusions can be drawn from the analysis:

(1)   The streamflow and water level complexity increases when the time-scale increases. The sample entropy of the streamflows increases when the time-scale increases from the daily to seasonal scale. The sample entropy of the water levels increases when the time-scale increases from the daily to monthly scale.

(2)   The complexity of the outflows is greater than that of the inflows. It may be caused by the complex river-lake interrelationships. The outflow processes of Poyang Lake are synthetically impacted by the inflow processes, lake regulation, and the streamflow processes of the Yangtze River.

(3) Significant upward trends can be detected in the sample entropy series, which are calculated using the streamflow and water level data, for most time-scales between the daily to monthly scale. The increased sample entropy indicates the enhanced streamflow and water level complexity, which may be caused by both climate change and anthropogenic activities. The mechanisms of the hydrological complexity changes will be studied in ongoing research.

Acknowledgments: This work is supported by the National Natural Science Foundation Projects of China (grant numbers 41401011, 51309131, 51679118, 41401010, and 41371098); Science and Technology Projects of Water Resources Department of Jiangxi Province (grant number KT201538); CRSRI Open Research Program (grant number CKWV2015237/KY); and Program for Changjiang Scholars and Innovative Research Team in University (grant number IRT13062). The authors acknowledge constructive comments from the editor and anonymous reviewers, which lead to improvement of the paper.

Author Contributions: The authors designed and performed the research together. Feng Huang wrote the draft of the paper. Xunzhou Chunyu, Yuankun Wang, Yao Wu, Bao Qian, Lidan Guo, Dayong Zhao and Ziqiang Xia made some comments and corrections. All authors have read and approved the final manuscript.

References

1. Sang, Y.F.; Wang, D.; Wu, J.C.; Zhu, Q.P.; Wang, L. Wavelet-Based Analysis on the Complexity of Hydrologic Series Data under Multi-Temporal Scales. *Entropy* **2011**, *13*, 195–210. [CrossRef]
2. Zhang, Q.; Zhou, Y.; Singh, V.P.; Chen, X.H. The influence of dam and lakes on the Yangtze River streamflow: long-range correlation and complexity analyses. *Hydrol. Process.* **2012**, *26*, 436–444. [CrossRef]
3. Zhou, Y.; Zhang, Q.; Li, K.; Chen, X.H. Hydrological effects of water reservoirs on hydrological processes in the East River (China) basin: Complexity evaluations based on the multi-scale entropy analysis. *Hydrol. Process.* **2012**, *26*, 3253–3262. [CrossRef]
4. Li, Z.W.; Zhang, Y.K. Multi-scale entropy analysis of mississippi river flow. *Stoch. Environ. Res. Risk Assess.* **2008**, *22*, 507–512. [CrossRef]
5. Huang, F.; Xia, Z.Q.; Zhang, N.; Zhang, Y.D.; Li, J. Flow-Complexity Analysis of the Upper Reaches of the Yangtze River, China. *J. Hydrol. Eng.* **2011**, *16*, 914–919. [CrossRef]
6. Liu, M.; Liu, D.; Liu, L. Complexity research of regional groundwater depth series based on multiscale entropy: a case study of Jiangsanjiang Branch Bureau in China. *Environ. Earth Sci.* **2013**, *70*, 353–361. [CrossRef]
7. Lai, X.J.; Liang, Q.H.; Jiang, J.H.; Huang, Q. Impoundment Effects of the Three-Gorges-Dam on Flow Regimes in Two China's Largest Freshwater Lakes. *Water Resour. Manag.* **2014**, *28*, 5111–5124. [CrossRef]
8. Cao, L.; Fox, A.D. Birds and people both depend on China's wetlands. *Nature* **2009**, *460*, 173. [CrossRef] [PubMed]
9. Zhao, G.J.; Hoermann, G.; Fohrer, N.; Zhang, Z.X.; Zhai, J.Q. Streamflow Trends and Climate Variability Impacts in Poyang Lake Basin, China. *Water Resour. Manag.* **2010**, *24*, 689–706. [CrossRef]
10. Ye, X.C.; Zhang, Q.; Liu, J.; Li, X.H.; Xu, C.Y. Distinguishing the relative impacts of climate change and human activities on variation of streamflow in the Poyang Lake catchment, China. *J. Hydrol.* **2013**, *494*, 83–95. [CrossRef]
11. Lai, X.J.; Huang, Q.; Zhang, Y.H.; Jiang, J.H. Impact of lake inflow and the Yangtze River flow alterations on water levels in Poyang Lake, China. *Lake Reserv. Manag.* **2014**, *30*, 321–330. [CrossRef]
12. Zhang, Z.X.; Chen, X.; Xu, C.Y.; Hong, Y.; Hardy, J.; Sun, Z.H. Examining the influence of river-lake interaction on the drought and water resources in the Poyang Lake basin. *J. Hydrol.* **2015**, *522*, 510–521. [CrossRef]
13. Guo, H.; Hu, Q.; Zhang, Q.; Feng, S. Effects of the Three Gorges Dam on Yangtze River flow and river interaction with Poyang Lake, China: 2003–2008. *J. Hydrol.* **2012**, *416*, 19–27. [CrossRef]
14. Dai, X.; Wan, R.R.; Yang, G.S. Non-stationary water-level fluctuation in China's Poyang Lake and its interactions with Yangtze River. *J. Geogr. Sci.* **2015**, *25*, 274–288. [CrossRef]
15. Chou, C.M. Complexity analysis of rainfall and runoff time series based on sample entropy in different temporal scales. *Stoch. Environ. Res. Risk Assess.* **2014**, *28*, 1401–1408. [CrossRef]

16. Costa, M.; Peng, C.K.; Goldberger, A.L.; Hausdorff, J.M. Multiscale entropy analysis of human gait dynamics. *Physica A* **2003**, *330*, 53–60. [CrossRef]

17. Richman, J.S.; Moorman, J.R. Physiological time-series analysis using approximate entropy and sample entropy. *Am. J. Physiol. Heart Circ. Physiol.* **2000**, *278*, H2039–H2049. [PubMed]

18. Costa, M.; Goldberger, A.L.; Peng, C.K. Multiscale entropy analysis of complex physiologic time series. *Phys. Rev. Lett.* **2002**, *89*, 068102. [CrossRef] [PubMed]

19. Wu, S.D.; Wu, C.W.; Lin, S.G.; Wang, C.C.; Lee, K.Y. Time Series Analysis Using Composite Multiscale Entropy. *Entropy* **2013**, *15*, 1069–1084. [CrossRef]

20. Kundzewicz, Z.W.; Robson, A.J. Change detection in hydrological records—A review of the methodology. *Hydrol. Sci. J.* **2004**, *49*, 7–19. [CrossRef]

21. Huang, J.; Sun, S.L.; Zhang, J.C. Detection of trends in precipitation during 1960–2008 in Jiangxi province, southeast China. *Theor. Appl. Clim.* **2013**, *114*, 237–251. [CrossRef]

22. Zhao, Q.H.; Liu, S.L.; Deng, L.; Dong, S.K.; Yang, J.J.; Wang, C. The effects of dam construction and precipitation variability on hydrologic alteration in the Lancang River Basin of southwest China. *Stoch. Environ. Res. Risk Assess.* **2012**, *26*, 993–1011. [CrossRef]

23. Huang, F.; Xia, Z.Q.; Li, F.; Guo, L.D.; Yang, F.C. Hydrological Changes of the Irtysh River and the Possible Causes. *Water Resour. Manag.* **2012**, *26*, 3195–3208. [CrossRef]

24. Yue, S.; Pilon, P.; Cavadias, G. Power of the Mann-Kendall and Spearman's rho tests for detecting monotonic trends in hydrological series. *J. Hydrol.* **2002**, *259*, 254–271. [CrossRef]

25. Chou, C.M. Applying Multiscale Entropy to the Complexity Analysis of Rainfall-Runoff Relationships. *Entropy* **2012**, *14*, 945–957. [CrossRef]

26. Costa, M.; Goldberger, A.L.; Peng, C.K. Multiscale entropy analysis of biological signals. *Phys. Rev. E* **2005**, *71*, 021906. [CrossRef] [PubMed]

27. Zhang, Q.; Li, L.; Wang, Y.G.; Werner, A.D.; Xin, P.; Jiang, T.; Barry, D.A. Has the Three-Gorges Dam made the Poyang Lake wetlands wetter and drier? *Geophys. Res. Lett.* **2012**, *39*, L20402. [CrossRef]

28. Xiao, Y.; Zhang, X.; Wan, H.; Wang, Y.Q.; Liu, C.; Xia, J. Spatial and temporal characteristics of rainfall across Ganjiang River Basin in China. *Meteorol. Atmos. Phys.* **2016**, *128*, 167–179. [CrossRef]

29. Tao, H.; Fraedrich, K.; Menz, C.; Zhai, J.Q. Trends in extreme temperature indices in the Poyang Lake Basin, China. *Stoch. Environ. Res. Risk Assess.* **2014**, *28*, 1543–1553. [CrossRef]

30. Ye, X.C.; Li, X.H.; Liu, J.; Xu, C.Y.; Zhang, Q. Variation of reference evapotranspiration and its contributing climatic factors in the Poyang Lake catchment, China. *Hydrol. Process.* **2014**, *28*, 6151–6162. [CrossRef]

31. Mei, X.F.; Dai, Z.J.; Du, J.Z.; Chen, J.Y. Linkage between Three Gorges Dam impacts and the dramatic recessions in China's largest freshwater lake, Poyang Lake. *Sci. Rep.* **2015**, *5*, 18197. [CrossRef] [PubMed]

32. Zhang, Q.; Liu, J.Y.; Singh, V.P.; Gu, X.H.; Chen, X.H. Evaluation of impacts of climate change and human activities on streamflow in the Poyang Lake basin, China. *Hydrol. Process.* **2016**, *30*, 2562–2576. [CrossRef]

33. Huang, F.; Chen, Q.H.; Li, F.; Zhang, X.; Chen, Y.Y.; Xia, Z.Q.; Qiu, L.Y. Reservoir-Induced Changes in Flow Fluctuations at Monthly and Hourly Scales: Case Study of the Qingyi River, China. *J. Hydrol. Eng.* **2015**, *20*, 05015008. [CrossRef]

34. Yang, T.; Zhang, Q.; Chen, Y.D.; Tao, X.; Xu, C.Y.; Chen, X. A spatial assessment of hydrologic alteration caused by dam construction in the middle and lower Yellow River, China. *Hydrol. Process.* **2008**, *22*, 3829–3843. [CrossRef]

35. Chen, Q.H.; Zhang, X.; Chen, Y.Y.; Li, Q.F.; Qiu, L.Y.; Liu, M. Downstream effects of a hydropeaking dam on ecohydrological conditions at subdaily to monthly time scales. *Ecol. Eng.* **2015**, *77*, 40–50. [CrossRef]

36. Wang, Y.K.; Wang, D.; Wu, J.C. Assessing the impact of Danjiangkou reservoir on ecohydrological conditions in Hanjiang river, China. *Ecol. Eng.* **2015**, *81*, 41–52. [CrossRef]

37. Wang, Y.K.; Rhoads, B.L.; Wang, D. Assessment of the flow regime alterations in the middle reach of the Yangtze River associated with dam construction: Potential ecological implications. *Hydrol. Process.* **2016**, *30*, 3949–3966. [CrossRef]

38. Serinaldi, F.; Zunino, L.; Rosso, O.A. Complexity-entropy analysis of daily stream flow time series in the continental United States. *Stoch. Environ. Res. Risk Assess.* **2014**, *28*, 1685–1708. [CrossRef]

39. Stosic, T.; Telesca, L.; Ferreira, D.V.D.; Stosic, B. Investigating anthropically induced effects in streamflow dynamics by using permutation entropy and statistical complexity analysis: A case study. *J. Hydrol.* **2016**, *540*, 1136–1145. [CrossRef]

5

Rainfall Network Optimization using Radar and Entropy

Hui-Chung Yeh [1], Yen-Chang Chen [2], Che-Hao Chang [2], Cheng-Hsuan Ho [2] and Chiang Wei [3,*]

[1] Department of Natural Resources, Chinese Culture University, Taipei 11114, Taiwan; xhz@faculty.pccu.edu.tw

[2] Department of Civil Engineering, National Taipei University of Technology, Taipei 10608, Taiwan; yenchen@ntut.edu.tw (Y.-C.C.); chchang@ntut.edu.tw (C.-H.C.); mazdaabuse@gmail.com (C.-H.H.)

[3] Experimental Forest, National Taiwan University, NanTou 55750, Taiwan

* Correspondence: d87622005@ntu.edu.tw

Abstract: In this study, a method combining radar and entropy was proposed to design a rainfall network. Owing to the shortage of rain gauges in mountain areas, weather radars are used to measure rainfall over catchments. The major advantage of radar is that it is possible to observe rainfall widely in a short time. However, the rainfall data obtained by radar do not necessarily correspond to that observed by ground-based rain gauges. The in-situ rainfall data from telemetering rain gauges were used to calibrate a radar system. Therefore, the rainfall intensity; as well as its distribution over the catchment can be obtained using radar. Once the rainfall data of past years at the desired locations over the catchment were generated, the entropy based on probability was applied to optimize the rainfall network. This method is applicable in remote and mountain areas. Its most important utility is to construct an optimal rainfall network in an ungauged catchment. The design of a rainfall network in the catchment of the Feitsui Reservoir was used to illustrate the various steps as well as the reliability of the method.

Keywords: entropy; information transfer; optimization; radar; rainfall network

1. Introduction

Rainfall data form the fundamental basis for hydraulic and hydrological engineering. Adequate and long-term rainfall data are essential in planning and management of water resources. The definition of rainfall is any product of atmospheric water that reaches the surface of Earth in the form of droplets of water [1]. Thus, rain gauges are the gold standard of precipitation measurement [2]. They are the principal source of rainfall data for rainfall network design. However, accurate and reliable rainfall data of catchments depend on well-designed rainfall networks. Ideally, a higher number of rainfall gauges in a catchment provides a clearer picture of the aerial distribution of the rainfall. Usually, network density and rainfall gauge distribution depend on the particular application. Many factors may affect the number and locations of rain gauges. However, there is no definite rule for constructing a rainfall network. The actual density of a rainfall network is significantly poorer than the values recommended by the World Meteorological Organization (WMO) [3]. Therefore, various methods have been used in the past to investigate the density of rainfall networks and the optimization of these networks. The WMO recommends certain densities of rain gauge stations for different types of catchments. In flat regions of temperate zones, 500 km^2 per station is recommended. For small mountainous islands with irregular precipitation, 25 km^2 per station is recommended [3]. Langbein [4] suggested that the densities of rain gauge stations are usually proportional to population density. The use of statistical characteristics is generally desirable in the design of rainfall networks; Rodriguez-Iturbe and Mejia [5] used a random process technique to develop design curves for estimating the mean of a rainfall event. Shih [6] introduced various steps based on a covariance factor among rain gauge stations to design a rainfall network. Patra [1] applied the coefficient of variance and allowable percentage of error to estimate the optimal number of rain gauge stations. Basalirwa et al. [7] attempted to design a minimum rainfall network by using principal component analysis. Likewise, geostatistics is frequently used in the design of rainfall networks. Kassim and Kottegoda [8] prioritized rain gauges with respect to their contribution in error reduction in the network through comparative Kriging methods. Chen et al. [9] developed a method by using Kriging and entropy that can determine the optimum number and spatial distribution of rain gauge stations in a catchment. Chebbi et al. [10] proposed an algorithm composed of a geostatistical variance-reduction method and simulated annealing to expand the existing rainfall network. Ridolfi et al. [11] introduce an entropy approach for evaluating the maximum information content achievable by an urban rainfall network. Shaghaghian and Abedini [12] selected an optimal subset of stations in the network by using Kriging, factor analysis, and clustering techniques to achieve the optimum rainfall network. Chebbi et al. [13] identified the optimal network using an intensity-duration-frequency curve and a variance-reduction method. Related works based on entropy since Krstanovic and Singh [14] is widen and undergoing [15–18]. Wei et al. [19] introduced entropy to evaluate the effect of spatiotemporal scaling on rainfall network design.

Two major scientific problems need to be addressed in studies on rainfall networks: the first problem is the number of rain gauge stations required to provide adequate representation of a catchment's rainfall characteristics, and the second problem is that the positioning of these rain gauge stations. These two issues are essential to the optimal network design and addressed by previous studies in river points [20], water level networks in polders [21,22], cross-section spacing for river modeling [23], groundwater quality monitoring [24,25], monitoring network design [26,27], and the homogeneity of the study region, also discussed by new clustering method based on entropy [28]. As these are interrelated problems, they need to be considered in conjunction during the design of rainfall networks. In this work, the data of Project Quantitative Precipitation Estimation and Segregation Using Multiple Sensors (QPESUMS) were used to estimate the spatial distribution of rainfall, whereas entropy was used to evaluate the uncertainty of each rain gauge station and to determine how the uncertainty and spatial distribution of rainfall interact with each other. For this purpose, a comprehensive evaluation of the Feitsui Reservoir's rain gauge station locations and distribution were performed using a three-step procedure: the first step was to use the currently available rainfall data to calibrate the radar parameters of the QPESUMS. The second step was to apply the radar to estimate the rainfall data of the candidate rain gauge stations. The third step was to use information entropy to determine the priority of the candidate rain gauge stations and to estimate the minimum required number of rain gauge stations. By combining the use of the radar of the QPESUMS and information entropy, we can determine the locations where new stations should be set up and the number of rain gauge stations required for a rainfall network. Therefore, during the determination of rain gauge station locations using this method, new rain gauge stations will be suggested if a rainfall network has fewer stations than the saturation number determined by our method, to provide an adequate quantity of catchment rainfall data. Conversely, if a rainfall network has a higher number of rain gauge stations than the saturation number, the rainfall data provided by the excess rain gauge stations will be limited, thus the removal of rain gauge stations would be suggested to improve the cost efficiency of hydrological information systems.

In this study, the Feitsui Reservoir's catchment was used to demonstrate how the proposed method can be used to construct an optimal rainfall network. As global climate change may have altered the hydrological characteristics of the Feitsui Reservoir's catchment, long-term hydrological data was used to assess whether an adjustment was needed for the rain gauge stations in this area, so that rain gauge stations may be added to ensure the operational safety of the reservoir. For this purpose, the measured rainfall data of the rain gauge stations were used to calibrate the parameters of the weather radar in this area, which was subsequently used to estimate the spatial distribution of rainfall in the reservoir catchment and the historical rainfall data of the candidate stations. In addition, the information transfer theories of information entropy were used to derive the importance of each candidate station and to construct an optimal rainfall network for the catchment after information was accumulated up to a specified saturation level. Finally, suggestions were provided on the appropriateness of rain gauge station additions.

2. Methodology

2.1. Radar Estimation of Rainfall

Radar was recognized for the measurement of precipitation in the late 1940s. An equation that relates the intensity of rainfall with radar echo factors was constructed after the Second World War, which gave birth to studies on the application of radar in the observation of rainfall. Radar observation is based on the scattering and reflection of high-power electromagnetic waves (emitted by radar antennas towards the atmosphere) when they encounter droplets of water or ice in clouds or raindrops. The energy of the reflected electromagnetic waves received by the radar antennas may then be used to estimate the quantity of rainfall.

The operation of radars is based on the Doppler effect, and radar is mainly used in meteorology to measure rainfall and track storms. Radars operate by performing 360° scans at different elevation angles (from the highest to lowest elevation angles), and the observation of precipitation may then be performed using the Doppler principle. As the topography of Taiwan is highly complex and mountainous, radar scans at low elevation angles are often blocked by mountains. The current solution for circumventing this issue is to use radar echo data from higher elevations to replace the data of topographically obstructed regions, i.e., by selecting echo factors with the lowest (unobstructed) elevation angle for an obstructed region, as shown in Figure 1.

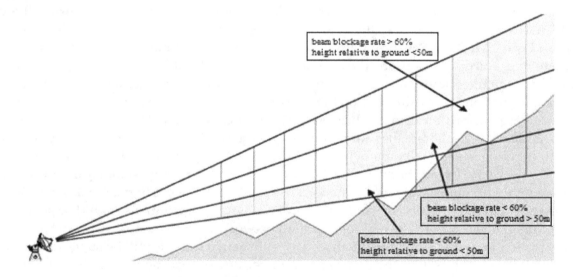

Figure 1. Schematic diagram of radar echoes with elevation (Zhang, 2006). When the beam blockage of a pulse volume's echo factor exceeds 60%, or if the central point of a pulse volume is less than 50 m above ground, its data is then replaced by the echo factor of the next (higher) elevation angle.

The basic meteorological radar equation for the quantification of rainfall is

$$P_r = \frac{\pi^3}{2^{10}\ln} \frac{P_t g^2 \theta h}{\lambda^2 r_0^2} |k_w|^2 Z \tag{1}$$

In this equation, P_r is the power of the echoes received by the radar antenna, P_t is the transmission power of the radar antenna, g is the radar antenna's gain, θ is the beam width, h is the spatial pulse length, k_w is the dielectric constant of the medium, r_0 is the distance from the radar to the area of precipitation, and Z is the radar reflectivity factor.

When the raindrops are very small in diameter and homogeneously distributed in space, the radar reflectivity factor and the raindrop diameter within a unit volume of the radar beam are proportionally related by a power of 6. Hence, the echo factor (the value of Z) may be expressed as

$$Z = \frac{1}{\Delta V} \sum_i D_i^6 = \int_0^\infty D^6 N(D) dD \tag{2}$$

In this equation, ΔV is the unit volume, D is the diameter of the raindrops, $N(D)$ is the raindrop diameter distribution function for a diameter, D (i.e., the raindrop density). The raindrop diameter distribution proposed by Marshall & Palmer [29] is

$$N(D) = N_0 e^{-\Lambda D} \tag{3}$$

Here, N_0 is a constant, and Λ is the rainfall rate function. Therefore, the function for the total number of raindrops may be expressed as

$$N_{total} = \int_0^\infty D^0 N(D) dD \tag{4}$$

The rainfall intensity (R) (mm/h) of the rainfall rate is the total rainfall on each unit of surface area per unit time. Hence, the relationship between rainfall intensity and raindrop diameter is

$$R = \frac{\pi}{6} \sigma_w \int_0^\infty D^3 N(D)(W_t - W) dD \tag{5}$$

In this equation, σ_w is the density of liquid water, W_t is the terminal velocity of the raindrops, and W is the flow rate of ascending airflows.

Equations (2) and (5) demonstrate that Z and R are both related to the raindrop density function; according to statistical analyses of the observed rainfall intensity and rainfall density data, the empirical function that relates rainfall intensity to the radar echo wave is

$$Z = aR^b \tag{6}$$

In this equation, a and b are parameters that may be derived from regression analysis with the rainfall data collected on the ground. In this study a and b are 32.5 and 1.65, respectively [30].

2.2. Information Transfer by Using Entropy

In 1948, Shannon proposed the probability-based concept of information entropy, which is quite different from thermodynamic entropy [31].

$$H(x) = -\sum_i p(x_i) \ln p(x_i) \tag{7}$$

In this equation, $H(x)$ is the entropy value. x represents an event, and $p(x)$ represents the probability of this event.

The data acquired by the rain gauge stations of a rainfall network may overlap with each other. If we treat the data of two rain gauge stations as two variables, x and y, the joint probability of x and y, p_{ij}, may then be expressed as

$$p_{ij} = p(x = x_i, y = y_j) \tag{8}$$

The total information content may be deduced from the joint entropy, which is

$$H(x,y) = -\sum_i \sum_j p_{ij} \ln\left(p_{ij}\right) \tag{9}$$

Equation (9) represents the uncertainty between two rain gauge stations. Like the characteristics of the joint probability distribution, the sum of the marginal entropies of x and y should be larger than, or equal to, the joint probability

$$H(x,y) \leq H(x) + H(y) \tag{10}$$

Similarly, the joint probability of three rain gauge stations (x, y, and z) is

$$H(x,y,z) = -\sum_i \sum_j \sum_k p_{ijk} \ln p_{ijk} \tag{11}$$

In this equation, p_{ijk} is the joint probability between rain gauge stations x, y and z.

When a rainfall signal is measured by station x, the residual uncertainty of station y may be expressed by the conditional entropy. The conditional probability for an event occurring at x when an event has occurred at y may be expressed as

$$p(x|y) = p_{i|j} = \frac{p_{ij}}{p_j} \tag{12}$$

Hence,

$$\begin{aligned}
H(x,y) \quad &= -\sum_i \sum_j p_{ij} \ln\left(p_{ij}\right) \\
&= -\sum_i \sum_j p(x|y)p(y) \ln(p(x|y)p(y)) \\
&= -\sum_i \sum_j p(x|y)p(y)[\ln(p(x|y)) + \ln(p(y))] \\
&= -\sum_j p(y)\sum_i p(x|y) \ln(p(x|y)) - \sum_j p(y) \ln(p(y))\sum_i p(x|y)
\end{aligned} \tag{13}$$

The first term of Equation (13) is the conditional entropy, $H(x|y)$, while the second term is the conditional probability; therefore, $\sum_i \sum_j p(x|y) = 1$ becomes the entropy value of $y.\sum_i \sum_j p(x|y) = 1$. Equation (12) may then be written as

$$H(x,y) = H(x|y) + H(y) \tag{14}$$

and

$$p_{ij} = p(x|y)p(y) = p(y|x)p(x) \tag{15}$$

Therefore,

$$H(x,y) = H(y|x) + H(x) \tag{16}$$

It may be inferred from Equations (13) and (16) that

$$H(x|y) \leq H(x) \tag{17}$$

Furthermore, the conditional entropy may be inferred from the equation below

$$H(x|y) = -\sum_j \sum_i p_{ij} \ln p_{i|j} \tag{18}$$

In the conditional entropy of a single rain gauge station, there will be no uncertainty. Hence, the conditional entropy of a single station is

$$H(x|x) = 0 \tag{19}$$

The calculation of transferable information may be used to determine whether two rain gauge stations will possess shared or redundant information, which would allow the rainfall at Station y to be deduced from the data of Station x. The equation for calculating the transferable information is

$$\begin{aligned} T(x,y) &= H(y) - H(y|x) \\ &= H(x) - H(x|y) \\ &= H(x) + H(y) - H(x,y) \end{aligned} \tag{20}$$

or

$$T(x,y) = \sum_i \sum_j p_{ij} \ln \frac{p_{ij}}{p_i p_h} \tag{21}$$

The importance of each rain gauge station in a rainfall network is described by its entropy value, and the priority of the rain gauge stations may be displayed by sorting the stations by entropy. The rain gauge station that has the largest entropy value will have the highest uncertainties, and it should be the first station selected for entry into the rainfall network. After the first rain gauge station has been determined, the rain gauge stations with the lowest quantity of redundant information should be systematically added to the rainfall network one after another, to reduce the uncertainty of the system. Hence, the criterion for determining the second most important rain gauge station for addition to the rainfall network is

$$Min\{H(x_1) - H(x_1|x_2)\} \tag{22}$$

This selects for the station with the highest $H(x_1 | x_2)$ value. The selection criterion for the j-th most important rain gauge station is then

$$Min\{H(x_1, x_2, \cdots, x_{j-1}) - H[(x_1, x_2, \cdots, x_{j-1}|x_j)]\} \tag{23}$$

The stations with the highest value of $H[(x_1, x_2, \cdots, x_{j-1}|x_j)]$ may then be selected from the calculations. This yields a ranking of all of the rain gauge stations in a rainfall network by the redundancy of their data, and the station with the highest redundancy will be the last station to be added to the network.

The ranking of rain gauge stations by importance derived from their entropy values may be used as an ordering for the removal of stations, and the increase in uncertainty may be used as a criterion for determining the removal of a station. After a certain number of stations have been added to the network, the value of $H[(x_1, x_2, \cdots, x_{j-1}|x_j)]$ will no longer increase or change significantly, and the information provided by further additions will be limited. Hence, an exponential model may be defined using the number of rain gauge stations and the value of $H[(x_1, x_2, \cdots, x_{j-1}|x_j)]$ to find the critical quantity of information and the required number of rain gauge stations, as shown below

$$H(n) = \omega \left[1 - \exp\left(\frac{-n}{c}\right)\right] \tag{24}$$

where ω and c are to-be-determined parameters.

If the number of rain gauge stations in a rainfall network is larger than the required number of stations, then the stations that rank lower than the required number may then be removed. Conversely, if the number of rain gauge stations in a rainfall network is lower than the required number of stations, more stations then need to be added.

The ranking of rain gauge stations by importance based on their entropy values may be used as an ordering for the removal of stations. The maximization of entropy is the objective of each station selection stage. The addition of each station should increase the joint entropy, but after a certain number of stations have been added to the network, it may be observed that the entropy value, $H(n)$ no longer increases or changes significantly, and converges to a fixed value instead. This indicates that all further additions to the system will only be able to provide a limited quantity of information. The index model of this study was used to plot the relationship between $H\left[(x_1, x_2, \cdots, x_{j-1}|x_j)\right]$ and the number of rain gauge stations, to find the critical quantity of information and the number of required stations. Here, we define a k_m coefficient, which represents the ratio between the entropy value of the m-th station added to the system and the total entropy of the study area. Hence, k_m may be used to represent the quantity of information provided by the m-th station added to the network.

Suppose that the study area has n measurement stations; after the base station has been selected, each subsequent addition is performed with the objective of maximizing entropy. The definition of k_m is then

$$k_m = \frac{H(x_1, x_2, \ldots, x_m)}{H(x_1, x_2, \ldots, x_m, \ldots, x_{n-1}, x_n)}, m < n \tag{25}$$

and $k_1, k_2, \ldots, k_m, \ldots, k_{n-1}, k_n < 1$.

When determining the number of rain gauge stations for an area, a threshold value, k_m^* also needs to be determined. When $k_m > k_m^*$, the number of rain gauge stations for a study area may then be obtained. The determination of the threshold value may be determined by the increase in efficacy, as revealed by the increase in k_m. The threshold is usually defined as $k_m = 0.95$, i.e., 95% of the information content. If the number of rain gauge stations in a rainfall network is larger than the required number of stations, the stations that rank lower than the required number may then be removed. Conversely, if the number of rain gauge stations in a rainfall network is smaller than the required number of stations, more stations then need to be added.

3. Study Area and Data Description

The Feitsui Reservoir is located southeast of Taipei. The main river of the Feitsui Reservoir's catchment is approximately 50 km long, with a drainage area of about 303 km^2 (Figure 2). The dendritic drainage system of this catchment includes four tributaries (the Daiyuku Creek, the Jingualiao Creek, the Houkengzi Creek, and the Huoshaozhang Creek) joined together into the main river, the Beishi River. The source of the Beishi River lies in the western slope at the northern end of the Xueshan Range, and it flows into Pinglin before feeding into the Daiyuku Creek and Jingualiao Creek. After this point, the Beishi River broadens and slows before it flows westward into the Feitsui Reservoir.

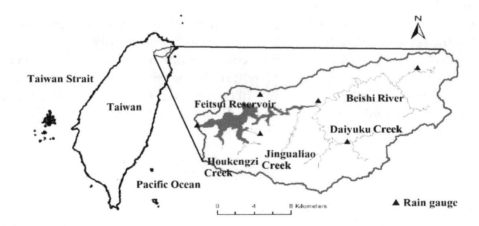

Figure 2. The Feitsui Reservoir's catchment and the locations of the rain gauge stations.

The Feitsui Reservoir catchment is within the subtropical climate zone. Cold and wet northeasterly monsoons prevail in winter, and the cold Arctic air will invade southwards from time to time. Cold snaps, low clouds and drizzles thus have a high probability of occurrence in winter. The southwesterly monsoons in summer have a minimal impact on this region, due to the obstruction of the Xueshan range. However, local showers often occur in the afternoon since solar radiation on the river valleys and hillsides has a significant impact on local convection. During the transition between summer and autumn, typhoons will bring about warm and highly humid airflows, and extremely heavy and intense rainfalls. The temporal and spatial distributions of rainfall were already taken into consideration during the initial design of the current rain gauge stations. For this reason, the Taipei Feitsui Reservoir Administration has established an integrated weather station for the Feitsui Reservoir in one location, and rain gauge stations in five locations, as shown in Figure 2.

The maximum, minimum and standard deviation of the monthly rainfall in the watershed are shown in Table 1. The rainy season is from August to November, and the variance in mean monthly rainfall often exceeds 100 mm. The rainfall tends to fluctuate from year to year; the annual rainfall ranges between 2520 and 5740 mm, and the mean annual rainfall is approximately 3760 mm. From August to October, this area is impacted by typhoons and storms that frequently bring about heavy rainfall, whereas the rainfall frequency peaks in the period between October and January, due to the impacts of the northeasterly monsoon. During 2006–2015, at least 30 typhoons invaded Taiwan. The rainfall data during 2006 and 2015, including the heavy rain coming with the typhoons, is used for this study.

Table 1. The maximum, minimum and standard deviation of the monthly rainfall of six rain gauges in Feitsui Reservoir's catchment.

Station	Taiping	Sirsangoo	Pingling	Feitsui	Geochungan	Beefu
Grid number	19	61	66	102	129	157
Maximum (mm/M)	2668.5	1723.0	2113.5	1902.0	1925.5	2330.0
Minimum (mm/M)	0.5	0.0	12.0	43.0	26.5	0.0
Mean (mm/M)	449.3	282.3	294.7	299.0	302.5	333.2
Std. Dev. (mm/M)	358.1	218.4	263.0	234.1	234.4	234.4

The weather radar station was installed by the Central Weather Bureau of Taiwan to provide real-time severe weather information. It is located around the most northeastern Taiwan. The radar was tested in July 1996, however it was destroyed by a typhoon in August 1996. The radar was repaired in 1998, and the data acquisition system was upgraded in 2006. In order to make the data consistent, only the data after 2006 is used in this study. The radar has a wavelength of 10 cm, a spatial resolution of $1.3 \times 1.3 \ km^2$, and makes rainfall observations once every 10 min. Therefore, the catchment of the Feitsui Reservoir may be divided into 217 grids according to the radar's spatial resolution, with the center of each grid being the location of a candidate rain gauge station, as shown in Figure 3. The 19th, 61st, 66th, 102nd, 129th, and 157th grids in Figure 3 correspond to the locations of the six currently existing rain gauge stations.

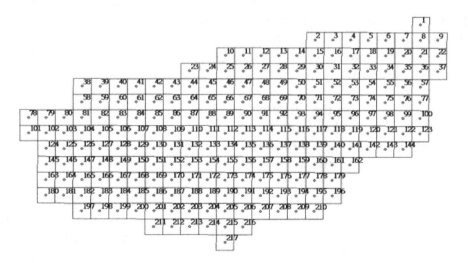

Figure 3. The positions and numbering of the candidate rain gauges in the Feitsui Reservoir's catchment.

Prior to this study, it has not been assessed whether increases in the number of rain gauge stations are necessary in light of alterations to the spatial and temporal distribution of rainfall induced by global climate changes, to ensure that the rainfall data being acquired is sufficient for the smooth operation of the reservoir. To address this shortfall, the mean monthly rainfall data from January 2006 up to December 2015 (120 months in total) was used to evaluate the efficacy of the current rainfall network. Figure 4 illustrates the monthly rainfall map of the Feitsui Reservoir's catchment from 2006 to 2015. Heavy rainfalls always occur in summer due to the arrival of typhoons, while the Meiyu front brings rain in spring, and the northeasterly monsoon also brings rain in winter, thus resulting in rainfall over long periods of the year. The minimum in rainfall usually occurs in autumn. The Feitsui Reservoir provides the water supply and usually operates a regulation line for 10-days. According to this, the rainfall temporal period for six-month/year seems too long for the reservoir operation and the analysis, hourly data may contribute to the flood control, while the month scale is reasonable for reservoir operation and also reflecting the variation during dry and wet seasons.

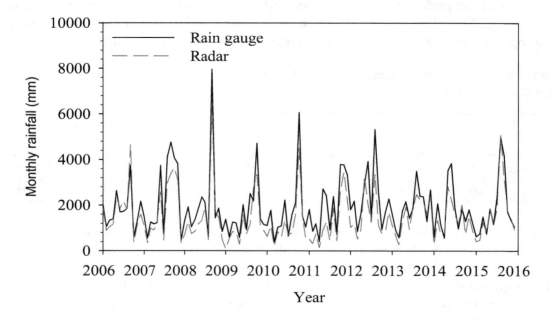

Figure 4. The monthly rainfall in the Feitsui Reservoir's catchment from 2006 to 2015.

4. Results and Discussion

As rainfall network evaluations will require considerations on the long-term temporal and spatial distributions of rainfall, the monthly rainfall data was selected for data analysis in this work. The arithmetic mean was used to estimate the mean rainfall of the catchment to minimize the impact of the spatial distribution of rainfall, and because the six currently existing rain gauge stations are spatially distributed in a uniform manner.

To estimate the rainfall data of the candidate rain gauge stations for the evaluation of the rainfall network, the observed mean monthly rainfall of the catchment was derived from the rainfall data of the six currently existing rain gauge stations, while the radar of QPESUMS was used to estimate the monthly rainfall at these six locations to obtain the radar-derived mean monthly rainfall. A linear regression was then used to probe the relationship between these quantities. Figure 5 displays the relationship between the rainfalls measured by the six currently existing rain gauge stations and the rainfalls estimated using the radar system. The horizontal axis (R_{QP}) indicates the monthly rainfall observed by the radar, while the vertical axis (R_G) represents the mean monthly rainfall of the catchment measured by the rain gauge stations. It was found that these quantities were related by $R_G = 1.1R_{QP} + 36.13$, with a correlation coefficient of 0.87. Since the correlation is quite strong, this equation and the radar's rainfall data may then be used to estimate the mean monthly rainfalls of all the candidate rain gauge stations in the 211 grids of the catchment, over the last 20 years.

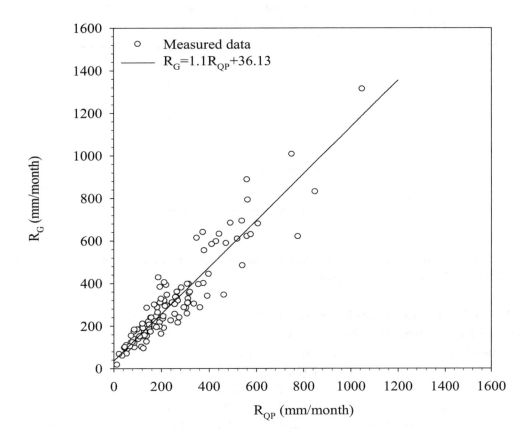

Figure 5. The relationship between the measured mean monthly rainfall and the estimated mean monthly rainfall.

In the study area, stations were selected according to transferable information calculations and joint entropy-based ordering. Hence, the entropy values of the selected stations in the study area were calculated using joint entropy and transferable information, and the stations were then ordered by sorting the calculated values. The ranking of rain gauge stations by importance based on their entropy values may be used as an ordering for the removal of stations. The maximization of entropy is the objective of each station selection stage. Each station addition should increase the joint entropy, but after the number of added stations has reached a certain value, it may be observed that the entropy value, $H(n)$, no longer increases or changes significantly, and converges to a fixed value; this indicates that all further additions to the system will only provide a limited quantity of information.

The study area has 217 grids. The grids corresponding to the six currently existing stations were selected as necessary locations for the reservoir's operation, while the remaining grids were added one after another, based on the principle of entropy maximization. A threshold value is usually defined to determine the number of stations required by some area. In this case, the threshold, k_m^*, was defined as 0.95, which corresponds to a threshold value of 95%. By doing so, almost all of the rainfall information of a region can be acquired using only a few rain gauge stations. Hence, when $k_m \geq k_m^*$, the number of stations that need to be added to the study area and the location of these stations may then be obtained. The relationship between the entropy value and the optimal number of stations is illustrated in Figure 6; an index function was used in this work (as shown in Equation (24)) to obtain an estimate for the optimal number of stations.

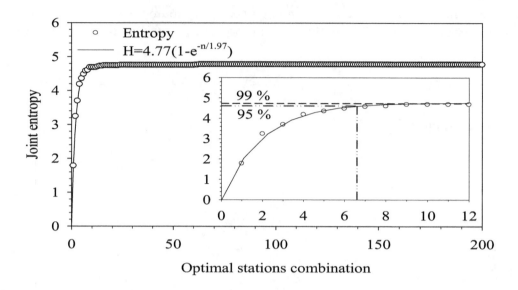

Figure 6. The relationship between joint entropy value and the optimal number of stations; all of the grids and the first 12 grids within the graph.

The grids labeled 1–6 in Figure 7 correspond to the locations of the six currently existing rain gauge stations in the catchment. The entropy value of these six stations is already ~94%, which indicates that these stations are sufficient for normal operation of the water reservoir. Nonetheless, if an improvement in the completeness of the data is desired, the addition of a station at Grid 7 will increase k_m from 0.938 to 0.958, which is larger than the $k_m^* = 0.95$ threshold. Therefore, we propose that one more rain gauge station should be added to the study area, in addition to the pre-existing rain gauge stations. The location of this station is indicated by Grid 7 in Figure 7, and it is located at the boundary of the catchment in the southeastern part of the study area.

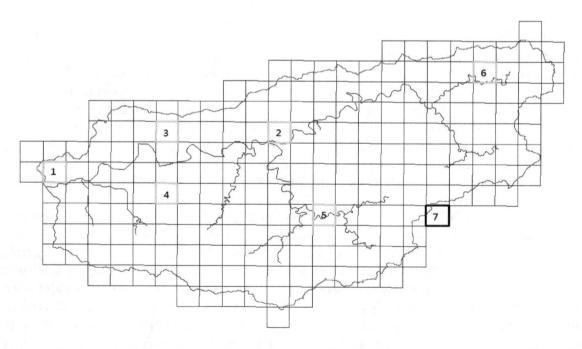

Figure 7. The position of the recommended rain gauge station (the bold grid) for addition to the rainfall network; Blue grids 1 to 6 are the existed rainfall stations.

5. Conclusions

In this study, surface and radar rainfall data were employed in unison to estimate the rainfall data of the candidate rain gauge sites, and information entropy was used to evaluate the information content and uncertainty of each rain gauge station. An optimal rainfall network may then be constructed by combining these methods. Unlike previous estimations of rainfall based on the Kriging method, the rainfall data obtained using radar is an actual measurement, whereas rainfalls estimated using statistical methods are at best, estimated values, which may not accurately reflect on the temporal and spatial distributions of rainfall in a catchment. Previously, it was impossible to obtain an exact answer for the location of rain gauge stations and the minimum required number of stations; with information entropy, it is now possible to simultaneously obtain an answer for both of these questions. This method will provide an important basis for the management of watersheds and the establishment of rain gauge stations. Furthermore, this method may also be used to assess the sufficiency of the rainfall data provided by currently existing rain gauge stations. More stations need to be added to a rainfall network if the data is insufficient, whereas stations need to be removed if the redundancy in data is too high. It is hoped that this method will be used to evaluate or adjust currently existing rainfall networks in the catchment.

To identify and test the robustness of this method of rainfall network design in catchments, the rainfall network at the Feitsui Reservoir's catchment was evaluated to highlight the applicability and reliability of our method. A grid was defined every 1.3 km in the study area, and actual radar data was used to reconstruct the historical rainfall data of these grids. Information entropy was then used to evaluate the spatial information content and the uncertainty of the information, and the estimated entropy values were used to add stations to the network one after another. It was shown that seven stations is the optimal number of rain gauge stations for the selected study area, as this is the number of stations required to obtain 95% of the study area's rainfall information, additional manpower and resources for initial establishing and maintaining more stations can be saved.

Acknowledgments: This paper is based on work partially supported by Ministry of Science and Technology, Taiwan (Grant No. MOST 106-2221-E-027-031-). The authors would like to thank the Taipei Feitsui Reservoir Administration for providing the fundamental, rainfall data and related assistance.

Author Contributions: Hui-Chung Yeh and Yen-Chang Chen conceived and designed the experiments; Che-Hao Chang analyzed the radar data; Cheng-Hsuan Ho performed the materials/analysis tools; Chiang Wei performed the analysis and wrote the manuscript. All authors have read and approved the final manuscript.

References

1. Patra, K.C. *Hydrology and Water Resources Engineering*; Alpha Science: Oxford, UK, 2010.
2. Strangeways, I. *Precipitation: Theory, Measurement and Distribution*; Cambridge University Press: Cambridge, UK, 2007.
3. WMO. *Guide to Hydrological Practices, WMO-164*; WMO: Geneva, Switzerland, 1994.
4. Langbein, W.B. Hydrologic data networks and methods of extrapolating or extending available hydrologic data. In *Hydrologic Networks and Method*; United Nations: Bangkok, Thailand, 1960.
5. Rodriguez-Iturbe, I.; Mejia, J.M. The design of rainfall networks in time and space. *Water Resour. Res.* **1974**, 23, 181–190. [CrossRef]
6. Shih, S.F. Rainfall variation analysis and optimization of gaging systems. *Water Resour. Res.* **1982**, 18, 1269–1277. [CrossRef]
7. Basalirwa, C.P.K.; Ogallo, L.J.; Mutua, F.M. The design of regional minimum rain gauge network. *Int. J. Water Resour. Dev.* **2007**, 9, 411–424. [CrossRef]
8. Kassim, A.H.M.; Kottegoda, N.T. Rainfall network design through comparative kriging methods. *Hydrol. Sci. J.* **1991**, 36, 223–240. [CrossRef]
9. Chen, Y.C.; Wei, C.; Yeh, H.C. Rainfall network design using kriging and entropy. *Hydrol. Process.* **2008**, 22, 340–346. [CrossRef]

10. Chebbi, A.; Bargaoui, Z.K.; Cunha, M.D.C. Optimal extension of rain gauge monitoring network for rainfall intensity and erosivity index interpolation. *J. Hydrol. Eng. ASCE* **2011**, *16*, 665–676. [CrossRef]

11. Ridolfi, E.; Montesarchio, V.; Russo, F.; Napolitano, F. An entropy approach for evaluating the maximum information content achievable by an urban rainfall network. *Nat. Hazards Earth Syst. Sci.* **2011**, *11*, 2075–2083. [CrossRef]

12. Shaghaghian, M.R.; Abedini, M.J. Rain gauge network design using coupled geostatistical and multivariate techniques. *Sci. Iran.* **2013**, *20*, 259–269. [CrossRef]

13. Chebbi, A.; Bargaoui, Z.K.; Cunha, M.D.C. Development of a method of robust rain gauge network optimization based on intensity-duration-frequency results. *Hydrol. Earth Syst. Sci.* **2013**, *17*, 4259–4268. [CrossRef]

14. Krstanovic, P.F.; Singh, V.P. Evaluation of rainfall networks using entropy: I. Theoretical development. *Water Resour. Manag.* **1992**, *6*, 279–293. [CrossRef]

15. Yoo, C.; Jung, K.; Lee, J. Evaluation of Rain Gauge Network Using Entropy Theory: Comparison of Mixed and Continuous Distribution Function Applications. *J. Hydrol. Eng.* **2008**, *13*, 226–235. [CrossRef]

16. Leach, J.M.; Kornelsen, K.C.; Samuel, J.; Coulibaly, P. Hydrometric network design using streamflow signatures and indicators of hydrologic alteration. *J. Hydrol.* **2015**, *529*, 1350–1359. [CrossRef]

17. Chacon-Hurtado, J.C.; Alfonso, L.; Solomatine, D.P. Rainfall and streamflow sensor network design: A review of applications, classification, and a proposed framework. *Hydrol. Earth Syst. Sci. Discuss.* **2017**, *21*, 3071–3091. [CrossRef]

18. Stosic, T.; Stosic, B.; Singh, V.P. Optimizing streamflow monitoring networks using joint permutation entropy. *J. Hydrol.* **2017**, *552*, 306–312. [CrossRef]

19. Wei, C.; Yeh, H.C.; Chen, Y.C. Spatiotemporal scaling effect on rainfall network design using entropy. *Entropy* **2014**, *16*, 4626–4647. [CrossRef]

20. Harmancioglu, N.; Yevjevich, V. Transfer of hydrologic information among river points. *J. Hydrol.* **1987**, *91*, 103–118. [CrossRef]

21. Alfonso, L.; Lobbrecht, A.; Price, R. Information theory-based approach for location of monitoring water level gauges in polders. *Water Resour. Res.* **2010**, *46*. [CrossRef]

22. Alfonso, L.; Lobbrecht, A.; Price, R. Optimization of water level monitoring network in polder systems using information theory. *Water Resour. Res.* **2010**, *46*, W12553. [CrossRef]

23. Ridolfi, E.; Alfonso, L.; Baldassarre, G.D.; Dottori, F.; Russo, F.; Napolitano, F. An entropy approach for the optimization of cross-section spacing for river modelling. *Hydrol. Sci. J.* **2013**, *59*, 126–137. [CrossRef]

24. Mogheir, Y.; Singh, V.P. Application of information theory to groundwater quality monitoring networks. *Water Resour. Manag.* **2002**, *16*, 37–49. [CrossRef]

25. Mogheir, Y.; de Lima, J.L.M.P.; Singh, V.P. Assessment of spatial structure of groundwater quality variables based on the entropy theory. *Hydrol. Earth Syst. Sci.* **2003**, *7*, 707–721. [CrossRef]

26. Alfonso, L.; Ridolfi, E.; Gaytan-Aguilar, S.; Napolitano, F.; Russo, F. Ensemble entropy for monitoring network design. *Entropy* **2014**, *16*, 1365–1375. [CrossRef]

27. Ridolfi, E.; Yan, K.; Alfonso, L.; Baldassarre, G.D.; Napolitano, F.; Russo, F.; Bates, P.D. An entropy method for floodplain monitoring network design. *AIP Conf. Proc.* **2012**, *1479*, 1780–1783. [CrossRef]

28. Ridolfi, F.; Rianna, E.; Trani, G.; Alfonso, L.; Baldassarre, G.D.; Napolitano, G.; Russo, F. A new methodology to define homogeneous regions through an entropy based clustering method. *Adv. Water Resour.* **2016**, *96*, 237–250. [CrossRef]

29. Marshall, J.S.; Palmer, W.M. The distribution of raindrops with size. *J. Meteorol.* **1948**, *5*, 165–166. [CrossRef]

30. Xin, L.; Reuter, G.; Larochelle, B. Reflectivity-rain rate relationships for convective rainshowers in Edmonton: Research note. *Atmos. Ocean* **1997**, *35*, 513–521. [CrossRef]

31. Shannon, C.E. A mathematical theory of communication. *Bell Syst. Tech. J.* **1948**, *27*, 623–656. [CrossRef]

6

Comparison of Two Entropy Spectral Analysis Methods for Streamflow Forecasting in Northwest China

Zhenghong Zhou [1], Juanli Ju [1,*], Xiaoling Su [1], Vijay P. Singh [2] and Gengxi Zhang [1]

[1] College of Water Resources and Architectural Engineering, Northwest A&F University, Yangling 712100, China; zzh199302@nwsuaf.edu.cn (Z.Z.); xiaolingsu@nwsuaf.edu.cn (X.S.); gengxizhang@nwsuaf.edu.cn (G.Z.)

[2] Department of Biological & Agricultural Engineering and Zachry Department of Civil Engineering, Texas A&M University, 2117 TAMU, College Station, TX 77843, USA; vsingh@tamu.edu

*   Correspondence: jujuanli@nwsuaf.edu.cn

Abstract: Monthly streamflow has elements of stochasticity, seasonality, and periodicity. Spectral analysis and time series analysis can, respectively, be employed to characterize the periodical pattern and the stochastic pattern. Both Burg entropy spectral analysis (BESA) and configurational entropy spectral analysis (CESA) combine spectral analysis and time series analysis. This study compared the predictive performances of BESA and CESA for monthly streamflow forecasting in six basins in Northwest China. Four criteria were selected to evaluate the performances of these two entropy spectral analyses: relative error (RE), root mean square error (RMSE), coefficient of determination (R^2), and Nash–Sutcliffe efficiency coefficient (NSE). It was found that in Northwest China, both BESA and CESA forecasted monthly streamflow well with strong correlation. The forecast accuracy of BESA is higher than CESA. For the streamflow with weak correlation, the conclusion is the opposite.

Keywords: monthly streamflow forecasting; Burg entropy; configurational entropy; entropy spectral analysis time series analysis

1. Introduction

Accurate streamflow forecasting is important for developing measures to flood control, river training, navigation, reservoir operation, hydropower generation plan and water resources management. Time series models, such as autoregressive (AR) or autoregressive moving average (ARMA) models, as proposed by Box and Jenkins [1], are generally used for monthly streamflow forecasting [2–4]. These models assume that streamflow time series is stochastic and are linear which limits their application [5]. Monthly streamflow time series not only exhibits stochastic characteristics but also seasonal and periodic patterns. Entropy spectral analysis can extract important information of time series, such as the periodic characteristics [6–11]. Therefore, combining entropy spectral theory with time series analysis provides a new way for streamflow forecasting. Considering frequency f as a random variable, Burg [12] defined entropy, called Burg entropy, and developed an algorithm for the estimation of spectral density function of time series using the principle of maximum entropy (POME). The algorithm is termed Burg entropy spectral analysis (BESA) and has been widely used for spectral analysis of geomagnetic series [13], climate indices [8,14], surface air temperature [15], tide levels [16], precipitation and runoff series [17], and flood stage [18]. BESA is recommended as better than traditional methods for long-term hydrological forecasting [19–23]. Huo et al. [24] applied BESA to simulate and predict groundwater in the west of Shandong province plain of the Yellow River downstream and achieved satisfactory results. Wang and Zhu [25] considered that implicit periodic components of monthly and annual hydrological time series were better identified by BESA. Shen et al. [26] proposed a more rigorous recursion algorithm for maximum entropy spectral estimation method. In addition to meeting forward and backward minimum error of BESA, the algorithm also needed to satisfy a condition that the optimal prediction error was orthogonal to the signal. It was considered that the spectral density resolution of this method was higher than that of BESA. Boshnakov and Lambert-Lacroix [27] proposed an extension of the periodic Levinson-Durbin algorithm which was considered more reliable. However, multi-peak spectral density is difficult to determine under non-stationary conditions. Hence, monthly streamflow that features strong seasonal and periodic characteristics cannot be well simulated [28].

Frieden [29] was the first to use configurational entropy in image reconstruction and Gull and Daniell [30] applied it to radio astronomy. Based on the finite length cepstrum model, Wu [31] deduced an explicit spectral density function estimation formula and solved the complex calculation problem of CESA. Nadeu [32] regarded that spectral estimation precision of CESA was higher than that of BESA for both ARMA and MA, while the corresponding precisions were quite similar for AR. Katsakos et al. [33] found that the precision was higher when the spectral density of white noise series was estimated. Based on the spectral density estimation formula constructed by Wu, Cui, and Singh [28] derived a single variable streamflow forecasting model and found that the forecasting accuracy of CESA was superior to BESA for 19 different rivers in the US. For monthly streamflow forecasting, resolution and reliability of CESA were better than those of BESA.

The objective of this paper therefore was to compare the forecast performances of BESA and CESA for monthly streamflow forecasting in Northwestern China. The paper is organized as follows. First, a brief introduction to streamflow forecasting is given. Second, a maximum entropy spectral analysis prediction model is derived and evaluation methods are discussed. Third, application to streamflow forecasting is discussed. Fourth, results are discussed. Finally, conclusions were given.

2. Derivation and Evaluation of Maximum Entropy Spectral Analysis Prediction Model

2.1. Maximum Entropy Model

Let streamflow time series frequency f be a random variable, and the normalized spectral density $P(f)$ be taken as the probability density function. Thus, the Burg entropy can be defined as

$$H_B(f) = -\int_{-W}^{W} \ln[P(f)]df \tag{1}$$

The configurational entropy is defined in the same form as the Shannon entropy and can be written as

$$H_C(f) = -\int_{-W}^{W} P(f) \ln[P(f)]df \tag{2}$$

where $W = 1/(2\Delta t)$ is the Nyquist fold-over frequency and f is the frequency that varies from $-W$ to W, Δt is the sampling period, $P(f)$ is the normalized spectral density of streamflow series.

2.2. Constraints for Model

For a given streamflow time series, the constraints can be formed from the relationship between the spectral density $P(f)$ and autocorrelation function $\rho(n)$, which can be written as

$$\rho(n) = \int_{-W}^{W} P(f)e^{i2\pi fn\Delta t}df, \quad -N \leq n \leq N \tag{3}$$

where Δt is the discretization or sampling interval, and $i = \sqrt{-1}$. N is normally taken from 1/4 up to 1/2 of the series length according to the periodicity of streamflow.

2.3. Determination of Spectral Density

To obtain the least biased spectral density $P(f)$ by entropy maximizing, one needs to maximize the Burg and configurational entropies. Entropy maximizing can be done by using the method of Lagrange multipliers in which the Lagrangian function for the Burg entropy and configurational entropies can be formulated as follows:

$$L_B(f) = -\int_{-W}^{W} \ln[P(f)]df - \sum_{n=-N}^{N} \lambda_n \left[\int_{-W}^{W} P(f)e^{i2\pi f n \Delta t}df - \rho(n) \right] \qquad (4)$$

$$L_C(f) = -\int_{-W}^{W} P(f) \ln[P(f)]df - \sum_{n=-N}^{N} \lambda_n \left[\int_{-W}^{W} P(f)e^{i2\pi f n \Delta t}df - \rho(n) \right] \qquad (5)$$

where λ_n, $n = 0, 1, 2, \ldots, N$, are the Lagrange multipliers. Taking the partial derivative of Equations (4) and (5) with respect to $P(f)$ and equating the derivative to zero, the least-biased spectral densities $P(f)$ obtained from the maximization of the Burg entropy and configurational entropy, respectively, are

$$P_B(f) = -\frac{1}{\sum\limits_{n=-N}^{N} \lambda_n \exp(-i2\pi f n \Delta t)} \qquad (6)$$

$$P_C(f) = \exp\left(-1 - \sum_{n=-N}^{N} \lambda_n e^{i2\pi f n \Delta t}\right) \qquad (7)$$

It can be seen from the above two equations that the spectral density derived from the Burg entropy is in the form of inverse of polynomials, while the one from the configurational entropy is in the exponential form, which is easier to manipulate. The form in Equation (6) suggests that BESA is related to a linear prediction process.

2.4. Solution of the BESA Model

The spectral density derived is defined in the same form as the autoregressive model. On the basis of minimum of the forward and backward prediction error, a method of parameter estimation was presented by Burg, which can be written as

$$a_k(i) = \begin{cases} a_{k-1}(i) + k_k a_{k-1}(k-i), & i = 1, 2, \ldots, k-1 \\ k_k, & i = k \end{cases} \qquad (8)$$

where $a_k(i)$ is the i-th parameter value of the k-order autoregressive model, and the parameter k_k is estimated by minimizing the forward and backward prediction error.

$$e_k^f(i) = e_{k-1}^f(i) + k_k e_{k-1}^b(i-1)$$
$$e_k^b(i) = e_{k-1}^b(i-1) + k_k e_{k-1}^f(i)$$
$$k_k = \frac{-2\sum\limits_{i=k}^{N-1} e_{k-1}^f(i)e_{k-1}^b(i-1)}{\sum\limits_{i=k}^{N-1} \left| e_{k-1}^f(i) \right|^2 + \sum\limits_{i=k}^{N-1} \left| e_{k-1}^b(i-1) \right|^2} \qquad (9)$$

where $e^f{}_0(t) = e^b{}_0(t) = x(t)$, $x(t)$ is the streamflow time series.

For configurational entropy, the Lagrange multipliers and the extension of autocorrelation function can be computed by cepstrum analysis. Then, Wu [31] deduced the explicit solution based on the maximization of configurational entropy. Taking the inverse Fourier transform of the log-magnitude of Equation (7), it becomes

$$\int_{-W}^{W}\{1 + \ln[P(f)]\}\exp(i2\pi k\Delta t)df = \int_{-W}^{W}\left[-\sum_{n=-N}^{N}\lambda_n\exp(i2\pi fn\Delta t)\right]\exp(i2\pi fk\Delta t)df \qquad (10)$$

where the second part of the left side of Equation (10) can be denoted as

$$e(k) = \int_{-W}^{W}\ln[P(f)]\exp(i2\pi fk\Delta t)df \qquad (11)$$

Doing the integration of both sides of Equation (10), one gets

$$\delta_k + e(k) = -\sum_{n=-N}^{N}\lambda_n\delta_{k-n} \qquad (12)$$

where δ_n is the Dirac delta function defined as

$$\delta_n = \begin{cases} 1, & n = 0 \\ 0, & n \neq 0 \end{cases} \qquad (13)$$

Equation (12) can be expanded as a set of N linear equations:

$$\begin{aligned} \lambda_0 &= -e(0) - 1 \\ \lambda_1 &= -e(1) \\ &\vdots \\ \lambda_N &= -e(N) \end{aligned} \qquad (14)$$

Equation (14) shows that the Lagrange multipliers can be determined from the values of cepstrum which entails the spectral density that is obtained from Equation (7). It is the main difference from Burg entropy.

For convenience of solving for the spectral density function, Nadeu [32] developed a simple method for computing cepstrum based on the use of the causal part of autocorrelation, where $\rho(n)$ is used only for $-N \leq n \leq N$. Thus, cepstrum can be estimated by the following recursive relation:

$$e(n) = \begin{cases} 2\left[\rho(n) - \sum_{k=1}^{n-1}\frac{k}{n}e(k)\rho(n-k)\right], & n > 0 \\ 0, & n \leq 0 \end{cases} \qquad (15)$$

On the other hand, for the configurational entropy, the autocorrelation is extended with the inverse relationship of Equation (15) using the autocepstrum as

$$\rho(n) = \frac{e(n)}{2} + \sum_{k=1}^{n-1}\frac{k}{n}e(k)\rho(n-k) \qquad (16)$$

Therefore, with model order m determined, the autocorrelation function can be estimated as

$$\rho(m) = \sum_{k=1}^{m}a_k\rho(m-k), \ m \leq N \qquad (17)$$

with extension coefficients $a_k = \frac{k}{m}e(k)$, and m is the model order.

Equation (17) extends the autocorrelation function with the configurational entropy maximized. Surprisingly, the autocorrelation extends with a linear combination of past lags, which is the same with the Burg entropy or the AR method. Thus, Equation (17) can be also written as

$$\rho(t) = \sum_{k=1}^{m} a_k \rho(t-k), \ t > T \tag{18}$$

where T is the total time period.

The extended autocorrelation in Equation (18) is a linear combination of the previous values weighted with coefficients a_k. Burg (1975) suggested weighing time series using the extension coefficients as

$$x(t) = \sum_{k=1}^{m} a_k x(t-k), \ t > T \tag{19}$$

Equation (19) represents the forecast using the entropy-based extended autocorrelation. It has been shown by Burg and Krstanovic and Singh [12,19–21] that Equation (19) satisfies the least squares prediction.

2.5. Determination of Model Order

The order of forecasting model m is identified by the Bayesian Information Criterion (BIC) [34]. BIC can reduce the order of the model by penalizing free parameters more strongly compared with AIC (Akaike information criterion).

$$BIC(m) = N \ln \sigma_\varepsilon^2 + m \ln N \tag{20}$$

where N is the length of time series and σ_ε^2 is the variance of residual.

2.6. Procedure for Streamflow Forecasting

The computation procedure for monthly streamflow forecasting is shown in Figure 1. The computation steps are as follows: (1) Streamflow data $x(t)$ are normalized with Equation (21); (2) The parameters in the model (BESA and CESA) are estimated and the cepstrum values are determined for computing the Lagrange multipliers; (3) The forecast order m is identified by the Bayesian Information Criterion (BIC) and monthly streamflow is forecasted; (4) The prediction results of streamflow series are obtained by inverse normalization and exponential transformation.

$$y(t) = zscore[\ln x(t)] \tag{21}$$

where $zscore$ is a standardized function and $y(t)$ is a logarithmic sequence minus the mean divided by the standard deviation of the original sequence.

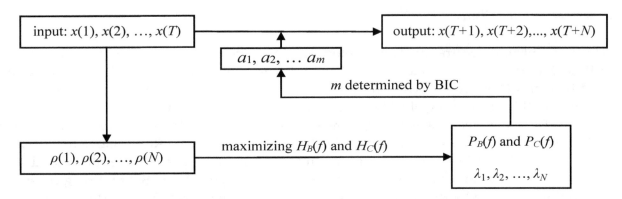

Figure 1. The computation procedure of entropy spectral analysis.

2.7. Evaluation of Model Forecast Performances

Four criteria were selected to evaluate the prediction model performance: relative error (RE), root mean square error (RMSE), coefficient of determination (R^2), and Nash–Sutcliffe efficiency coefficient

(NSE). The relative error provides the average magnitude of differences between observed values and predicted values relative to observed values. RMSE also represents the difference between observed and predicted values, however, it is scale-dependent. The coefficient of determination is defined as the square of the coefficient of correlation. It ranges between 0 and 1, and its higher values indicate better prediction. The Nash–Sutcliffe efficiency coefficient, defined by Nash and Sutcliffe [35], ranges from negative infinity to 1. Higher values of NSE represent more agreement between model predictions and observations, and negative values indicate that the model is worse than the mean value as a predictor.

$$RE = \frac{1}{N} \sum_{i=1}^{N} \left| \frac{Q_f(i) - Q_o(i)}{Q_o(i)} \right| \tag{22}$$

$$RMSE = \sqrt{\frac{\sum_{i=1}^{N} \left(Q_o(i) - Q_f(i)\right)^2}{N - 1}} \tag{23}$$

$$R^2 = \left\{ \frac{\sum_{i=1}^{N} \left(Q_o(i) - \overline{Q_o}\right)\left(Q_f(i) - \overline{Q_f}\right)}{\left[\sum_{i=1}^{N} \left(Q_o(i) - \overline{Q_o}\right)^2\right]^{0.5} \left[\sum_{i=1}^{N} \left(Q_f(i) - \overline{Q_f}\right)^2\right]^{0.5}} \right\}^2 \tag{24}$$

$$NSE = 1 - \frac{\sum_{i=1}^{N} \left|Q_o(i) - Q_f(i)\right|^2}{\sum_{i=1}^{N} \left|Q_o(i) - \overline{Q_o}\right|^2} \tag{25}$$

where N is the number of observed streamflow data, $Q_o(i)$ is the i-th observed streamflow, $Q_f(i)$ is the i-th forecasted streamflow, $\overline{Q_o}$ and $\overline{Q_f}$ are the average values of observed and forecasted streamflow, respectively.

3. Application to Streamflow Forecasting

3.1. Observed Data and Characteristics

The two entropy spectral analysis methods, BESA and CESA, were testedusing observed streamflow data from six river sites on the Yellow River, Heihe River, Zamu River, Xiying River, Datong River, and Daxia River. The Yellow River has a large drainage area of 752,443 km^2, with an average monthly streamflow of 633 m^3/s. Datong River and Daxia River are tributaries of the Yellow River. These two rivers have drainage areas of 151,33 km^2 and 7154 km^2, with average monthly streamflow of 88 m^3/s and 27 m^3/s, respectively. Zamu River and Xiying River belong to the Shiyang River watershed, with drainage areas of 851 km^2 and 1120 km^2. The Heihe River is the second largest interior river in Northwest China, with a drainage area of 130,000 km^2. Six hydrological stations selected in this paper are located in the Yellow River, Heihe River and Shiyang River, respectively. Tangnaihai station is located on the mainstream of Yellow River, while Xiangtang and Zheqiao stations are located on the tributary of Yellow River, Zamusi and Jiutiaoling stations are situated on the Shiyang River. Yingluoxia station is located in the Heihe River and it marks the boundary between the upstream and middle reaches. The location and basic information of each station are shown in Figure 2 and Table 1.

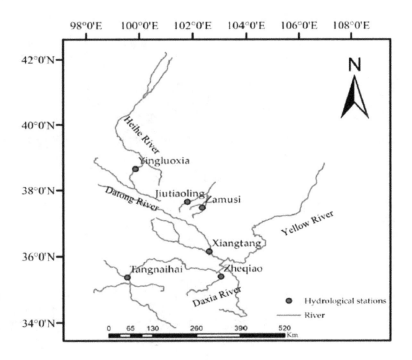

Figure 2. The location of selected stations.

Table 1. Basic information of steamflow data for selected stations.

No.	Station	Longitude	Latitude	River	Basin Area (km²)	Catchment Area (km²)	Record Length	Average (m³/s)	Peak (m³/s)
1	Xiangtang	102°51′E	36°22′N	Datong	15133	15,126	1950–2016	88	506
2	Yingluoxia	100°11′E	38°48′N	Heihe	130,000	10,009	1954–2012	51	214
3	Zamusi	102°34′E	37°42′N	Zamu	851	851	1952–2010	8	58.2
4	Jiutiaoling	102°03′E	37°52′N	Xiying	1120	1077	1972–2010	10	43.7
5	Tangnaihai	100°09′E	35°30′N	Yellow	752,443	121,972	1956–2016	633	3550
6	Zheqiao	103°16′E	35°38′N	Daxia	7154	6843	1963–2016	27	210

Monthly streamflow box-plots with all available data are presented in Figure 3. The bottom (Q1) and top (Q3) of the box are the first and third quartiles of streamflow, and the band inside the box is the median of streamflow. The inter quartile range (IQR) is equal to the difference between first and third quartiles. The limit of whiskers is called the inner fence which is 1.5 IQR from the quartile and the outer fence is 3 IQR from the quartile. Outliers are points that fall outside the limits of whiskers. + represents mild outliers which are between an inner and outer fence. × represents the extreme outliers which are beyond one of the outer fences. As shown in Figure 3, streamflow is concentrated during the flood season (June–September), and it drops down in the non-flood season. Because precipitation is the most important streamflow supply and the precipitation in these basins is concentrated during June–September.

There are many mild and extreme outliers for monthly streamflow data during the flood season at Xiangtang and Zheqiao stations, respectively. This is mainly due to poor vegetation coverage and barren hills in Datong River (Xiangtang station) downstream regions. Meanwhile, rainfall is mainly concentrated from June to September and mainly consists of heavy rain. Daxia River (Zheqiao station) upstream and downstream flow through the rocky mountainous region and loess plateau, separately. Serious soil erosion, heavy rain, mudslides, and landslides are frequent there. Streamflow during the flood season has many positive outliers for every station (Figure 3), and logarithmic processing is able to reduce the skewness of positive outliers in Section 2.7.

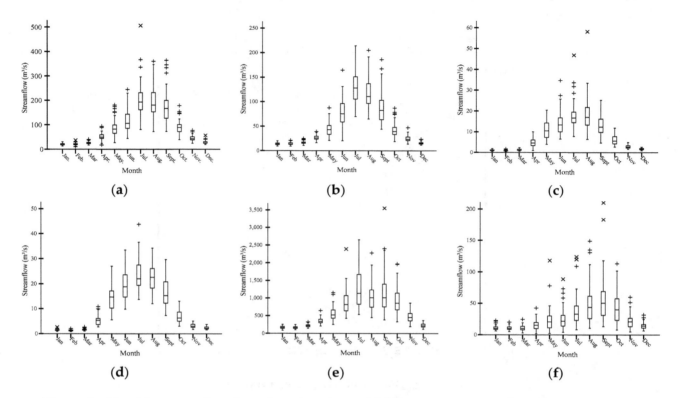

Figure 3. Monthly streamflow for selected stations. (**a**) Xiangtang station; (**b**) Yingluoxia station; (**c**) Zamusi station; (**d**) Jiutiaoling station; (**e**) Tangnaihai station; (**f**) Zheqiao station.

3.2. Comparison the Results of BESA and CESA

It is shown in a previous study that with the increase of the training time, both the accuracy of the training time and the precision of the lead time do not increase. Streamflow is forecasted by the two entropy spectral analysis methods with a five year training time (2003–2007) and a three year lead time (2008–2010) for representative stations. The simulated values and observed values for the two entropy spectrum models during the training period are shown in Figure 4. Both models were capable of simulating preferably streamflow variations at all stations. However, simulation results were better for the short leading time than that of the long leading time. The error between observed and simulated values was increasing with the lead time extension. The simulation values were better in drought seasons than in flood seasons, which mainly reflected the peak position and peak values. The maximum discharge during the flood seasons for six rivers appeared in different months for every year. This may lead to one month in advance or delay for the simulated values than the observed values for both methods. The predicted streamflow in the flood seasons was lower than observed streamflow for some stations. It was mostly at Yingluoxia, Tangnaihai and Zheqiao stations. Compared with CESA, the predicted and observed values were closer in flood seasons for BESA model. Overall, the simulation of streamflow time series at the above stations was superior for BESA to CESA.

The forecasted values and observed values for two entropy spectral models in the lead time are shown in Figure 5. For Xiangtang, Yingluoxia, Zamusi, and Jiutiaoling stations, both models satisfactorily forecasted streamflow. In the first lead year, two models accurately forecasted the time of maximum monthly streamflow at Xiangtang, Yingluoxia, and Jiutiaoling stations. Nevertheless, the predicted maximum streamflow for the last two years appeared one month earlier or later. At Zamusi station, BESA accurately forecasted the bi-modal values of the flood season for the lead time, while CESA did not. BESA forecasted the number of peaks in the following two years, but the peak position appeared one month earlier or later. At Tangnaihai and Zheqiao stations, the difference between the predicted and observed values for BESA model was large. However, CESA still did not forecast the multimodal pattern of partial flood season, while uni-modal year of streamflow in the flood season had better forecast results.

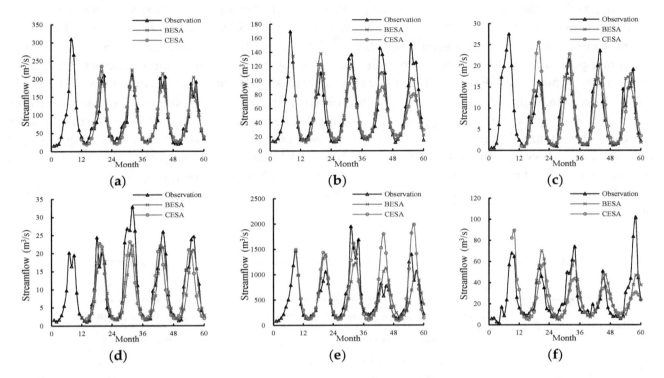

Figure 4. Streamflow forecasted using entropy spectral analysis of representative stations in training time. (**a**) Xiangtang station; (**b**) Yingluoxia station; (**c**) Zamusi station; (**d**) Jiutiaoling station; (**e**) Tangnaihai station; (**f**) Zheqiao station.

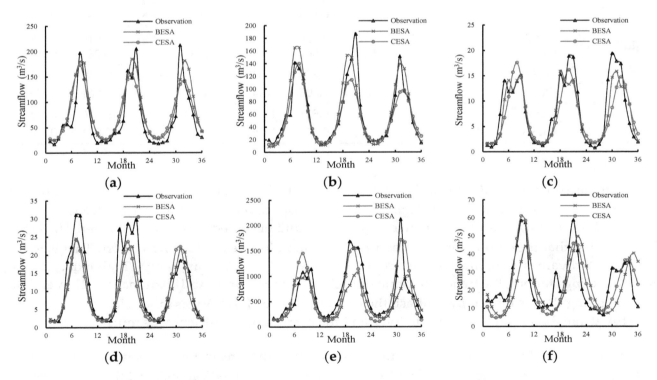

Figure 5. Streamflow forecasted using entropy spectral analysis of representative stations in lead time. (**a**) Xiangtang station; (**b**) Yingluoxia station; (**c**) Zamusi station; (**d**) Jiutiaoling station; (**e**) Tangnaihai station; (**f**) Zheqiao station.

The performance metrics of the models are shown in Table 2. It can be seen that the optimum order of BESA and CESA models range from 8 to 16 and 8 to 13, respectively. The R^2 and NSE values at Xiangtang, Yingluoxia, Zamusi, and Jiutiaoling stations during the training period were relatively high, with the values of over 0.86 and 0.70, respectively. The R^2 and NSE values at Tangnaihai and Zheqiao stations were lower than at the former four stations. The simulation of streamflow time series during the training period for BESA was better than that of CESA for the former four stations, and performance metrics were superior to CESA. For the other stations, the simulation results were equivalent for the two models. Streamflow forecasting by the two entropy spectrum models was good at Xiangtang, Yingluoxia, Zamusi, and Jiutiaoling stations during the verification period. The corresponding R^2 and NSE values were all higher than 0.88 and 0.70, respectively. However, streamflow forecasting at Tangnaihai and Zheqiao stations was relatively poor. Although the R^2 values were more than 0.76, the NSE values were between 0.48 and 0.49. BESA performed better at Xiangtang, Yingluoxia, Zamusi, and Jiutiaoling stations while CESA was more suitable for forecasting streamflow at Tangnaihai and Zheqiao stations.

Table 2. Results of forecasting at representative stations by two entropy methods.

Station	Model	Model Order	Training Time (2003–2007)				Lead Time (2008–2010)			
			RE	RMSE	R^2	NSE	RE	RMSE	R^2	NSE
xiangtang	BESA	16	0.169	19.4	0.953	0.904	0.376	27.3	0.913	0.773
	CESA	11	0.231	24.5	0.920	0.840	0.405	28.1	0.890	0.760
yingluoxia	BESA	8	0.202	17.3	0.915	0.834	0.235	21.2	0.912	0.798
	CESA	10	0.267	22.8	0.874	0.708	0.185	21.1	0.917	0.801
zamusi	BESA	14	0.207	2.7	0.913	0.832	0.268	2.5	0.925	0.838
	CESA	12	0.316	3.7	0.864	0.690	0.382	3.5	0.840	0.684
jiutiaoling	BESA	11	0.229	4.5	0.920	0.768	0.245	4.9	0.915	0.755
	CESA	13	0.346	5.1	0.868	0.697	0.318	5.4	0.886	0.707
tangnaihai	BESA	13	0.312	303.3	0.750	0.548	0.295	354.8	0.759	0.482
	CESA	8	0.335	340.3	0.805	0.447	0.360	273.0	0.861	0.693
zheqiao	BESA	15	0.362	19.0	0.621	0.255	0.291	9.1	0.876	0.618
	CESA	8	0.466	17.8	0.640	0.365	0.369	8.6	0.843	0.659

Comparison of monthly streamflow estimated by BESA and CESA and observed values during the verification period is shown in Figures 6 and 7, respectively. The slope of the trend line was closer to 1, indicating that the bias between predicted and observed values was smaller. The larger R^2 suggested that the correlation between predicted and observed values was better. That is to say, the predicted values were much closer to the observed values. At Xiangtang, Yingluoxia, Zamusi, and Jiutiaoling stations, the trend line slope of BESA was much closer to 1 than that of CESA and the corresponding R^2 was higher. By contrast, the trend line slope of CESA was much closer to 1 than that of BESA at Tangnaihai and Zheqiao stations, and the correlation coefficient was much higher.

Above all, the fitness of BESA for simulating the observed streamflow sequence was better than that of CESA. The forecast accuracy of BESA at Xiangtang, Yingluoxia, Zamusi, and Jiutiaoling stations was better than that of CESA. Nevertheless, it was the opposite at Tangnaihai station. Neither model made better forecasts at Zheqiao station.

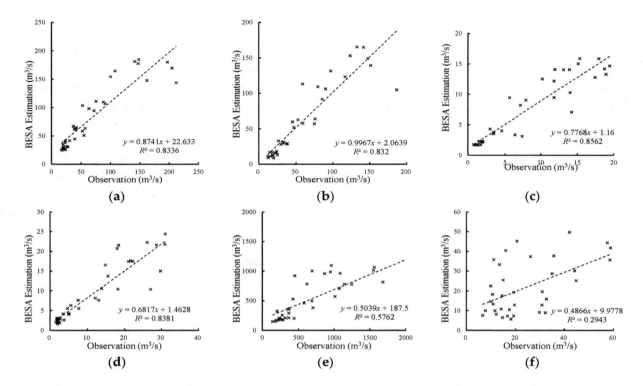

Figure 6. Forecasted values of Burg entropy spectral analysis (BESA) related to observed values in the lead time. (**a**) Xiangtang station; (**b**) Yingluoxia station; (**c**) Zamusi station; (**d**) Jiutiaoling station; (**e**) Tangnaihai station; (**f**) Zheqiao station.

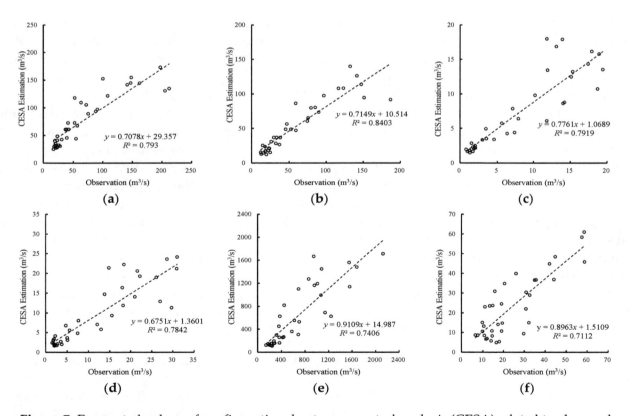

Figure 7. Forecasted values of configurational entropy spectral analysis (CESA) related to observed values in the lead time. (**a**) Xiangtang station; (**b**) Yingluoxia station; (**c**) Zamusi station; (**d**) Jiutiaoling station; (**e**) Tangnaihai station; (**f**) Zheqiao station.

3.3. Comparison with Other Autocorrelation Models

The autoregressive coefficients of BESA and CESA are obtained by maximizing Burg entropy and Configurational entropy, respectively. In order to demonstrate the improved accuracy of the predictions, we performed comparison with two other autocorrelation models. The first one is the AR model, and its coefficients are calculated by Yule-Walker function. The second one is the seasonal autoregressive model (SAR), and it rearranges the streamflow by month to avoid the seasonality. The comparison of the performance metrics is shown in Table 3. It can be seen that the BESA and CESA performed better than the AR and SAR models. This is because the BESA and CESA combine the maximum entropy principle and spectral analysis. The model estimated by maximum entropy principle is unbiased with all available data and no further hypothesis are needed. The spectral analysis can detect the periodical pattern of time series. Thus, BESA and CESA is more accurate and reliable than AR and SAR model.

Table 3. Comparison of the performance metrics by four models.

Station	BESA		CESA		AR		SAR	
	NSE of TT	NSE of LT	NSE of TT	NSE of LT	NSE of TT	NSE of LT	NSE of TT	NSE of LT
xiangtang	0.904	0.773	0.840	0.760	0.686	0.666	0.564	0.631
yingluoxia	0.834	0.798	0.708	0.801	0.612	0.755	0.429	0.624
zamusi	0.832	0.838	0.690	0.684	0.644	0.524	0.479	0.540
jiutiaoling	0.768	0.755	0.697	0.707	0.585	0.511	0.456	0.466
tangnaihai	0.548	0.482	0.447	0.693	0.495	0.34	0.403	0.373
zheqiao	0.255	0.618	0.365	0.659	0.485	0.144	0.364	0.728

Notes: TT represents training time and LT represents lead time. AR: autoregressive; SAR: seasonal autoregressive.

4. Discussion

Despite the fact that the six hydrological stations are in the same area, they differ in factors such as the type of river, control catchment area, vegetation condition, and human activities. Datong River (Xiangtang station), Daxia River (Zheqiao station), and the Yellow River (Yingluoxia station) belong to outflow rivers, whereas Heihe River (Yingluoxia station), Xiying River (Jiutiaoling station), and Zamu River (Zamusi station) belong to interior rivers. As the upstream of Yellow River, the catchment area of Tangnaihai station is the largest, which is about 120,000 km^2. Both Xiangtang and Zheqiao stations are located on the tributary of Yellow River, while Zamusi and Jiutiaoling stations are situated on the Shiyang River. Although four hydrological stations are located on the downstream, the catchment area are all less than 20,000 km^2. Yingluoxia station is located on the upstream Heihe River, with a catchment area of 10,009 km^2. For Xiangtang, Zamusi and Jiutiaoling stations, the upper reaches of piedmont watershed scale have good vegetation coverage and little human activity. By comparison, vegetation coverage is slightly poorer in the middle and lower reaches. Yingluoxia station is located on the Heihe upstream, but its catchment area and vegetation coverage are similar to the upstream of the watershed mentioned above. The impact of human activity is small above Tangnaihai station. In addition, industrial and agricultural water use is rare. Since the 1990s, the grassland has a tendency to gradually degradation. The upstream piedmont of Daxia River (i.e., Zheqiao station-owned) stony mountainous area covered with pasture except for few woods. Its downstream flow through loess plateau with ravines crossbar, poor vegetation and serious soil erosion and concentrated human activities.

Whether outliers, control catchment area, vegetation condition, and human activities, all reflect correlation of streamflow data. The autocorrelation of the six stations is shown in Figure 8. It can be seen that the autocorrelation of Tangnaihai and Zheqiao stations is relatively low (i.e., the autocorrelation coefficient is less than 0.5 at the 12th lag, and other stations are higher than 0.6). Based on the level of autocorrelation, the six stations can be grouped into two categories. The autocorrelation in the first category (Xiangtang, Yingluoxia, Zamusi, and Jiutiaoling stations) is higher than the second category (Tangnaihai and Zheqiao stations). As streamflow forecasting is based on the autocorrelation with the past series, the streamflow series with strong correlation will be more reliable forecasted. Thus, streamflow forecasting effects of the first category are better than the second category. BESA fitness may be better and unbiased because of the difference methods of the two forecast models. For BESA, the autoregressive coefficients are calculated by Levinson–Burg algorithm, which is developed from the AR model. For CESA, the autoregressive coefficients are calculated by cepstrum estimation. Therefore, entropy spectrum analysis methods need to be chosen carefully according to the situation of the study area.

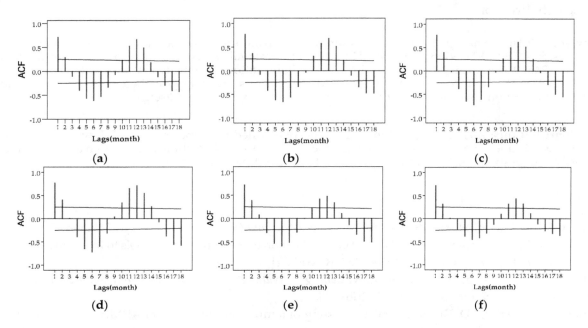

Figure 8. Autocorrelation plot of representative stations. (**a**) Xiangtang station; (**b**) Yingluoxia station; (**c**) Zamusi station; (**d**) Jiutiaoling station; (**e**) Tangnaihai station; (**f**) Zheqiao station.

The six stations selected in this paper are all located in Northwest China. Streamflow is principally composed of precipitation. Precipitation is the main recharge source of streamflow among them. Annual precipitation mainly occurs during June to September. All six rivers originate from alpine regions where April and May are spring flood periods, and flood is mainly formed by snow melt. Although annual precipitation mainly occurs during June to September, rainfall is mostly heavy rain and the month with maximum precipitation is not fixed. Hence, the maximum monthly streamflow is also unset. The input of autoregressive model is only previous monthly streamflow data, which may influence the forecast accuracy of the autoregressive model in the flood season. Therefore, adding precipitation as a predictor, selecting one or more models with high accuracy in the flood season, and using the entropy spectrum model and its combination (such as combined streamflow forecasting based on cross entropy [36]) to forecast can be used as the next research direction.

5. Conclusions

Two entropy spectral analysis methods (Burg entropy and configurational entropy) are mainly developed for streamflow forecasting in Northwest China. The following conclusions are drawn from this study:

1. The autoregressive coefficients obtained by maximizing Burg entropy and configurational entropy leads to more reliable than those by Levinson-Durbin algorithm. So, the streamflow forecasted by BESA and CESA is more accurate than that of the AR and SAR models.

2. For the streamflow with strong correlation, both BESA and CESA forecast monthly streamflow well. The R^2 and NSE were over 0.84 and 0.68, respectively. The forecast accuracy of BESA is higher than that of CESA. For the streamflow with weak correlation, the conclusion is the opposite.

3. The time of peak flow forecasted by both models (BESA and CESA) may be either earlier or later than observed. The peak flow is generally underestimated by both models. BESA accurately forecasted the bi-modal values of the flood season for the lead time, while CESA had better forecast results for the streamflow data with weak correlation.

4. In Northwest China, streamflow in the flood periods is principally composed of precipitation. The month with maximum precipitation is not fixed. Hence, the study of streamflow characteristics and spectral pattern associated with the precipitation can be used as the next research direction.

Acknowledgments: We are grateful for the grant support from the National Natural Science Fund in China (Project Nos. 91425302, 51279166 and 51409222), the Fundamental Research Funds for the Central University (Project No. 2452016069), and the Doctoral Program Foundation of Northwest A & F University (Project No. 2452015290). We wish to thank the editor and anonymous reviewers for their valuable comments and constructive suggestions, which were used to improve the quality of the manuscript.

Author Contributions: Z.H.Z. designed the study and performed the experiments. J.L.J. wrote the paper and reviewed it. X.L.S. and V.P.S. reviewed and edited the manuscript. G.X.Z. drew the research area map and gave some feasible suggestions. All authors have read and approved the final manuscript.

References

1. Box, G.E.; Jenkins, G.M. *Time Series Analysis: Forecasting and Control*; Holden-Day: Oakland, CA, USA, 1976.

2. Carlson, R.F.; Maccormick, A.J.A.; Watts, D.G. Application of linear random models to four annual streamflow series. *Water Resour. Res.* **1970**, *6*, 1070–1078. [CrossRef]

3. Hipel, K.W.; McLeod, A.I. *Time Series Modelling of Water Resources and Environmental Systems*; Elsevier: Amsterdam, The Netherlands, 1994; Volume 45.

4. Haltiner, J.P.; Salas, J.D. Development and testing of a multivariate, seasonal ARMA (1,1) model. *J. Hydrol.* **1988**, *104*, 247–272. [CrossRef]

5. Elshorbagy, A.; Simonovic, S.; Panu, U. Noise reduction in chaotic hydrologic time series: Facts and doubts. *J. Hydrol.* **2002**, *256*, 147–165. [CrossRef]

6. Singh, V.P.; Cui, H. Entropy theory for streamflow forecasting. *Environ. Process.* **2015**, *2*, 449–460. [CrossRef]

7. Fleming, S.W.; Marsh Lavenue, A.; Aly, A.H.; Adams, A. Practical applications of spectral analysis to hydrologic time series. *Hydrol. Process.* **2002**, *16*, 565–574. [CrossRef]

8. Ghil, M.; Allen, M.; Dettinger, M.; Ide, K.; Kondrashov, D.; Mann, M.; Robertson, A.W.; Saunders, A.; Tian, Y.; Varadi, F. Advanced spectral methods for climatic time series. *Rev. Geophys.* **2002**, *40*. [CrossRef]

9. Labat, D. Recent advances in wavelet analyses: Part 1. A review of concepts. *J. Hydrol.* **2005**, *314*, 275–288. [CrossRef]

10. Labat, D.; Ronchail, J.; Guyot, J.L. Recent advances in wavelet analyses: Part 2—Amazon, Parana, Orinoco and Congo discharges time scale variability. *J. Hydrol.* **2005**, *314*, 289–311. [CrossRef]

11. Marques, C.; Ferreira, J.; Rocha, A.; Castanheira, J.; Melo-Goncalves, P.; Vaz, N.; Dias, J. Singular spectrum analysis and forecasting of hydrological time series. *Phys. Chem. Earth Parts A/B/C* **2006**, *31*, 1172–1179. [CrossRef]

12. Burg, J.P. Maximum entropy spectral analysis. In Proceedings of the 37th Annual International Meeting, Oklahoma, OK, USA, 31 October 1967.

13. Currie, R.G. Geomagnetic line spectra-2 to 70 years. *Astrophys. Space Sci.* **1973**, *21*, 425–438. [CrossRef]

14. Pardo-Igúzquiza, E.; Rodríguez-Tovar, F.J. Maximum entropy spectral analysis of climatic time series revisited: Assessing the statistical significance of estimated spectral peaks. *J. Geophys. Res.* **2006**, *111*. [CrossRef]

15. Hasanean, H. Fluctuations of surface air temperature in the eastern mediterranean. *Theor. Appl. Climatol.* **2001**, *68*, 75–87. [CrossRef]

16. Wang, D.; Chen, Y.-F.; Li, G.-F.; Xu, Y.-H. Maximum entropy spectral analysis for annual maximum tide levels time series of the changjiang river estuary. *J. Coast. Res.* **2004**, *43*, 101–108.

17. Dalezios, N.R.; Tyraskis, P.A. Maximum entropy spectra for regional precipitation analysis and forecasting. *J. Hydrol.* **1989**, *109*, 25–42. [CrossRef]

18. Huang, Z.S. The application of spectrum analysis method in hydrogy. *Hydrology* **1983**, *3*, 101–108.
19. Krstanovic, P.; Singh, V.P. A univariate model for long-term streamflow forecasting. *Stoch. Hydrol. Hydraul.* **1991**, *5*, 173–188. [CrossRef]
20. Krstanovic, P.; Singh, V.P. A real-time flood forecasting model based on maximum-entropy spectral analysis: I. Development. *Water Resour. Manag.* **1993**, *7*, 109–129. [CrossRef]
21. Krstanovic, P.; Singh, V.P. A real-time flood forecasting model based on maximum-entropy spectral analysis: II. Application. *Water Resour. Manag.* **1993**, *7*, 131–151. [CrossRef]
22. Singh, V.P. *Entropy Theory and Its Application in Environmental and Water Engineering*, 1st ed.; Wiley: Hoboken, NJ, USA, 2013.
23. Singh, V.P. *Entropy Theory in Hydrologic Science and Engineering*; McGraw-Hill: New York, NY, USA, 2015.
24. Huo, C. Use of auto-regression model of time series in the simulation and forecast of groundwater dynamic in irrigation areas. *Geotech. Investig. Surv.* **1990**, *1*, 36–38.
25. Wang, D.; Zhu, Y.S. Research on cryptic period of hydrologic time series based on mem1spectral analysis. *Hydrology* **2002**, *2*, 19–23.
26. Shen, H.F.; Li, M.S.; Luo, F. Strict maximum entropy spectral estimation based on recursive algorithm. *Radar Sci. Technol.* **2008**, *4*, 288–291.
27. Boshnakov, G.N.; Lambert-Lacroix, S. A periodic levinson–durbin algorithm for entropy maximization. *Comput. Stat. Data Anal.* **2012**, *56*, 15–24. [CrossRef]
28. Cui, H.; Singh, V.P. Configurational entropy theory for streamflow forecasting. *J. Hydrol.* **2015**, *521*, 1–17. [CrossRef]
29. Frieden, B.R. Restoring with maximum likelihood and maximum entropy. *J. Opt. Soc. Am.* **1972**, *62*, 511–518. [CrossRef] [PubMed]
30. Gull, S.F.; Daniell, G.J. Image reconstruction from incomplete and noisy data. *Nature* **1978**, *272*, 686–690. [CrossRef]
31. Wu, N.L. An explicit solution and data extension in the maximum entropy method. *IEEE Trans. Acoust. Speech Signal Process.* **1983**, *31*, 486–491.
32. Nadeu, C. Finite length cepstrum modelling—A simple spectrum estimation technique. *Signal Process.* **1992**, *26*, 49–59. [CrossRef]
33. Katsakos-Mavromichalis, N.; Tzannes, M.; Tzannes, N. Frequency resolution: A comparative study of four entropy methods. *Kybernetes* **1986**, *15*, 25–32. [CrossRef]
34. Schwarz, G. Estimating the dimension of a model. *Ann. Stat.* **1978**, *6*, 461–464. [CrossRef]
35. Nash, J.E.; Sutcliffe, J.V. River flow forecasting through conceptual models partI—A discussion of principles. *J. Hydrol.* **1970**, *10*, 282–290. [CrossRef]
36. Men, B.; Long, R.; Zhang, J. Combined forecasting of streamflow based on cross entropy. *Entropy* **2016**, *18*, 336. [CrossRef]

7

Application of Entropy Ensemble Filter in Neural Network Forecasts of Tropical Pacific Sea Surface Temperatures

Hossein Foroozand [1], Valentina Radić [2] and Steven V. Weijs [1,*]

[1]   Department of Civil Engineering, University of British Columbia, Vancouver, BC V6T 1Z4, Canada;
      hosseinforoozand@civil.ubc.ca

[2]   Department of Earth, Ocean and Atmospheric Sciences, University of British Columbia, Vancouver,
      BC V6T 1Z4, Canada; vradic@eoas.ubc.ca

*   Correspondence: steven.weijs@civil.ubc.ca

Abstract: Recently, the Entropy Ensemble Filter (EEF) method was proposed to mitigate the computational cost of the Bootstrap AGGregatING (bagging) method. This method uses the most informative training data sets in the model ensemble rather than all ensemble members created by the conventional bagging. In this study, we evaluate, for the first time, the application of the EEF method in Neural Network (NN) modeling of El Nino-southern oscillation. Specifically, we forecast the first five principal components (PCs) of sea surface temperature monthly anomaly fields over tropical Pacific, at different lead times (from 3 to 15 months, with a three-month increment) for the period 1979–2017. We apply the EEF method in a multiple-linear regression (MLR) model and two NN models, one using Bayesian regularization and one Levenberg-Marquardt algorithm for training, and evaluate their performance and computational efficiency relative to the same models with conventional bagging. All models perform equally well at the lead time of 3 and 6 months, while at higher lead times, the MLR model's skill deteriorates faster than the nonlinear models. The neural network models with both bagging methods produce equally successful forecasts with the same computational efficiency. It remains to be shown whether this finding is sensitive to the dataset size.

Keywords: entropy ensemble filter; ensemble model simulation criterion; EEF method; bootstrap aggregating; bagging; bootstrap neural networks; El Niño; ENSO; neural network forecast; sea surface temperature; tropical Pacific

1. Introduction

Most data-mining algorithms require proper training procedures [1–14] to learn from data. The Bootstrap AGGregatING (bagging) method is a commonly used tool in the machine learning methods to increase predictive accuracy. Despite its common application, the bagging method is considered to be computationally expensive, particularly when used to create new training data sets out of large volumes of observations [15–17]. To improve the computational efficiency, Wan et al. [15] proposed a hybrid artificial neural network (HANN), while Kasiviswanathan et al. [17] combined the bagging method with the first order uncertainty analysis (FOUA). The combined method reduced the computational time of simulation for uncertainty analysis with limited statistical parameters such as mean and variance of the neural network weight vectors and biases. Wang et al. [16] showed that sub-bagging (SUBsample AGGregatING) gives similar accuracy but is computationally more efficient than bagging. This advantage is highlighted for Gaussian process regression (GPR) since its computational time increases in cubic order with the increase of data. Yu and Chen [18] compared different machine learning techniques and found that the fully Bayesian regularized artificial neural network (BANN) methods are much more time consuming than support vector machine (SVM) and maximum likelihood estimation (MLE)-based Gaussian process (GP) model. Table 1 provides a brief overview of studies that applied the bagging method in a range of different machine learning algorithms.

The Entropy Ensemble Filter (EEF) method, as a modified bagging procedure, has been proposed recently by Foroozand and Weijs [19]. The EEF method uses the most informative training data sets in the ensemble rather than all ensemble members created by the conventional bagging method. The EEF method achieved a reduction of the computational time of simulation by around 50% on average for synthetic data simulation, showing a potential for its application in computationally demanding environmental prediction problems.

In this paper, the first application of the EEF method on real-world data simulation is presented. We test its application in forecasting the tropical Pacific sea surface temperatures (SST) anomalies based on the initially proposed neural-network model of Wu et al. [20]. We chose this particular application due to the numerous studies of the El Nino-southern oscillation (ENSO) phenomenon and its use for water resources management. The ENSO is the strongest climate fluctuation on time scales ranging from a few months to several years and is characterized by inter-annual variations of the tropical Pacific sea surface temperatures, with warm episodes called El Niño, and cold episodes, La Niña. As ENSO affects not only the tropical climate but also the extra-tropical climate [21,22], the successful prediction of ENSO is of great importance. One example is the Pacific Northwest of North America, where water management operations depend on the accuracy of seasonal ENSO forecasts. For the Columbia River hydropower system, the use of ENSO information, in combination with adapted operating policies, could lead to an increase of $153 million in expected annual revenue [23]. Successful long-term forecasts of ENSO indices themselves could increase forecast lead-times, potentially further increasing benefits from hydropower operations. Vu et al. [24] recently argued that information entropy suggests stronger nonlinear links between local hydro-meteorological variables and ENSO, which could further strengthen its predictive power. Also, recent drought in coastal British Columbia, Canada, has increased the need for reliable seasonal forecasts to aid water managers in, for example, anticipating drinking water supply issues.

Since the early 1980s, much effort has been allocated to forecasting the tropical Pacific SST anomalies with the use of dynamical, statistical and hybrid models [21,25,26]. Because of ENSO's nonlinear features [20,21,27–30], many studies applied nonlinear statistical models such as a neural network (NN) model. Detailed comparisons between linear and nonlinear models in ENSO forecasts have been conducted in [2,20,30]. Wu et al. [20] developed a multi-layer perceptron (MLP) NN approach, where sea level pressure (SLP) field and SST anomalies over Tropical Pacific were used to predict the five leading SST principal components at lead times from 3 to 15 months. The performance of the MLP model, when compared to the multiple-linear regression (MLR) models, showed higher correlation skills and lower root mean square errors over most Nino domains. In this study, we incorporate the EEF method in both MLP and MLR models and evaluate their performance and computational efficiency relative to the original models. In addition to the original MLP model that uses Bayesian neural network (BNN), henceforth labeled as BNN model, we also test the MLP model that applies a cross-validation with Levenberg-Marquardt optimization algorithm [31–33], henceforth labeled as NN model. The main difference between BNN and NN models is their procedure to prevent overfitting. The NN model splits the provided data into training and validation and uses the early stop training procedure to prevent overfitting, while the BNN model uses all of the provided data points for training and uses weight penalty function (complexity penalization) to prevent overfitting (see [2,31,32,34] for details).

This paper is structured as follows: in Section 2 we give a brief explanation of the EEF method and model structures, followed by a description of data, predictors, and predictands in Section 3. In Section 4 we present and discuss the results of the three models (MLR, BNN, NN) run with the conventional bagging method in comparison to the runs with the EEF method. Finally, a conclusion and outlook are presented in Section 5.

Table 1. Examples of studies on machine learning algorithms with bagging methods and summary of their discussion on computational efficiency.

Authors (Year)	Machine Learning Method *	Computational Efficiency of the Bagging Method
Wan et al. (2016) [15]	HANN and BBNN	The bootstrapped NN training process is extremely time-consuming. The HANN approach is nearly 200 times faster than the BBNN approach with 10 hours runtime.
Liang et al. (2016) [35]	BNN and BMH	The bootstrap sample cannot be very large for the reason of computational efficiency.
Zhu et al. (2016) [27]	BNN	The proposed improvement in accuracy comes at the cost of time-consumption during the network training.
Gianola et al. (2014) [36]	GBLUP	Bagging is computationally intensive when one searches for an optimum value of BLUP-ridge regression because of the simultaneous bootstrapping.
Faridi et al. (2013) [37]	ANN	Each individual network is trained on a bootstrap re-sampling replication of the original training data.
Wang et al. (2011) [16]	ANN and GPR	Subagging gives similar accuracy but requires less computation than bagging. This advantage is especially remarkable for GPR since its computation increases in cubic order with the increase of data.
Mukherjee and Zhang (2008) [38]	BBNN	Dividing the batch duration into fewer intervals will reduce the computation effort in network training and batch optimisation. However, this may reduce the achievable control performance . . .
Yu and Chen (2005) [18]	BNN, SVM, and MLE-GP	Fully Bayesian methods are much more time consuming than SVM and MLE- GP.
Rowley et al. (1998) [39]	ANN	To improve the speed of the system different methods have been discussed, but this work is preliminary and is not intended to be an exhaustive exploration of methods to optimize the execution time.

* ANN (artificial neural network), BNN (Bayesian neural network), HANN (hybrid artificial neural network), BBNN (bootstrap-based neural network), BMH (bootstrap Metropolis-Hastings), GPR (Gaussian process regression), GBLUP (genomic best linear unbiased prediction), MLE-GP (maximum likelihood estimation-based Gaussian process), SVM (support vector machine) and V-SVM (virtual support vector machine).

2. Methods

2.1. Entropy Ensemble Filter

The EEF method is a modified bagging procedure to improve efficiency in ensemble model simulation (see [19] for details). The main novelty and advantages of the EEF method are rooted in using the self-information of a random variable, defined by Shannon's information theory [40,41] for selection of most informative ensemble models which are created by conventional bagging method [42]. Foroozand and Weijs [19] proposed that an ensemble of artificial neural network models or any other machine learning technique can use the most informative ensemble members for training purpose rather than all bootstrapped ensemble members. The results showed a significant reduction in computational time without negatively affecting the performance of simulation. Shannon information theory quantifies information content of a dataset based on calculating the smallest possible number of bits, on average, to convey outcomes of a random variable, e.g., per symbol in a message [40,43–47]. The Shannon entropy H, in units of bits (per symbol), of ensemble member M in a bagging dataset, is given by:

$$H_M(X) = -\sum_{k=1}^{K} p_{x_k} \log_2 p_{x_k},$$

(1)

where p_{x_k} is the probability of occurrence outcome k of random variable X within ensemble member M. This equation calculates the Shannon entropy in the units of "bits" because logarithm's base is 2. H gives the information content of each ensemble member in a discretized space, where the bootstrapped members are processed using K bins of equal bin-size arranged between the signal's minimum and maximum values. These bin sizes are user-defined and chosen to strike a balance between having enough data points per bin and keeping enough detail in representing the data distribution of the time series. In this case, we chose K = 10.

Figure 1 illustrates the flowchart of the EEF procedure as applied in this study. The EEF method will assess and rank the ensemble members, initially generated by the bagging procedure, to filter and select the most informative ones for the training of the NN model. As it is expected in machine learning, the overall computational time depends roughly linearly on the number of retained ensemble members which potentially leads to significant time savings.

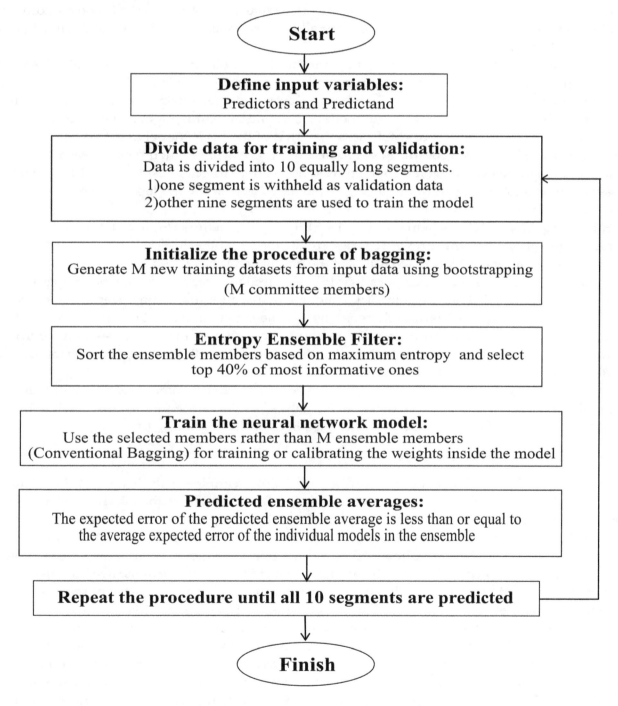

Figure 1. The flowchart of Entropy Ensemble Filter (EEF) method applied in the study.

2.2. Models

Following Wu et al. [20], we adopt the same structure for the NN models used with both the conventional bagging and the EEF method. For details on the model's training and cross-validation procedures to prevent overfitting, we refer to that paper [20]. The original model, a standard feed-forward multilayer perceptron neural network model with Bayesian regularization (BNN model) is run in MATLAB using the 'trainbr.m' function of the Neural Network Toolbox [34]. We introduce an additional NN model that applies the Levenberg-Marquardt optimization algorithm instead of Bayesian regularization and is run using the 'trainlm.m' MATLAB function (NN model) [32,48]. Finally, we run a multiple linear regression (MLR) model using the same 12 PCs as predictors. Following the recommendation of Wu et al. [20], in both neural network models, we optimize the number of hidden neurons (m) which is varying from 1 to 8 (the network architectures of 12-m-1) based on the correlation skill for each predictand out of five SST PCs. For all models we apply the following three schemes in order to produce the model ensemble runs:

(1) MLR, NN, BNN scheme (following the original method in Wu et al [20]): Using the conventional bagging method, training data from the nine segments is randomly drawn multiple times in order to train the BNN and NN model separately. In each 'bagging' draw we include the model run into the ensemble only if the model's correlation skill is higher than that of the MLR model, and its mean-square-error (MSE) less than that of the MLR model; if otherwise, the model run is rejected. This entire procedure is repeated until 30 ensemble members for each NN model are accepted. The ensemble average is then used as the final forecast result. With 10 segments for cross-validation and each segment producing 30 models, a total of 300 models are used for forecasting over the whole record.

(2) MLR_E, NN_E, BNN_E scheme: Using the EEF bagging method (Figure 1), training data from the nine segments is randomly drawn 12 times, producing 12 model ensemble for each data segment. We chose the model ensemble size to be 40% of the original one above, i.e., 12 out of 30, following the recommendations for EEF method application [19]. No selection criteria involving the comparison with the MLR model is applied here. A total of 120 models are used for forecasting over the whole record.

(3) MLR_{rand}, NN_{rand}, BNN_{rand} scheme: This scheme is the same as (2) except the conventional bagging scheme is used instead of the EEF method. This scheme mainly serves as a control run, i.e., for direct comparison of its performance with the scheme (2), both yielding the same total amount of 120 models over the whole period.

The above procedures are repeated until all five SST PCs at all lead times (3, 6, 9, 12 and 15 months) are predicted. For each lead time, we therefore have the ensemble-mean forecast from each of the nine models (MLR, MLR_E, MLR_{rand}, NN, NN_E, NN_{rand}, BNN, BNN_E, BNN_{rand}).

3. Data, Predictors, and Predictands

Monthly SST and SLP data in this study are from European Re-Analysis Interim (ERA-Interim) which is a global atmospheric reanalysis dataset [49] downloaded from the ECMWF website (https://www.ecmwf.int/en/forecasts/datasets/reanalysis-datasets/era-interim) for the tropical Pacific region (124° E–90° W, 20° N–20° S) at 0.75° × 0.75° for the period January 1979 to August 2017. Anomalies in both variables are calculated by subtracting their monthly climatology based on the 1979–2017 period. Following Wu et al. [20], we define the predictand as one of the five leading principal components (PCs) of the SST anomalies over the whole spatial domain, i.e., each of the five SST PCs is predicted separately. The corresponding spatial patterns of the eigenvectors (also called empirical orthogonal functions, EOFs) of the predictands, together explaining 80% of the total variance of the SST anomalies, are displayed in Figure 2. Note that our eigenvectors and SST PCs are somewhat different from those in Wu et al. [20] since their study used different reanalysis data and different time period (1948–2004). For the predictors, after applying a 3-month running mean to the gridded anomaly data of SLP and SST, a separate principal component analysis (PCA) is performed over the whole spatial domain with 7 SLP PCs and 9 SST PCs retained. These retained PCs are then separately normalized by dividing them by the standard deviation of their first PC. To set up the predictors' structure, the 7 SLP PCs supplied at time leads of 0, 3, 6, and 9 months and the 9 SST PCs at time leads of 0 months are stacked together, altogether yielding 37 PC time series (4 × 7 SLP PCs and 9 SST PCs). Finally, another PCA is performed on the 37 PC time series to yield 12 final PCs that are used as the predictors in the models. As in Wu et al. [20], the lead time is defined as the time from the center of the period of the latest predictors to the center of the predicted period. The data record was partitioned into 10 equal segments (Figure 1), each 44 months long; one segment is withheld to provide independent data for testing the model forecasts, while the other nine segments are used to train the model. By repeating this procedure until all 10 segments are used for testing, we provide the forecast over the whole period. Following the recommendation in [20] for model evaluation criteria, we then calculate the correlation and root mean square error (RMSE) between the predicted SST anomalies and the corresponding target data (ERA-Interim) over the whole record. This is consistent with the Gaussianity assumptions underlying PCA and routinely employed in forecasting SST anomalies. Mean squared error (MSE) has been used inside of both neural network optimization procedures.

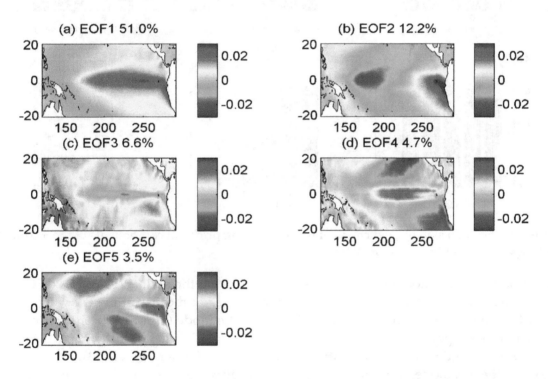

Figure 2. Spatial patterns (eigenvectors) of the first five PCA modes for the SST anomaly field. The percentage variance explained by each mode is given in the panel titles.

4. Results and Discussion

We first inter-compare the performance of the nine modelling schemes across all five SST PCs and all five lead times by looking at the correlation between modelled and observed predictands (Figure 3). Note that, relative to Wu et al. [20] all our correlations are higher. It is important to note, though, that different time periods were used for the forecast, due to data availability. The pattern across PCs is the following: for the first PC, which carries most of the variance and spatially best resembles the ENSO pattern, all models perform equally well at all lead times with correlation coefficient greater than 0.9 for the lead times of 3–9 months. As we move to higher PCs in our experiments, with the exception of PC4, the neural network models out-perform the MLR model, especially at the lead times 12 and 15 months. The better performance of neural network models is particularly striking for the PC2 with lead times 12 and 15 months where the correlation of MLR model substantially drops from the value greater than 0.8 at the 3-month lead time to the value less than 0.2 at the 15-month lead time.

This result indicates the importance of using nonlinear models for higher lead times, corroborating the findings in Wu et al. [20]. The BNN and NN models perform similarly well for PC1, PC2 and PC4, while the BNN model scores higher for PC3 and lower for PC5. The variation in skill between different methods seems to increase with increasing lead time and with higher PC modes. In general, the expected decrease of skill at higher lead times is visible throughout all PCs, except for PC4 where the predictability peaks at 9-months lead time.

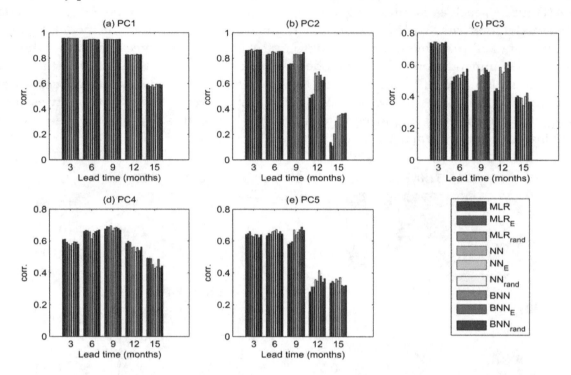

Figure 3. Correlation skill of predictions of the five leading principal components of the SST fields at lead times from 3 to 15 months for all 9 models.

To assess the overall model performance, i.e., combining the skill across all five PCs, we derive the weighted mean correlation across all the PCs, for each lead time, assigning the weights to each PC mode according to the amount of variance explained by the mode (Figure 4a). The following patterns emerge: (1) all models perform equally well for the lead time 3 and 6 months; (2) MLR model's skill drops at 9 months lead time more substantially than the skill of other models; (3) at the lead times of 9 and 15 months, the NN models outperform MLR models by roughly 0.05 difference in mean correlation (the difference between these models' correlations is statistically significant, Steiger's $Z = 4.174$ [50], $p < 0.01$ and Steiger's $Z = 1.73$, $p < 0.05$ at the lead times of 9 and 15 months respectively); (4) the NN models all produce correlations very close to each other (within 0.01 difference in correlation); (5) at the 15-month lead time the BNN model is the best performing model; and (6) overall, EEF method achieved a significant reduction in computational time and performed well especially in the forecast at the first 3 lead times. However, its underperformance at 12 and 15 lead time can be regarded as a compromise on computational time-saving.

Next, we look into the computational time for the models to produce the forecast for each PC mode (Figure 4b). The assessment is performed on the computer with 8 parallel quad-core processors (Intel® Core™ i7-4790 CPU @ 3.60GHz × 8). As expected, the BNN simulation is the most computationally expensive with 30 hours runtime in total. Considering the similar performance among all the models at the lead time of 3 and 6 months, the faster algorithms (e.g., MLR model, NN model with EEF method or with random selection) have the computational advantage over the BNN model. It appears that the original Wu et al. [20] model, which relies on the selection criteria, i.e., the inclusion of ensemble members that perform better than their MLR equivalents, does not have an advantage over the modeling scheme without this selection criteria. Finally, the EEF method in BNN and NN models performs equally well and is equally computationally efficient as the application of random reduced ensemble selection, i.e., the conventional bagging. As expected, computation time is mainly driven by ensemble size, as training the models is the most computationally expensive step of the forecasting procedure.

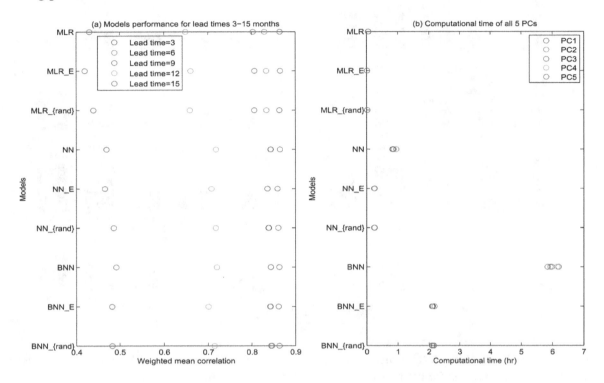

Figure 4. Weighted mean correlation and computational time for all models.

Regional Forecast

To estimate the regional forecast skills over the whole tropical Pacific, SST anomaly fields were reconstructed from five predicted PCs multiplied by their corresponding EOF spatial patterns. We spatially averaged the reconstructed SST anomalies over the Niño 4 (160° E–150° W, 5° S–5° N), Niño 3.4 (170° W–120° W, 5° S–5° N), Niño 3 (150° W–90° W, 5° S–5° N), and Niño 1+2 (90° W–80° W, 10° S–0°) regions, and then computed the correlation skills and root mean squared error (Figure 5). We focus on the difference among the model performance with the EEF method (MLR$_E$, NN$_E$, and BNN$_E$). Overall, their performances are at the same level at lead times of 3–12 months. The BNN model provides a better correlation skill than the other two models at 15 months lead time for all regions and especially for Niño 1+2 domain. We also look into the correlation skills spatially across each domain, i.e., correlation between reconstructed modeled and observed SST anomalies for each grid cell in the domain (Figure 6). At 15 months lead time, the large part of the domain is best simulated with the BNN model. There are significant variations between BNN and NN forecast skill at different lead times in central- northern equatorial region of the tropical Pacific. We also compared the time series of modeled vs observed SST anomalies averaged within the Niño 3 and Niño 1+2 regions at lead times of 3–15 months for the three models with the EEF method (see Appendix A). As expected, all models produced successful forecasts for the major El Niño and La Niña episodes at 3 months lead time. The BNN model outperformed other models at higher lead times, especially in Niño 1+2 region.

The spatial distribution of prediction skill was tested and compared for the various models at different lead times. Note that the performance of the BNN model relative to the NN model changes within the same region at different lead times (Figure 6). The same is true for the linear versus the nonlinear model. For example, for a lead time of 12 months, the nonlinear model outperforms the linear model in the Western Pacific, but not near Middle America, while the results for 15 month lead time are opposite.

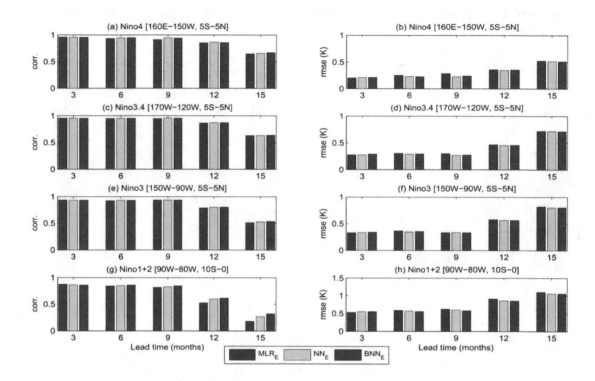

Figure 5. Correlation skills (**left column**) and RMSE scores (**right column**) of SST anomaly forecasts at lead times of 3–15 months for the Niño 4, Niño 3.4, Niño 3 and Niño 1+2 regions.

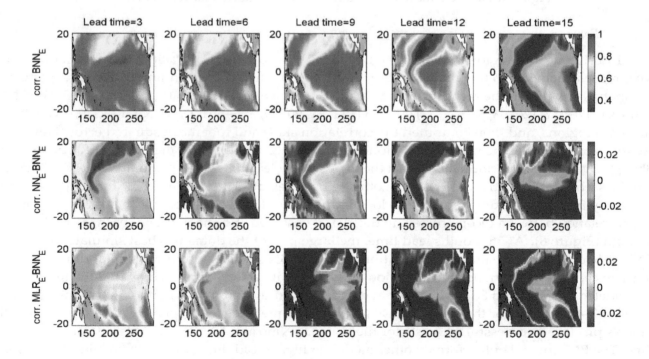

Figure 6. Forecast performance (correlation) per pixel of the forecast reconstructed from 5 leading principal components at lead times of 3–15 months for the period 1979–2017. Top row: BNN$_E$ model, middle and bottom rows: improvement of performance of NN$_E$ and MLR$_E$ over BNN$_E$.

5. Conclusions

In this study, we performed sea-surface temperature (SST) forecasts over the tropical Pacific using both linear and nonlinear (neural network) models with different training and ensemble generation schemes. In addition to the conventional bagging scheme (randomly generated samples for model training), we applied the ensemble entropy filter (EEF) method. This method reduces the original model ensemble size, and thus computation time, while trying to maintain prediction quality by prioritizing the retention of the most informative training data sets in the ensemble. We incorporated the EEF method in a multiple-linear regression (MLR) model and two neural network models (NN and BNN) and evaluated their performance and computational efficiency relative to the same models when conventional bagging is used. The predictands were the principal components (PCs) of the first five modes of SST monthly anomaly fields over the tropical Pacific for the period 1979–2017. The models' skills were tested for five different lead times: from 3 to 15 months, with 3 months increment.

We show that all models perform equally well at the lead time of 3 and 6 months, while a significant drop in MLR model's skill occurs at 9 month lead time and progressively deteriorates at 12 and 15 months lead time. At the higher lead times, the NN models outperform MLR models, while at the 15 months lead time the BNN model is the best performing model. Models with the EEF method perform equally well as the same models with the conventional bagging method with larger and equal ensemble sizes. Although the EEF method does not improve the correlation skill in the nonlinear forecast, it does not deteriorate it either. Considering that the EEF method selects only a portion of the data to be used in the forecasting, the improvement of computational efficiency with a minimum reduction in model skill makes this method attractive for the application on big datasets. For this particular case, however, the conventional bagging draws random ensembles that closely resemble the optimal ensembles from the EEF method. Thus, the neural network model with both bagging methods produced equally successful forecasts with the same computational efficiency. It remains to be shown, however, whether this finding is sensitive to the number of observations in the dataset.

A limitation of this study is that we deal with deterministic predictions only. Using an ensemble of models in prediction would, in theory, lend itself to generate probabilistic forecasts if an appropriate post-processing scheme can be developed. This would then allow for information-theoretical analysis of prediction skill. Also, a more detailed analysis of variations in predictability could be undertaken, investigating links with slower modes of large scale circulation patterns, such as the Pacific Decadal Oscillation (PDO). This is best investigated in a practical prediction context.

Future work will focus on further exploring the possibilities for forecasting streamflows in the Pacific Northwest, where seasonal forecasts can provide significant benefits for hydropower production, prediction of drinking water shortages and water management in general. Reducing computation time would enable smaller individual organizations to use seasonal forecasts tailored to their specific river basins and water resources management problems, rather than using agency issued ENSO index forecasts that may not always exploit the maximum information in the teleconnection to their most important variables. Also, the use of the ensemble prediction techniques to produce uncertainty estimates with the forecasts is a promising area of research that can inform risk-based decision making for water resources.

Acknowledgments: This research was supported by funding from H. Foroozand's NSERC scholarship and S.V. Weijs's NSERC discovery grant.

Author Contributions: Hossein Foroozand, Valentina Radic, and Steven V. Weijs designed the method and experiments; Hossein Foroozand, Valentina Radic, and Steven V. Weijs performed the experiments; Hossein Foroozand, Valentina Radic, and Steven V. Weijs wrote the paper.

Appendix A. Results of Regional Forecast

In this appendix, the results are shown for predictions of various regional averages that are often used SST indices. The figures show forecasts for various lead times, reconstructed for the 1979–2017 period. Figure A1 shows the Niño 3 SST anomalies, while Figure A2 shows the Niño 1+2 SST anomalies.

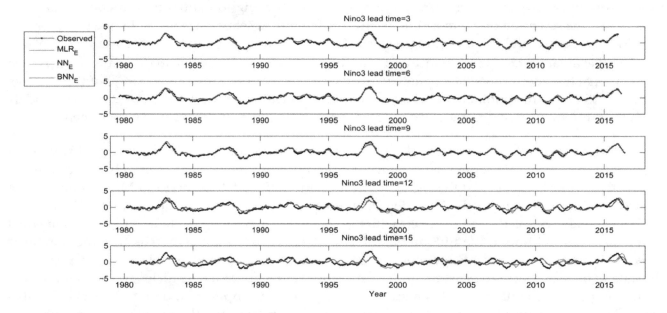

Figure A1. Time observed series of the Niño 3 SST anomalies as well as those predicted by different models at lead times of 3–15 months for the period 1979–2017.

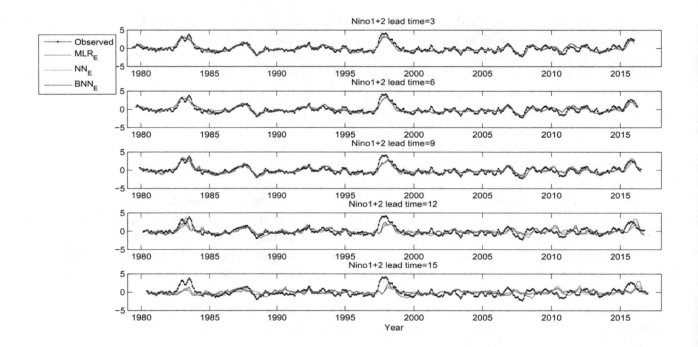

Figure A2. Time observed series of the Niño 1+2 SST anomalies as well as those predicted by different models at lead times of 3–15 months for the period 1979–2017.

References

1. Chau, K. Use of meta-heuristic techniques in rainfall-runoff modelling. *Water* **2017**, *9*, 186. [CrossRef]
2. Hsieh, W.W. *Machine Learning Methods in the Environmental Sciences: Neural Networks and Kernels*; Cambridge University Press: Cambridge, UK, 2009.
3. Lazebnik, S.; Raginsky, M. Supervised learning of quantizer codebooks by information loss minimization. *IEEE Trans. Pattern Anal. Mach. Intell.* **2009**, *31*, 1294. [CrossRef] [PubMed]
4. Zaky, M.A.; Machado, J.A.T. On the formulation and numerical simulation of distributed-order fractional optimal control problems. *Commun. Nonlinear Sci. Numer. Simul.* **2017**, *52*, 177. [CrossRef]
5. Ghahramani, A.; Karvigh, S.A.; Becerik-Gerber, B. HVAC system energy optimization using an adaptive hybrid metaheuristic. *Energy Build.* **2017**, *152*, 149. [CrossRef]
6. Foroozand, H.; Afzali, S.H. A comparative study of honey-bee mating optimization algorithm and support vector regression system approach for river discharge prediction case study: Kashkan river basin. In Proceedings of the International Conference on Civil Engineering Architecture and Urban Infrastructure, Tabriz, Iran, 29–30 July 2015.
7. Niazkar, M.; Afzali, S.H. Parameter estimation of an improved nonlinear muskingum model using a new hybrid method. *Hydrol. Res.* **2017**, *48*, 1253. [CrossRef]
8. Sahraei, S.; Alizadeh, M.R.; Talebbeydokhti, N.; Dehghani, M. Bed material load estimation in channels using machine learning and meta-heuristic methods. *J. Hydroinformatics* **2018**, *20*, 100. [CrossRef]
9. Nikoo, M.R.; Kerachian, R.; Alizadeh, M. A fuzzy KNN-based model for significant wave height prediction in large lakes. *Oceanologia* **2017**. [CrossRef]
10. Sivakumar, B.; Jayawardena, A.W.; Fernando, T.M.K.G. River flow forecasting: Use of phase-space reconstruction and artificial neural networks approaches. *J. Hydrol.* **2002**, *265*, 225. [CrossRef]
11. Moosavian, N.; Lence, B.J. Nondominated sorting differential evolution algorithms for multiobjective optimization of water distribution systems. *J. Water Resour. Plan. Manag.* **2017**, *143*, 04016082. [CrossRef]
12. Moosavian, N.; Jaefarzadeh, M.R. Hydraulic analysis of water distribution network using shuffled complex evolution. *J. Fluid.* **2014**, *2014*, 979706. [CrossRef]
13. Chen, X.Y.; Chau, K.W. A hybrid double feedforward neural network for suspended sediment load estimation. *Water Resour. Manag.* **2016**, *30*, 2179. [CrossRef]
14. Olyaie, E.; Banejad, H.; Chau, K.W.; Melesse, A.M. A comparison of various artificial intelligence approaches performance for estimating suspended sediment load of river systems: A case study in United States. *Environ. Monit. Assess.* **2015**, *187*, 189. [CrossRef] [PubMed]
15. Wan, C.; Song, Y.; Xu, Z.; Yang, G.; Nielsen, A.H. Probabilistic wind power forecasting with hybrid artificial neural networks. *Electr. Power Compon. Syst.* **2016**, *44*, 1656. [CrossRef]
16. Wang, K.; Chen, T.; Lau, R. Bagging for robust Non-Linear Multivariate Calibration of Spectroscopy. *Chemom. Intell. Lab. Syst.* **2011**, *105*, 1. [CrossRef]
17. Kasiviswanathan, K.S.; Sudheer, K.P. Quantification of the predictive uncertainty of artificial neural network based river flow forecast models. *Stoch. Environ. Res. Risk Assess.* **2013**, *27*, 137. [CrossRef]
18. Yu, J.; Chen, X.W. Bayesian neural network approaches to ovarian cancer identification from high-resolution mass spectrometry data. *Bioinformatics* **2005**, *21*, 487. [CrossRef] [PubMed]
19. Foroozand, H.; Weijs, S.V. Entropy ensemble filter: A modified bootstrap aggregating (Bagging) procedure to improve efficiency in ensemble model simulation. *Entropy* **2017**, *19*, 520. [CrossRef]
20. Wu, A.; Hsieh, W.W.; Tang, B. Neural network forecasts of the tropical Pacific sea surface temperatures. *Neural Netw.* **2006**, *19*, 145. [CrossRef] [PubMed]
21. Aguilar-Martinez, S.; Hsieh, W.W. Forecasts of tropical Pacific sea surface temperatures by neural networks and support vector regression. *Int. J. Oceanogr.* **2009**, *2009*, 167239. [CrossRef]
22. Wallace, J.M.; Rasmusson, E.M.; Mitchell, T.P.; Kousky, V.E.; Sarachik, E.S.; von Storch, H. On the structure and evolution of ENSO-related climate variability in the tropical Pacific: Lessons from TOGA. *J. Geophys. Res. Oceans* **1998**, *103*, 14241. [CrossRef]
23. Hamlet, A.F.; Huppert, D.; Lettenmaier, D.P. Economic value of long-lead streamflow forecasts for columbia river hydropower. *J. Water Resour. Plan. Manag.* **2002**, *128*, 91. [CrossRef]
24. Vu, T.M.; Mishra, A.K.; Konapala, G. Information entropy suggests stronger nonlinear associations between hydro-meteorological variables and ENSO. *Entropy* **2018**, *20*, 38. [CrossRef]

25. Goddard, L.; Mason, S.J.; Zebiak, S.E.; Ropelewski, C.F.; Basher, R.; Cane, M.A. Current approaches to seasonal to interannual climate predictions. *Int. J. Climatol.* **2001**, *21*, 1111. [CrossRef]

26. Barnston, A.G.; Tippett, M.K.; L'Heureux, M.L.; Li, S.; DeWitt, D.G. Skill of real-time seasonal ENSO model predictions during 2002–11: Is our capability increasing? *Bull. Am. Meteorol. Soc.* **2011**, *93*, 631. [CrossRef]

27. Zhu, L.; Jin, J.; Cannon, A.J.; Hsieh, W.W. Bayesian neural networks based bootstrap aggregating for tropical cyclone tracks prediction in south China sea. In Proceedings of the 23rd International Conference ICONIP, Kyoto, Japan, 16–21 October 2016.

28. Tangang, F.T.; Hsieh, W.W.; Tang, B. Forecasting the equatorial Pacific sea surface temperatures by neural network models. *Clim. Dyn.* **1997**, *13*, 135. [CrossRef]

29. Cannon, A.J.; Hsieh, W.W. Robust nonlinear canonical correlation analysis: Application to seasonal climate forecasting. *Nonlinear Process. Geophys.* **2008**, *15*, 221. [CrossRef]

30. Tang, B.; Hsieh, W.W.; Monahan, A.H.; Tangang, F.T. Skill comparisons between neural networks and canonical correlation analysis in predicting the equatorial Pacific sea surface temperatures. *J. Clim.* **2000**, *13*, 287. [CrossRef]

31. Levenberg, K. A method for the solution of certain non-linear problems in least squares. *Q. Appl. Math.* **1944**, *2*, 164. [CrossRef]

32. Marquardt, D. An algorithm for least-squares estimation of nonlinear parameters. *J. Soc. Ind. Appl. Math.* **1963**, *11*, 431. [CrossRef]

33. Taormina, R.; Chau, K.W.; Sivakumar, B. Neural network river forecasting through baseflow separation and binary-coded swarm optimization. *J. Hydrol.* **2015**, *529*, 1788. [CrossRef]

34. MacKay, D.J.C. Bayesian Interpolation. *Neural Comput.* **1992**, *4*, 415. [CrossRef]

35. Liang, F.; Kim, J.; Song, Q. A bootstrap metropolis-hastings algorithm for Bayesian analysis of big data. *Technometrics* **2016**, *58*, 304. [CrossRef] [PubMed]

36. Gianola, D.; Weigel, K.A.; Krämer, N.; Stella, A.; Schön, C.C. Enhancing genome-enabled prediction by bagging genomic BLUP. *PLoS ONE* **2014**, *9*, 91693. [CrossRef] [PubMed]

37. Faridi, A.; Golian, A.; Mousavi, A.H.; France, J. Bootstrapped neural network models for analyzing the responses of broiler chicks to dietary protein and branched chain amino acids. *Can. J. Anim. Sci.* **2013**, *94*, 79. [CrossRef]

38. Mukherjee, A.; Zhang, J. A reliable multi-objective control strategy for batch processes based on bootstrap aggregated neural network models. *J. Process Control* **2008**, *18*, 720. [CrossRef]

39. Rowley, H.A.; Baluja, S.; Kanade, T. Neural network-based face detection. *IEEE Trans. Pattern Anal. Mach. Intell.* **1998**, *20*, 23. [CrossRef]

40. Shannon, C.E. A mathematical theory of communication. *Bell Syst. Tech. J.* **1948**, *27*, 379. [CrossRef]

41. Singh, V.P.; Byrd, A.; Cui, H. Flow duration curve using entropy theory. *J. Hydrol. Eng.* **2014**, *19*, 1340. [CrossRef]

42. Breiman, L. Bagging predictors. *Mach. Learn.* **1996**, *24*, 123. [CrossRef]

43. Cover, T.M.; Thomas, J.A. *Elements of Information Theory*; Wiley-Interscience: Hoboken, NJ, USA, 2006.

44. Shannon, C.E. Communication in the Presence of Noise. *Proc. IRE* **1949**, *37*, 10. [CrossRef]

45. Weijs, S.V.; van de Giesen, N. An information-theoretical perspective on weighted ensemble forecasts. *J. Hydrol.* **2013**, *498*, 177. [CrossRef]

46. Weijs, S.V.; van de Giesen, N.; Parlange, M.B. HydroZIP: How hydrological knowledge can be used to improve compression of hydrological data. *Entropy* **2013**, *15*, 1289–1310. [CrossRef]

47. Cui, H.; Singh, V.P. Maximum entropy spectral analysis for streamflow forecasting. *Phys. Stat. Mech. Its. Appl.* **2016**, *442*, 91. [CrossRef]

48. Hagan, M.T.; Menhaj, M.B. Training feedforward networks with the Marquardt algorithm. *IEEE Trans. Neural Netw.* **1994**, *5*, 989. [CrossRef] [PubMed]

49. Dee, D.P.; Uppala, S.M.; Simmons, A.J.; Berrisford, P.; Poli, P.; Kobayashi, S.; Andrae, U.; Balmaseda, M.A.; Balsamo, G.; Bauer, P.; et al. The ERA-interim reanalysis: Configuration and performance of the data assimilation system. *Q. J. R. Meteorol. Soc.* **2011**, *137*, 553. [CrossRef]

50. Steiger, J.H. Tests for comparing elements of a correlation matrix. *Psychol. Bull.* **1980**, *87*, 245. [CrossRef]

8

An Extension to the Revised Approach in the Assessment of Informational Entropy

Turkay Baran, Nilgun B. Harmancioglu *, Cem Polat Cetinkaya and Filiz Barbaros

Faculty of Engineering, Civil Engineering Department, Dokuz Eylul University, Tinaztepe Campus, Buca, 35160 Izmir, Turkey; turkay.baran@deu.edu.tr (T.B.); cem.cetinkaya@deu.edu.tr (C.P.C.); filiz.barbaros@deu.edu.tr (F.B.)
* Correspondence: nilgun.harmancioglu@deu.edu.tr

Abstract: This study attempts to extend the prevailing definition of informational entropy, where entropy relates to the amount of reduction of uncertainty or, indirectly, to the amount of information gained through measurements of a random variable. The approach adopted herein describes informational entropy not as an absolute measure of information, but as a measure of the variation of information. This makes it possible to obtain a single value for informational entropy, instead of several values that vary with the selection of the discretizing interval, when discrete probabilities of hydrological events are estimated through relative class frequencies and discretizing intervals. Furthermore, the present work introduces confidence limits for the informational entropy function, which facilitates a comparison between the uncertainties of various hydrological processes with different scales of magnitude and different probability structures. The work addresses hydrologists and environmental engineers more than it does mathematicians and statisticians. In particular, it is intended to help solve information-related problems in hydrological monitoring design and assessment. This paper first considers the selection of probability distributions of best fit to hydrological data, using generated synthetic time series. Next, it attempts to assess hydrometric monitoring duration in a netwrok, this time using observed runoff data series. In both applications, it focuses, basically, on the theoretical background for the extended definition of informational entropy. The methodology is shown to give valid results in each case.

Keywords: uncertainty; information; informational entropy; variation of information; continuous probability distribution functions; confidence intervals

1. Introduction

The concept of entropy has its origins in classical thermodynamics and is commonly known as "thermodynamic entropy" in relation to the second law of thermodynamics. Such a non-probabilistic definition of entropy has been used widely in physical sciences, including hydrology and water resources. Typical examples on the use of "thermodynamic entropy" in water resources involve problems associated with river morphology and river hydraulics [1,2].

Boltzmann's definition of entropy as a measure of disorder in a system was given in probabilistic terms and constituted the basis for statistical thermodynamics [3–5]. Later, Shannon [6] followed up on Boltzmann's definition, claiming that the entropy concept could be used to measure disorder in systems other than thermodynamic ones. Shannon's entropy is what is known as "informational entropy", which measures uncertainty (or, indirectly, information) about random processes. As uncertainty and information are the two most significant yet the least clarified problems in hydrology and water resources, researchers were intrigued by the concept of informational entropy. Thus, it has found a large number of diverse applications in water resources engineering.

Within a general context, the entropy principle is used to assess uncertainties in hydrological variables, models, model parameters, and water-resources systems. In particular, versatile uses of the concept range from specific problems, such as the derivation of frequency distributions and parameter estimation, to broader cases such as hydrometric data network design. The most distinctive feature of entropy in these applications is that it provides a measure of uncertainty or information in quantitative terms [7–19].

On the other hand, researchers have also noted some mathematical difficulties encountered in the computation of various informational entropy measures. The major problem is the controversy associated with the mathematical definition of entropy for continuous probability distribution functions. In this case, the lack of a precise definition of informational entropy leads to further mathematical difficulties and, thus, hinders the applicability the concept in hydrology. This problem needs to be resolved so that the informational entropy concept can be set on an objective and reliable theoretical basis and thereby achieve widespread use in the solution of water-resources problems based on information and/or uncertainty.

Some researchers [20, 21] attempted to revise the prevailing definition of informational entropy, where entropy relates to the amount of reduction of uncertainty, or indirectly to the amount of information gained through measurements of a random variable. The study presented extends on the revised definition of Jaynes [20] and Guiasu [21] to describe informational entropy, not as an absolute measure of information, but as a measure of the variation of information. The mathematical formulation developed herein does not depend on the use of discretizing intervals when discrete probabilities of hydrological events are estimated through relative class frequencies and discretizing intervals. This makes it possible to obtain a single value for the variation of information instead of several values that vary with the selection of the discretizing interval. Furthermore, the extended definition introduces confidence limits for the entropy function, which facilitates a comparison between the uncertainties of various hydrological processes with different scales of magnitude and different probability structures.

It must be noted that the present work is intended for hydrologists and environmental engineers more than for mathematicians and statisticians. In particular, entropy measures have been used to help solve information-related problems in hydrological monitoring design and assessment. These problems are manifold, ranging from the assessment of sampling frequencies (both temporal and spatial) and station discontiuance to statistical analyses of observed data. For the latter, this paper considers the selection of probability distributions of best fit to hydrological data. Hence, the informational entropy concept is used here only in the temporal domain. To test another feature of entropy measures, the present work also attempts to assess hydrometric monitoring duration in a gauging network, this time using observed runoff data series. In both applications, the paper focuses, basically, on the theoretical background for the extended definition of informational entropy, and the results are shown to give valid results.

2. Mathematical Difficulties Associated with Informational Entropy Measures

Entropy is a measure of the degree of uncertainty of random hydrological processes. It is also a quantitative measure of information contained in a series of data since the reduction of uncertainty equals the same amount of gain in information [7,22]. Within the scope of Mathematical Communication Theory, later known as Information Theory, Shannon [6] and later Jaynes [23] defined informational entropy as the expectation of information or, conversely, as a measure of uncertainty. If S is a system of events, $E_1, E_2, \ldots , E_n$, and $p(E_k) = p_k$ the probability of the k-th event recurring, then the entropy of the system is:

$$H(S) = -\sum_{k=1}^{n} p_k \ln p_k \tag{1}$$

With,

$$\sum_{k=1}^{n} p_k = 1$$

Shannon's entropy as given in Equation (1) is originally formulated for discrete variables and always assumes positive values. Shannon extended this expression to the continuous case by simply replacing the summation with an integral equation as:

$$H(X) = -\int_{-\infty}^{+\infty} f(x) \cdot \ln f(x) \cdot dx \tag{2}$$

With,

$$\int_{-\infty}^{+\infty} f(x) \cdot dx = 1$$

For the random variable $X \in (-\infty, +\infty)$, and where $H(X)$ is denoted as the marginal entropy of X, i.e., the entropy of a univariate process. Equation (2) is not mathematically justified, as it is not valid under the assumptions initially made in defining entropy for the discrete case. What researchers proposed for solving this problem has been to approximate the discrete probabilities p_k by $f(x)\Delta x$, where $f(x)$ is the relative class frequency and Δx, the size of class intervals. Under these conditions, the selection of Δx becomes a crucial problem, such that each specified class interval size gives a different reference level of zero uncertainty with respect to which the computed entropies are measured. In this case, various entropy measures become relative to the discretizing interval Δx and change in value as Δx changes. The unfavorable result here is that the uncertainty of a random process may assume different values at different selected values of Δx for the same variable and the same probability distribution function. In certain cases, the entropy of a random variable even becomes negative [16,17,22,24–27], a situation which contradicts Shannon's definition of entropy as the selection of particular Δx values produces entropy measures varying within the interval $(-\infty, +\infty)$. On the contrary, the theoretical background for the random variable X, $H(X)$ defines the condition:

$$0 \leq H(X) \leq \ln N \tag{3}$$

where N is the number of events X assumes. The condition above indicates that the entropy function has upper (In N) and lower (0 when X is deterministic) bounds, assuming positive values in between [6,8,10–13,16,17,22,24–28]. The discrepancies encountered in practical applications of the concept essentially result from the above errors in the definition of entropy for continuous variables.

Another significant problem is the selection of the probability distribution function to be used in the definition of entropy, as in Equation (2). The current expression for continuous entropy produces different values when different distribution functions are assumed for the same variable. In this case, there is the need for a proper selection of the distribution function which best fits the process analyzed. One may consider here a valid criterion in the form of confidence limits to assess the suitability of the selected distribution function for entropy computations.

Further problems are encountered when the objective is to compare the uncertainties of two or more random variables with widely varying means and thus with different scales of magnitude. For instance, if entropy values are computed, using the same discretizing interval Δx, for two variables with means of 100 units and 1 unit, respectively, the results become incomparable due to the improper selection of the reference level of zero uncertainty for each variable. Such a problem again stems from the inclusion of the discretizing interval Δx in the definition of entropy for continuous variables. Comparison of uncertainties of different variables is an important aspect of entropy-based hydrometric network design procedures, where the aforementioned problem leads to subjective evaluations of information provided by the network [7,19].

It follows from the above discussion that the main difficulty associated with the applicability of the informational entropy concept in hydrology is the lack of a precise definition for the case of the continuous variables. It is intended in this study to resolve this problem by extending the revised approach proposed by Guiasu [21] so that the informational entropy can be set on an objective and reliable theoretical basis in order to discard subjective assessments of information conveyed by hydrological data or of the uncertainty of hydrological processes.

3. The Revised Definition of Informational Entropy for Continuous Variables

To solve the difficulties associated with the informational entropy measure in the continuous case, some researchers have proposed the use of a function $m(x)$ such that the marginal entropy of a continuous variable X is expressed as:

$$H(X) = -\int_{-\infty}^{+\infty} f(x) \cdot ln\left[\frac{f(x)}{m(x)}\right] \cdot dx \qquad (4)$$

"where $m(x)$ is an 'invariant measure' function, proportional to the limiting density of discrete points" [20]. The approach seemed to be statistically justified; however, it still remained uncertain what the $m(x)$ function might represent in reality. Jaynes [20] also discussed that it could be an a priori probability distribution function, but there were then controversies over the choice of a priori distribution such that the problem was unresolved [8].

In another study, Guiasu [21] referred to Shannon's definition of the informational entropy for the continuous case. He considered that the entropy $\{H_S\}$ for the continuous variable X within an interval $[a, b]$ is:

$$H_S = -\int f(x) \cdot \ln f(x) \cdot dx \qquad (5)$$

When the random variable assumes a uniform probability distribution function as:

$$f(x) = \frac{1}{(b-a)} \; x \in [a, b] \qquad (6)$$

Then the informational entropy H_S for the continuous case within this interval can be expressed as:

$$H_S = \ln(b-a) \qquad (7)$$

If the interval $[a, b]$ is discretized into N equal intervals, the variable follows a discrete uniform distribution and its entropy $\{H_N\}$ can be expressed as:

$$x \in [a, b] \qquad (8)$$

When N goes to infinity, H_N will also approach infinity. In this case, Guiasu [21] claims that, although H_S and H_N are similarly defined, H_S will not approach H_N when $N \to \infty$. Accordingly, Guiasu [21] proposed an expression similar to that of Jaynes [20] for informational entropy in the continuous case as:

$$H(X/X^*) = -\int f(x) \cdot \ln\left[\frac{f(x)}{m(x)}\right] \cdot dx \qquad (9)$$

which he called as the variation of information. In Equation (9), X^* represents a priori information (i.e., information available before making observations on the variable X) and X is the a posteriori information (i.e., information obtained by making observations). Similarly, $m(x)$ is the a priori and $f(x)$ the a posteriori probability density function for the random variable X.

In previous studies by the authors [8,10–13], informational entropy has been defined as the variation of information, which indirectly equals the amount of uncertainty reduced by making observations. To develop such a definition, two measures of probability, p and q with (p and $q \in K$), are considered in the probability space (Ω, K). Here, q represents a priori probabilities (i.e., probabilities prior to making observations). When a process is defined in such a probability space, the information conveyed when the process assumes a finite value A $\{A \in K\}$ in the same probability space is:

$$I = -\ln\left(\frac{p(A)}{q(A)}\right) \tag{10}$$

The process defined in Ω can assume one of the finite and discrete events $(A_1, \ldots, A_n) \in K$; thus, the entropy expression for any value A_n can be written as:

$$H(p/q) = -\ln\left(\frac{p(A_n)}{q(A_n)}\right)(n = 1, \ldots, N) \tag{11}$$

The total information content of the probability space (Ω, K) can be defined as the expected value of the information content of its elementary events:

$$H(p/q) = -\sum p(A_n) \cdot \ln\left(\frac{p(A_n)}{q(A_n)}\right) \tag{12}$$

Similarly, the entropy $H(X/X^*)$ of a random process X defined in the same probability space can be defined as:

$$H(X/X^*) = -\sum p(x_n) \cdot \ln\left(\frac{p(x_n)}{q(x_n)}\right) \tag{13}$$

where, $H(X/X^*)$ is in the form of conditional entropy, i.e., the entropy of X conditioned on X^*. Here, the condition is represented by an a priori probability distribution function, which can be described as the reference level against which the variation of information in the process can be measured.

Let us assume that the a priori $\{q(x)\}$ and a posteriori $\{p(x)\}$ probability distribution functions of the random variable X are known. If the ranges of possible values of the continuous variable X are divided into N discrete and infinitesimally small intervals of width Δx, the entropy expression for this continuous case can be given as:

$$H(X/X^*) = -\int p(x) \cdot \ln\left(\frac{p(x)}{q(x)}\right) \cdot dx \tag{14}$$

The above expression describes the variation of information (or, indirectly, the uncertainty reduced by making observations) to replace the absolute measure of information content given in Equation (2). This definition is essentially in conformity with those given by Jaynes [20] and Guiasu [21] for continuous variables. When the same infinitesimally small class interval Δx is used for the a priori and a posteriori distribution functions, the term Δx drops out in the mathematical expression of marginal entropy in the continuous case. Thus, this approach eliminates the problems pertaining to the use of Δx discretizing class intervals involved in the previous definitions of informational entropy [8,10–13].

At this point, the most important issue is the selection of a priori distribution. In case the process X is not observed at all, no information is available about it so that it is completely uncertain. In probability terms, this implies the selection of the uniform distribution. In other words, when no information exists about the variable X, the alternative events it may assume may be represented by equal probabilities or simply by the uniform probability distribution function.

If the a priori $\{q(x)\}$ is assumed to be uniform, and a posteriori $\{p(x)\}$ distribution of X is assumed to be normal, the informational entropy $H(X/X^*)$ can be expressed as:

$$H(X/X^*) = \ln\sqrt{2\pi} + \ln\sigma + \frac{1}{2} - \ln(b - a) \tag{15}$$

By integrating Equation (14). The first three terms in this equation represent the marginal entropy of X and the last term stands for the maximum entropy. Accordingly, the variation of information can be expressed simply as:

$$H(X/X^*) = H(X) - H_{max} \tag{16}$$

If the a posteriori distribution of X is assumed to be lognormal, the informational entropy $H(X/X^*)$ becomes:

$$H(X/X^*) = \ln \sqrt{2\pi} + \ln \sigma_y + \mu_y + \frac{1}{2} - \ln(b - a) \tag{17}$$

with and μ_y and σ_y being the mean and standard deviation of $y = \ln x$.

If the a posteriori distribution of X is assumed to be 2-parameter gamma distribution with parameters α and β,

$$f_{(x)} = \frac{1}{\beta^\alpha \, \Gamma_{(\alpha)}} \, x^{\alpha - 1} \, e^{\frac{-x}{\beta}} \; x \geq 0 \tag{18}$$

The informational entropy $H(X/X^*)$ becomes:

$$H(X/X^*) = \ln[\beta \cdot \Gamma(\alpha)] + \mu_x/\beta - (\alpha - 1) \cdot \Phi(\alpha) - \ln(b - a) \tag{19}$$

where, μ_x is the mean of the series, α the shape parameter, and β the scale parameter.

In the above, entropy as the variation of information measures the amount of uncertainty reduced by making observations when the a posteriori distribution is estimated.

The maximum amount of information gained about the process X defined within the interval $[a, b]$ is H_{max}. Thus, the expression in Equation (16) will assume negative values. However, since $H(X/X^*)$ describes entropy as the variation of information, it is possible to consider the absolute value of this measure.

When the a posteriori probability distribution function is correctly estimated, the information gained about the random variable will increase as the number of observations increases. Thus, when this number goes to infinity, the entropy $H(X/X^*)$ will approach zero. In practice, it is not possible to obtain an infinite number of observations; rather, the availability of sufficient data is important. By using the entropy measure $H(X/X^*)$, it possible to evaluate the fitness of a particular distribution function to the random variable and to assess whether the available data convey sufficient information about the process.

4. Mathematical Interpretation of the Revised Definition of Informational Entropy

4.1. The Distance between Two Continuous Distribution Functions as Defined by the Euclidian Metric

The approach used to obtain Equation (16) is essentially a means of measuring the distance between the points in probability space, described by the a priori $\{q(x)\}$ and a posteriori $\{p(x)\}$ distribution functions. The distance between these two functions can be determined by different measures like the metric concept, which enables one to see whether the two functions coincide.

According to the Euclidian metric, the distance between $p(x)$ and $q(x)$ functions defined in the same probability space (Ω, K) is:

$$I = \int [p(x) - q(x)]^2 dx \tag{20}$$

If $p(x)$ is the standard normal, and $q(x)$, the standard uniform distribution function, one obtains:

$$\Phi(x) = \frac{1}{\sqrt{2\pi}} \left\{ \sqrt{1 + \frac{3\sqrt{\pi}}{x} \left[1 - \frac{4}{3} F(x) \right]} \right\} \tag{21}$$

By integrating Equation (20) to obtain the difference function $\Phi(x)$. The $F(x)$ function in Equation (21) represents the cumulative probabilities for the standard normal distribution. When the above difference function is equal to zero, $p(x)$ and $q(x)$, which are described as two points in the (Ω, K) probability space, will coincide at the same point. When the difference function assumes a minimum value, this will indicate a point of transition between $p(x)$ and $q(x)$, where the two functions can be expressed in terms of each other. The same point also refers to a minimum number of observations required to produce information about the process X. When the difference function is described as in Figure 1, the presence of such a minimum value can be observed. The difference function $\Phi(x)$ decreases until $x = x_0$, where it passes through a minimum value. At point x_0, the two functions $p(x)$ and $q(x)$ approach each other until the distance between them is approximated by a constant C. After this point, when x approaches infinity, the difference function gradually increases; and finally, the difference between $p(x)$ and $q(x)$ approaches zero at infinity. One may define x_0 as the point where the two probability functions can be used interchangeably with an optimum number of observations.

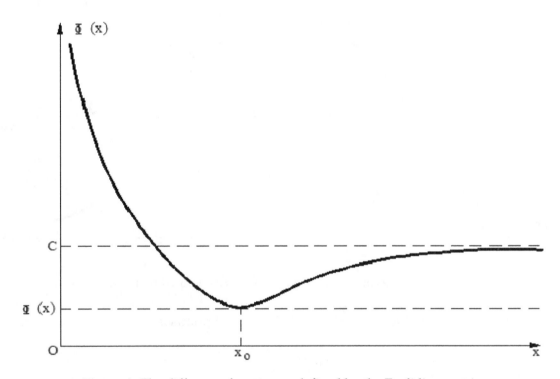

Figure 1. The difference function as defined by the Euclidian metric.

The purpose of observing the random variable X is to obtain a realistic estimate of its population parameters and to achieve reliable information about the process to allow for correct decisions in planning. On the other hand, each observation entails a cost factor; therefore, planners are interested in delineating how long the variable X has to be observed. Equation (16) is significant from this point of view. By defining the variation of information as the reduction of uncertainty via sampling, the point where no more increase or change in variation of information is obtained actually specifies the time point when sampling can be stopped. This is a significant issue which may be employed in considerations of gauging station discontinuance.

4.2. The Distance between Two Continuous Distribution Functions as Defined by Max-Norm

The max-norm can also be used to measure the distance between two functions defined in the probability space and to assess whether these two functions approach each other. According to the max-norm, the distance between two functions $p(x)$ and $q(x)$ is defined as:

$$\Delta(p, m) = \sup_{-\infty < x < +\infty} |p(x) - q(x)| \tag{22}$$

When $p(x)$ is used to represent the standardized normal and $q(x)$, the standardized uniform distribution functions, the difference function $\{h(x)\}$ will be:

$$h(x) = p(x) - q(x) \tag{23}$$

It may be observed in Figure 2 that, the critical points of the difference function are at h_0, h_1, and h_2 so that the difference between the two functions $\{\Delta(p,q)\}$ can be expressed as:

$$\Delta(p,q) = \max\{h_0, h_1, h_2\} \tag{24}$$

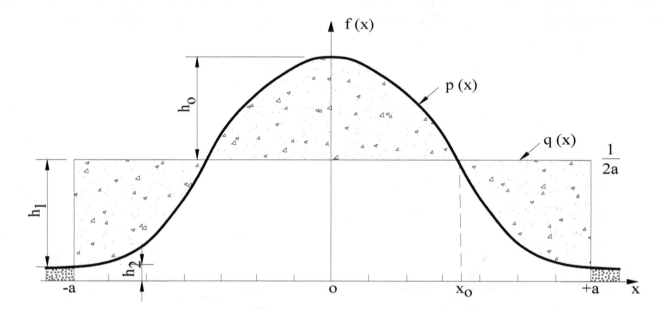

Figure 2. Critical values of the difference function as defined by the max-norm.

Based on the half-range value "a" in Figure 2, the critical points h_0, h_1, and h_2 can be obtained as:

$$h_0 = \frac{1}{\sqrt{2\pi}}\frac{1}{2a} \tag{25}$$

$$h_1 = \frac{1}{\sqrt{2\pi}}\frac{1}{2a}e^{(-a^2/2)} \tag{26}$$

$$h_2 = \frac{1}{\sqrt{2\pi}}e^{(-a^2/2)} \tag{27}$$

The problem here is then to find the distance between the two functions as the half-range value "a" which minimizes $\Delta(p,q)$ of Equation (24). The critical half-range value "a" that satisfies this supremum is:

$$a = \frac{3}{4}\sqrt{2\pi} \tag{28}$$

At the above critical half-range value "a", which is obtained by the max-norm, it is possible to use the two functions $p(x)$ and $q(x)$ interchangeably with an optimum number of observations.

When two points represented by the a posteriori and a priori distribution functions, $p(x)$ and $q(x)$, respectively, in the same probability space approach each other, this indicates, in information terms, an information increase about the random process analyzed. The case when the two points coincide represents total information availability about the process. Likewise, when $H(X/X^*)$ of Equation (16) approaches zero in absolute terms, this indicates a gain of total information about the process X defined within the interval $[a, b]$. One obtains sufficient information about the process when the variation information, as described by the Euclidian metric, approaches a constant value.

4.3. Asymptotic Properties of Shannon's Entropy

Vapnik [29] analyzed and provided proofs for some asymptotic properties of Shannon's entropy of the set of events on the sample size N. He used these properties to prove the necessary and sufficient conditions of uniform convergence of the frequencies of events to their probabilities.

In the work of Vapnik [23], it is shown that the sequence:

$$\frac{H(S)}{N}, N = 1, 2, \ldots, \tag{29}$$

has a limit c, when N goes to infinity. The lemma:

$$\lim_{N \to \infty} \frac{H(S)}{N} = c \ 0 \le c \le 1, \tag{30}$$

was proved by Vapnik [29] and was claimed to "repeat the proof of the analogous lemma in information theory for Shannon's entropy". Vapnik [29] also proved that, for any N, the sequence of Equation (29) is an upper bound for limit of Equation (30).

Vapnik [29] proved the above lemmas for Shannon's entropy, based on the discrete case of Equation (1). However, they are also valid for the continuous case as described by the Euclidian metric. Thus, it is possible to restate, using Vapnik's proofs, that the upper bound H_{max} of Shannon's entropy will be reached as the number of observations increases to approach the range of the population ($N \to \infty$) and that the variation of information of Equation (16) approaches a constant value "c".

In the next section, the derivation of the constant "c" is demonstrated for the case when the a priori distribution function is assumed to be uniform and the a posteriori function to be normal. These assumptions comply with the limits ($0 \le c \le 1$) defined for the discrete case as in Equation (30).

5. Further Development of the Revised Definition of the Variation of Information

If the observed range $[a, b]$ of the variable X is considered also as the population value of the range, R, of the variable, the maximum information content of the variable may be described as:

$$H_{max} = \ln R \tag{31}$$

With;

$$R = b - a \ a < x < b \tag{32}$$

When the a posteriori distribution of the variable is assumed to be normal, the marginal entropy of X becomes:

$$H(X) = \ln \sqrt{2\pi} + \ln \sigma + 1/2 \tag{33}$$

If the variable is actually normally distributed and if a sufficient number of observations are obtained, the entropy of Equation (16) will approach a value which can be considered to be within an acceptable region. This is the case where one may infer that sufficient information has been gained about the process.

When sufficient information is made available about X, it will be possible to make the best estimates for the mean (μ), variance (σ), and the range (R) of X. For this purpose, the variable has to be analyzed as an open series in the historic order. According to the approach used, the information gained about the process will continuously increase as the number of observations increase. Similarly, H_{max} and $H(X)$ will also increase, while $H(X/X^*)$ will decrease. When the critical point is reached, where the variable can be described by its population parameters, H_{max} will approach a constant value; $H(X)$ will also get closer to this value with $H(X/X^*)$ approaching a constant value of "c" as in Figure 3.

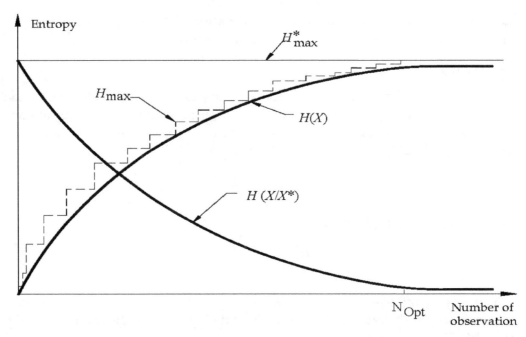

Figure 3. Maximum entropy $\{H_{max}\}$, marginal entropy $\{H(X)\}$ and entropy as the variation of information $\{H(X/X^*)\}$ versus the number of observations.

Determination on Confidence Limits for Entropy Defined by Variation of Information

The confidence limits (*acceptable region*) of entropy can be determined by using the a posteriori probability distribution functions. If the normal $\{N(0,1)\}$ probability density function is selected, the maximum entropy for the standard normal variable z is;

$$H_{max}(z) = \ln R_z, \tag{34}$$

with the range of z being,

$$R_z = 2a \tag{35}$$

Here, the value a describes the half-range of the variable. Then, the maximum entropy for variation x with $N(\mu,\sigma)$ is;

$$H_{max}(x) = \ln(R_z \sigma) \tag{36}$$

If the critical half-range value is foreseen as:

$$a = 4\sigma, \tag{37}$$

then the area under the normal curve may be approximated to be 1.

For the half-range value, replacing the appropriate values in Equation (16), one obtains the acceptable entropy value for the normal probability density function as:

$$H(X/X^*)_{cr} = 0.6605, \tag{38}$$

using natural logarithms. When the entropy $H(X/X^*)$ of the variable which is assumed to be normal remains below the above value, one may decide that the normal probability density function is acceptable and that a sufficient amount of information has been collected about the process.

If the a posteriori distribution function is selected as lognormal $LN(\mu_y, \sigma_y)$, the variation of information for the variable x can be determined as:

$$H(X/X^*) = \ln\left[2Sinh(a\sigma_y)\right] - \ln \sigma_y - 1.4189 \tag{39}$$

Here, since lognormal values will be positive, one may consider $0 \leq x \leq \infty$. Then the acceptable value of $H(X/X^*)$ for the lognormal distribution function will be;

$$H(X/X^*) = a\sigma_y - \ln \sigma_y - 1.4189 \tag{40}$$

According to Equation (40), no single constant value exists to describe the confidence limit for lognormal distribution. Even if the critical half-range is determined, the confidence limits will vary according to the variance of the variable. However, if the variance of x is known, the confidence limits can be computed.

6. Application

6.1. Application to Synthetic Series to Test the Fit of Probability-Distribution Functions

It is often difficult in practice to find long series of complete hydrological data. Thus, it is preferred here to test the above methodology on synthetically generated data for the purposes of evaluating the fit of different probability distribution functions. For this purpose, normal $\{N\,(\mu, \sigma)\}$ and lognormal $\{LN\,(\mu_y, \sigma_y)\}$ distributed time series are produced, using uniformly distributed series derived by the Monte Carlo method. Ten-year time series are obtained with normal $\{N\,(\mu, \sigma)\}$ and lognormal $\{LN\,(\mu_y, \sigma_y)\}$ distributions, respectively. Each series covered a period of (10×365) days with cumulative data for each year as $(i \times 365;$ where $i = 1, ..., N)$.

To test the methodology, $N(8, 10)$ distributed 3650 synthetic data are divided into subgroups with 365 data in each. First, maximum informational entropy $(H_{\max})$ is determined, using Equation (31) and the whole time series. Assuming that the a posteriori distribution is normal, marginal entropies $(H(X))$ and, finally, the informational entropy values $(H(X/X^*)$ are computed for the normal distribution using Equation (15). Consecutive values of these entropy measures are computed first for 365 generated data, next for 2×365 data, and for the last year 10×365 data. The confidence limits for the case of a posteriori normal distribution is determined by Equation (38). Figure 4 shows the results of this application. If a lognormal posteriori distribution is assumed for this series, which is actually normally distributed, this assumption is rejected on the basis of the computed confidence limits for normal distribution. Otherwise, the assumption is accepted. In Figure 4, the $H(X/X^*)$ values fall below the confidence level determined for normal distribution so that the assumption of a posteriori lognormal distribution is rejected.

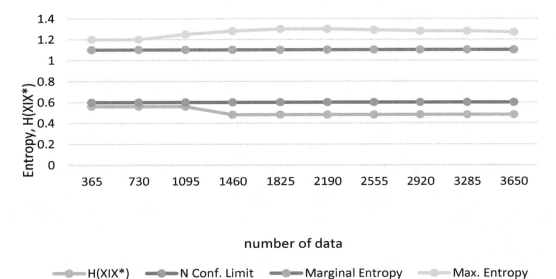

Figure 4. Normal distributed synthetic series, by the assumption of a posteriori normal distributed probability function (where; N Conf. Limit is the confidence limit for normal distribution).

If the same application is repeated by using the confidence limit for lognormal distribution, as in Figure 5, the assumption of a posterior lognormal distribution is rejected as the $H(X/X^*)$ values stay above the confidence level determined for lognormal distribution.

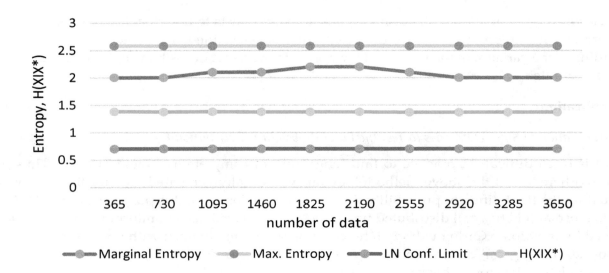

Figure 5. Normal distributed synthetic series, by the assumption of a posteriori lognormal distributed probability function (where; LN Conf. Limit is the limit of confidence for lognormal distribution).

Similar exercises may be run by generating lognormal distributed synthetic series and assuming the posteriori distribution first as lognormal (Figure 6) and then as normal distribution (Figure 7).

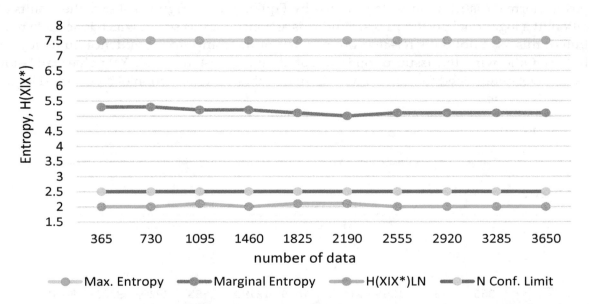

Figure 6. Lognormal distributed synthetic series, by the assumption of a posteriori lognormal distributed probability function (where; N Conf. Limit is the confidence limit for normal distribution).

The above exercises show that comparisons between assumptions of a posteriori normal and lognormal distributions on the basis of entropy-based confidence limits for each distribution give valid results by checking how the variation of information values behave with respect to the confidence limits.

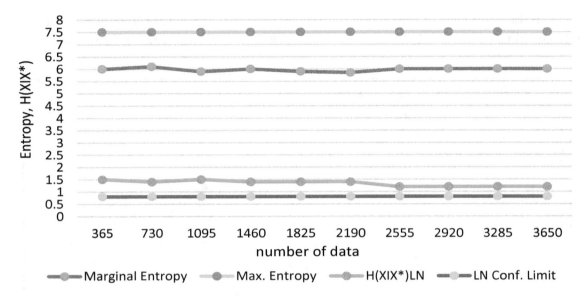

Figure 7. Lognormal distributed synthetic series, by the assumption of a posteriori normal distributed probability function (where; LN Conf. Limit is the limit of confidence for lognormal distribution).

6.2. Application to Runoff Data for Assessment of Sampling Duration

An important question regarding hydrometric data-monitoring networks is how long the observations should be continued. Considering the "data rich, information poor" data networks of our times, researchers and decisionmakers have wondered whether monitoring could be discontinued at certain sites, as data observation is a cost and labor-consuming activity [30,31]. To date, none of the approaches proposed for the problem of station discontinuance have found universal acceptance. Entropy measures as described in this work may as well be employed when a monitoring activity reaches an optimal point in time after which any new data does not produce new information. This feature of entropy measures is shown in Figure 3, where the marginal entropy of the process $H(X)$ approaches the total uncertainty Hmax as the number of observations (N) increase. Finally, a point is reached where H_{max} and $H(X)$ coincide after a certain number of observations, which can be defined as N_{opt}. After this point on, observed data do not produce new information, and thus monitoring can be discontinued. Certainly, the probability distribution of best fit to observed series must be selected first to evaluate this condition. This is an important feature of entropy measures as they can be used to infer about station discontinuance, based also on the selection of the appropriate distribution functions.

To test the above aspect of the entropy concept, observed runoff data at two monitoring stations (Kuskayasi and Akcil) in the Ceyhan river basin in Turkey are employed (Figure 8). The Ceyhan basin has been subject to several investigations and projects for the development of water schemes; thus, it is intended here to evaluate the monitoring activities in the basin in terms of entropy measures. Although there are other gauging stations along the river, their data are not homogeneous due to already-built hydraulic structures. Kuskayasi and Akcil are the two stations where natural flows are observed, although their common observation periods cover only 8 years.

The observations at Kuskayasi were discontinued after 1980 and Akcil after 1989. Thus, for the purposes of this application their common period between 1973 and 1980 is selected. Daily data for the observation period of 8 years are used, where the mean daily runoff at Kuskayasi is 10.8 m^3/s and that at Akcil is 27.18 m^3/s. The standard deviations are 11.77 m^3/s and 22.48 m^3/s, respectively.

Next, the fits of normal and lognormal distributions are tested at both stations again with the entropy concept. This analysis is followed by the computation of marginal entropies ($H(X)$, H_{max} and the variation of information $H(X)/H^*$) for these two distribution functions. The computations are carried out in a successive manner, using the first year's 365 data, the second year's 720, and so on until the total number of 2920 data are reached. Certainly, H_{max} changes with the total of data observed from the beginning of the observation period, assuming a ladder-like increase as in Figure 3, where H^* is used to represent $H_{\max}$ for the total observation period of 2920 daily data.

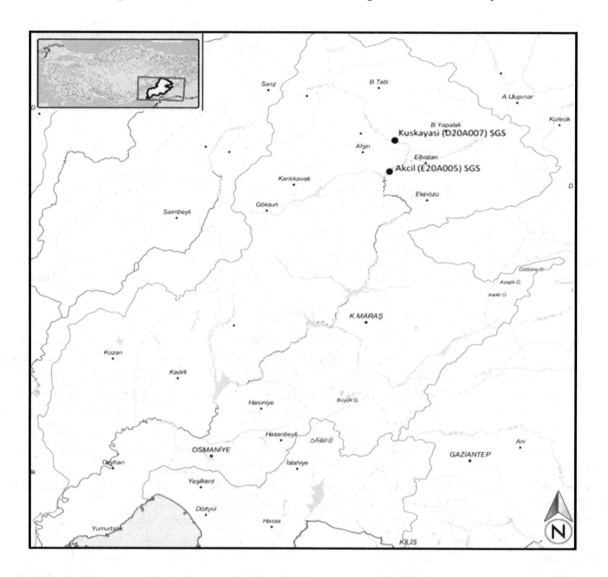

Figure 8. Ceyhan river basin in the south of Turkey and selected monitoring sites (Kuskayasi and Akcil).

Figures 9 and 10 show figures similar to Figure 3 under the assumption of normal and lognormal distributions fit to daily data for 8 years. Although both distributions seem to be sufficient, normal distribution shows more distinctively how $H(X)$ approaches the total entropy H^*. It may seem unusual for an upstream station with daily observations to reflect a normal distribution; yet this is physically due to karstic contributions to runoff, which stabilize the flows.

Whether the normal or lognormal distributions are selected, it can be observed in Figures 9 and 10 that 2920 observations are not sufficient to reach H^*. Although $H(X)$ approaches H^*, the optimal number of observations is not yet reached with only 8 years of observations.

Results for the downstream Akcil station are shown in Figures 11 and 12. Here, again, normal distribution appears to give a better fit to observed data. As can be observed especially in Figure 11, $H(X)$ closely approaches H^* for 8 years of data. If observations could be continued after 8 years of 2920 data, most probably the optimum number of observations would be reached.

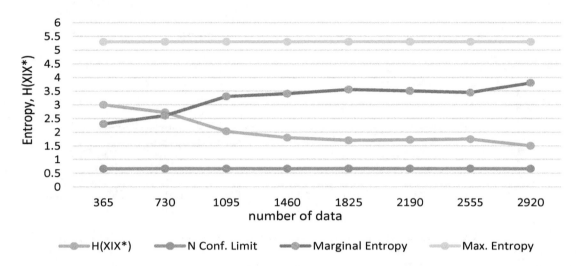

Figure 9. Kuskayasi (1973–1980), by the assumption of a posteriori normal distribution function (where; N Conf. Limit is the limit for normal distribution).

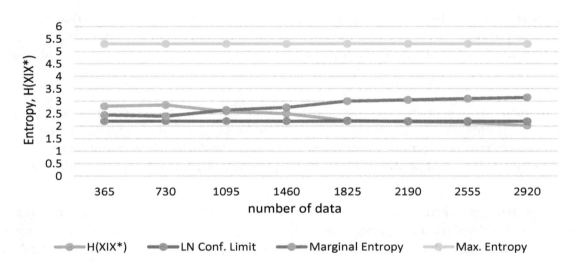

Figure 10. Kuskayasi (1973–1980), by the assumption of a posteriori lognormal distribution function (where; LN Conf. Limit is the limit of confidence for lognormal distribution).

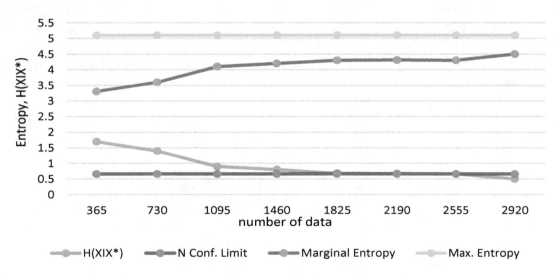

Figure 11. Akcil (1973–1980), by the assumption of a posteriori normal distribution function (where; N Conf. Limit is the limit of confidence for normal distribution).

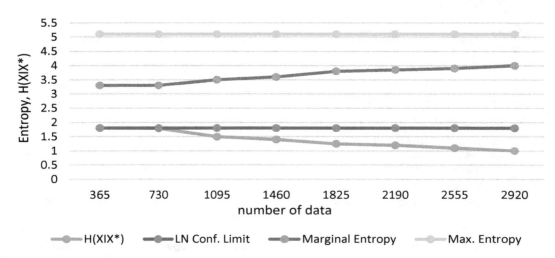

Figure 12. Akcil (1973–1980), by the assumption of a posteriori lognormal distribution function (where; LN Conf. Limit is the limit of confidence for lognormal distribution).

It is concluded on the basis of results obtained through the above application that, if sufficiently long observed time series are available, the entropy principle can be effectively used to infer on an important feature of hydrometric monitoring, i.e., sampling duration or station discontinuance.

7. Conclusions

The extension to the revised definition of informational entropy developed in this paper resolves further major mathematical difficulties associated with the assessment of uncertainty, and indirectly of information, contained in random variables. The description of informational entropy, not as an absolute measure of information but as a measure of the variation of information, has the following advantages:

- It eliminates the controversy associated with the mathematical definition of entropy for continuous probability distribution functions. This makes it possible to obtain a single value for the variation of information instead of several entropy values that vary with the selection of the discretizing interval when, in the former definitions of entropy for continuous distribution functions, discrete probabilities of hydrological events are estimated through relative class frequencies and discretizing intervals.
- The extension to the revised definition introduces confidence limits for the entropy function, which facilitates a comparison between the uncertainties of various hydrological processes with different scales of magnitude and different probability structures.
- Following from the above two advantages, it is further possible through the use of the concept of the variation of information to:

  ○ determine the contribution of each observation to information conveyed by data;
  ○ determine the probability distribution function which best fits the variable;
  ○ make decisions on station discontinuance.

The present work focuses basically on the theoretical background for the extended definition of informational entropy. The methodology is then tested via applications to synthetically generated data and observed runoff data and is shown to give valid results. For real-case observed data, long duration series with sufficient length and quality are needed. Currently, studies are being continued by the authors on long series of runoff, precipitation and temperature data.

It follows from the above discussions that the use of the concept of variation of information and of confidence limits makes it possible to:

- *determine the contribution of each observation to information conveyed by data;*
- *calculate the cost factors per information gained;*
- *determine the probability distribution function which best fits the variable;*
- *select the model which best describes the behavior of a random process;*
- *compare the uncertainties of variables with different probability density functions;*
- *make decisions on station discontinuance.*

The above points are different problems to be solved by the concept of entropy, and further extensions of the methodology are required to address each of them.

Acknowledgments: We gratefully acknowledge the support received from the authors' EU Horizon2020 Project entitled FATIMA (FArming Tools for external nutrient Inputs and water Management, Grant No. 633945) for providing the required funds to cover the costs towards publishing in open access.

Author Contributions: Turkay Baran and Nilgun B. Harmancioglu conceived and designed the experiments; Turkay Baran and Filiz Barbaros performed the experiments; Cem P. Cetinkaya and Filiz Barbaros analyzed the data; Turkay Baran, Nilgun B. Harmancioglu and Cem P. Cetinkaya contributed reagents/materials/analysis tools; Nilgun B. Harmancioglu wrote the paper.

References

1. Singh, V.P.; Fiorentino, M. (Eds.) A Historical Perspective of Entropy Applications in Water Resources. In *Entropy and Energy Dissipation in Water Resources*; Water Science and Technology Library; Kluwer Academic Publishers: Dordrecht, The Netherlands, 1992; Volume 9, pp. 155–173.
2. Fiorentino, M.; Claps, P.; Singh, V.P. An Entropy-Based Morphological Analysis of River Basin Networks. *Water Resour. Res.* **1993**, *29*, 1215–1224. [CrossRef]
3. Wehrl, A. General Properties of Entropy. *Rev. Mod. Phys.* **1978**, *50*, 221–260. [CrossRef]
4. Templeman, A.B. Entropy and Civil Engineering Optimization. In *Optimization and Artificial Intelligence in Civil and Structural Engineering*; NATO ASI Series (Series E: Applied Sciences); Topping, B.H.V., Ed.; NATO: Washington, DC, USA, 1989; Volume 221, pp. 87–105, ISBN 978-94-017-2490-6.
5. Schrader, R. On a Quantum Version of Shannon's Conditional Entropy. *Fortschr. Phys.* **2000**, *48*, 747–762. [CrossRef]
6. Shannon, C.E. A Mathematical Theory of Information. In *The Mathematical Theory of Information*; The University of Illinois Press: Urbana, IL, USA, 1948; Volume 27, pp. 170–180.
7. Harmancioglu, N.; Alpaslan, N. Water Quality Monitoring Network Design: A Problem of Multi-Objective Decision Making. *JAWRA* **1992**, *28*, 179–192. [CrossRef]
8. Harmancioglu, N.; Singh, V.P.; Alpaslan, N. Versatile Uses of The Entropy Concept in Water Resources. In *Entropy and Energy Dissipation in Water Resources*; Singh, V.P., Fiorentino, M., Eds.; Kluwer Academic Publishers: Dordrecht, The Netherlands, 1992; Volume 9, pp. 91–117, ISBN 978-94-011-2430-0.
9. Harmancioglu, N.; Alpaslan, N.; Singh, V.P. Assessment of Entropy Principle as Applied to Water Quality Monitoring Network Design. In *Stochastic and Statistical Methods in Hydrology and Environmental Engineering*; Water Science and Technology Library; Hipel, K.W., McLeod, A.I., Panu, U.S., Singh, V.P., Eds.; Springer: Dordrecht, The Netherlands, 1994; Volume 3, pp. 135–148, ISBN 978-94-017-3083-9.
10. Harmancioglu, N.B.; Yevjevich, V.; Obeysekera, J.T.B. Measures of information transfer between variables. In Proceedings of the Fourth International Hydrology Symposium—Multivariate Analysis of Hydrologic Processes, Fort Collins, CO, USA, 15–17 July 1985; Colorado State University: Fort Collins, CO, USA, 1986; pp. 481–499.
11. Harmancioglu, N.B.; Yevjevich, V. Transfer of hydrologic information among river points. *J. Hydrol.* **1987**, *91*, 103–118. [CrossRef]
12. Harmancioglu, N.; Cetinkaya, C.P.; Geerders, P. Transfer of Information among Water Quality Monitoring Sites: Assessment by an Optimization Method. In Proceedings of the EnviroInfo Conference 2004, 18th International Conference Informatics for Environmental Protection, Geneva, Switzerland, 21–23 October 2004; pp. 40–51.

13. Baran, T.; Bacanli, Ü.G. An Entropy Approach for Diagnostic Checking in Time Series Analysis. *SA Water* **2007**, *33*, 487–496.

14. Singh, V.P. The Use of Entropy in Hydrology and Water Resources. *Hydrol. Process.* **1997**, *11*, 587–626. [CrossRef]

15. Singh, V.P. The Entropy Theory as a Decision Making Tool in Environmental and Water Resources. In *Entropy Measures, Maximum Entropy Principle and Emerging Applications. Studies in Fuzziness and Soft Computing*; Karmeshu, Ed.; Springer: Berlin, Germany, 2003; Volume 119, pp. 261–297, ISBN 978-3-540-36212-8.

16. Harmancioglu, N.; Singh, V.P. Entropy in Environmental and Water Resources. In *Encyclopedia of Hydrology and Water Resources*; Herschy, R.W., Fairbridge, R.W., Eds.; Kluwer Academic Publishers: Dordrecht, The Netherlands, 1998; Volume 5, pp. 225–241, ISBN 978-1-4020-4497-7.

17. Harmancioglu, N.; Singh, V.P. Data Accuracy and Data Validation. In *Encyclopedia of Life Support Systems (EOLSS)*; Knowledge for Sustainable Development, Theme 11 on Environmental and Ecological Sciences and Resources, Chapter 11.5 on Environmental Systems; Sydow, A., Ed.; UNESCO Publishing-Eolss Publishers: Oxford, UK, 2002; Volume 2, pp. 781–798, ISBN 0 9542989-0-X.

18. Harmancioglu, N.B.; Ozkul, S.D. Entropy-based Design Considerations for Water Quality Monitoring Networks. In *Technologies for Environmental Monitoring and Information Production*; Nato Science Series (Series IV: Earth and Environmental Sciences); Harmancioglu, N.B., Ozkul, S.D., Fistikoglu, O., Geerders, P., Eds.; Springer: Dordrecht, The Netherlands, 2003; Volume 23, pp. 119–138, ISBN 978-94-010-0231-8.

19. Ozkul, S.; Harmancioglu, N.B.; Singh, V.P. Entropy-Based Assessment of Water Quality Monitoring Networks. *J. Hydrol. Eng.* **2000**, *5*, 90–100. [CrossRef]

20. Jaynes, E.T. *E. T. Jaynes: Papers on Probability, Statistics and Statistical Physics*; Rosenkrantz, R.D., Ed.; Springer: Dordrecht, The Netherlands, 1983; ISBN 978-94-009-6581-2.

21. Guiasu, S. *Information Theory with Applications*; Mc Graw-Hill: New York, NY, USA, 1977; 439p, ISBN 978-0070251090.

22. Harmancioglu, N.B. Measuring the Information Content of Hydrological Processes by the Entropy Concept. *J. Civ. Eng. Fac. Ege Univ.* **1981**, 13–88.

23. Jaynes, E.T. Information Theory and Statistical Mechanics. *Phys. Rev.* **1957**, *106*, 620–630. [CrossRef]

24. Harmancioglu, N. Entropy concept as used in determination of optimal sampling intervals. In Proceedings of the Hydrosoft 1984, International Conference on Hydraulic Engineering Software. Interaction of Computational and Experimental Methods, Portorož, Yugoslavia, 10–14 September 1984; Brebbia, C.A., Maksimovic, C., Radojkovic, M., Eds.; Editions du Tricorne: Geneva, Switzerland; pp. 99–110.

25. Harmancioglu, N.B. An Entropy-based approach to station discontinuance. In *Stochastic and Statistical Methods in Hydrology and Environmental Engineering*; Time Series Analysis and Forecasting; Hipel, K.W., McLeod, I., Eds.; Kluwer Academic Publishers: Dordrecht, The Netherlands, 1994; Volume 3, pp. 163–176.

26. Harmancioglu, N.B.; Alpaslan, N. *Basic Approaches to Design of Water Quality Monitoring Networks*; Water Science and Technology; Elsevier: Amsterdam, The Netherlands, 1994; Volume 30, pp. 49–56.

27. Harmancioglu, N.B.; Cetinkaya, C.P.; Barbaros, F. Environmental Data, Information and Indicators for Natural Resources Management. In *Practical Environmental Statistics and Data Analysis*; Rong, Y., Ed.; ILM Publications: Buchanan, NY, USA, 2011; Chapter 1, pp. 1–66.

28. Schultze, E. Einführung in die Mathematischen Grundlagen der Informationstheorie. In *Lecture Notes in Operations Research and Mathematical Economics*; Springer: Berlin, Germany, 1969; 116p, ISBN 978-3-642-86515-2.

29. Vapnik, V.N. *Statistical Learning Theory*; Wiley Interscience: New York, NY, USA, 1998; 736p, ISBN 978-0-47-03003-4.

30. Harmancioglu, N.B.; Singh, V.P.; Alpaslan, N. *Environmental Data Management*; Water Science and Technology Library; Kluwer Academic Publishers: Dordrecht, The Netherlands, 1998; 298p, ISBN 0792348575.

31. Harmancioglu, N.B.; Fistikoglu, O.; Ozkul, S.D.; Singh, V.P.; Alpaslan, N. *Water Quality Monitoring Network Design*; Water Science and Technology Library; Kluwer Academic Publishers: Dordrecht, The Netherlands, 1999; 290p, ISBN 978-94-015-9155-3.

9

Scaling-Laws of Flow Entropy with Topological Metrics of Water Distribution Networks

Giovanni Francesco Santonastaso [1,2], **Armando Di Nardo** [1,2], **Michele Di Natale** [1,2], **Carlo Giudicianni** [1] **and Roberto Greco** [1,2,*]

[1]   Dipartimento di Ingegneria Civile, Design, Edilizia e Ambiente, Università degli Studi della Campania "Luigi Vanvitelli", via Roma 29, 81031 Aversa, Italy; giovannifrancesco.santonastaso@gmail.com (G.F.S.); armando.dinardo@unicampania.it (A.D.N.); michele.dinatale@unicampania.it (M.D.N.); carlo.giudicianni@gmail.com (C.G.)

[2]   Action Group CTRL+SWAN of the European Innovation Partnership on Water, EU, B-1049 Brussels, Belgium

[*]   Correspondence: roberto.greco@unicampania.it

Abstract: Robustness of water distribution networks is related to their connectivity and topological structure, which also affect their reliability. Flow entropy, based on Shannon's informational entropy, has been proposed as a measure of network redundancy and adopted as a proxy of reliability in optimal network design procedures. In this paper, the scaling properties of flow entropy of water distribution networks with their size and other topological metrics are studied. To such aim, flow entropy, maximum flow entropy, link density and average path length have been evaluated for a set of 22 networks, both real and synthetic, with different size and topology. The obtained results led to identify suitable scaling laws of flow entropy and maximum flow entropy with water distribution network size, in the form of power–laws. The obtained relationships allow comparing the flow entropy of water distribution networks with different size, and provide an easy tool to define the maximum achievable entropy of a specific water distribution network. An example of application of the obtained relationships to the design of a water distribution network is provided, showing how, with a constrained multi-objective optimization procedure, a tradeoff between network cost and robustness is easily identified.

Keywords: scaling laws; power laws; water distribution networks; robustness; flow entropy

1. Introduction

The topology of water distribution networks (WDN) is being deeply studied with respect to its relationship with their robustness, i.e., their capability of effectively delivering the demanded flows to the users with the required pressure under unfavorable operating conditions [1]. In fact, evaluating the performance of a WDN requires the complex calibration of a hydraulic model of the network, and often a number of time-consuming simulations. Hence, establishing relationships, linking topological metrics of a WDN, easily achievable from the mere knowledge of the network layout, with its hydraulic behavior, would represent a powerful tool for the design, rehabilitation and management of WDN. In this respect, aiming at quantitative comparison of different network layouts, it is important to understand how topological metrics change with the size of the considered network.

In fact, the size variation of a system can cause changes in the order of predominance of physical phenomena; this is called scaling effect, and the laws that govern such an effect are called scaling laws. The scaling laws are relationships linking any parameter associated with an object (or system) with its length scale [2]. They constitute a very useful tool to predict the behavior and the properties of a large system by experimenting on a small-sized scale model, since the characteristics of a system can be expressed through various parameters in such a way that any change in size (i.e., scale) does not affect the magnitudes of these quantities. Scaling laws represent useful tools for understanding the interplay among various physical phenomena and geometric characteristics of complex systems, and often it happens that simple scaling laws can provide clues to some fundamental aspects of the system. In many fields, scaling laws have been identified. For example, scaling laws have been experimentally determined over a huge range of scales in probability distributions describing river basin morphology [3], whose geometrical description is of great importance for a deeper understanding of how some related natural events occur. The existence of a scaling law relating point precipitation depth records to duration has been known for at least 60 years through published tabulations of data and the associated graphs [4,5], even if there is no explanation of the mechanism underlying this remarkably robust relationship, making it even more tantalizing [6]. Scaling laws have been also identified in fluid mechanics, to describe turbulent energy distribution across scales [7,8], and in meteorology, to describe scaling of clouds [9], atmospheric variability [10], and fluctuations of Arctic sea ice [11]. In the field of network topology, it has been found that many real networks exhibit power–law shaped node degree distribution, where the degree is the number of connected links to each node. Such networks have been named scale-free networks [12], because power–laws have the property of retaining the same functional form at all scales. These networks result in the simultaneous presence of a few nodes (the hubs) linked to many other nodes, and a large number of poorly connected elements [13]. The World Wide Web (WWW) is one of the most famous scale-free networks. It is formed by the hyperlinks between different Web pages, and, with more than 10^8 nodes, it is the largest network ever studied.

Differently, water distribution networks (WDN) do not present hubs, as each node is connected only to a few nodes located in its immediate surroundings. The connections between nodes in a WDN ensure multiple possible flow paths, so to cope with abnormal working conditions, such as unexpected water requests by the users and failure of some elements [14]. In this respect, several topological metrics aimed at quantifying WDN connectivity have been proposed as proxies for network robustness and reliability [15].

Reliability, in a WDN, can be defined as the probability of the system being capable of supplying the water demands both under normal and abnormal conditions [16,17]. The assessment of reliability is influenced by many factors: spatial and temporal demand distribution, possible failure of one or more components, pressure-flow relationship, connectivity of the network, etc. Therefore, there is not an established measure of WDN reliability, and a review of different methods to evaluate it can be found in [18]. Reliability measures are categorized into three groups: topological, hydraulic and entropic. Topological reliability is based on the probability of node connectivity/reachability [19]; hydraulic reliability is focused on the probability of delivering design water demands, (e.g., [17]); and the last category adopts the informational entropy as a surrogate of the reliability [18].

The concept of informational entropy [20] has been widely applied in hydraulics and hydrology (i.e., to estimate velocity distribution in open channels, suspended sediment concentration profile, suspended sediment discharge, or precipitation variability, moisture profiles, etc.) [21]. In the field of WDN, Shannon's entropy has been proposed as a measure of connectivity and so as a proxy for reliability [22].

The adoption of entropy as a surrogate for network reliability was investigated by several authors [23–26]. The basic idea is that entropy is a measure of the uniformity of pipe flow rate [27], thus it is related to looped network redundancy, which makes it potentially more capable of facing unfavorable working conditions, such as concentrated peaks of demand or failure of pipes (e.g., [1]). Hence, redundancy increases network robustness, and so, indirectly, its reliability.

Hence, many studies [28–30] have proposed multi-objective optimization for water distribution network design or rehabilitation based on minimizing costs for construction, operation, and maintenance, coupled with the maximization of the entropy as a measure of robustness.

Traditionally, the robustness of water distribution networks was assured by means of densely looped layouts, so to provide alternative paths for each demand node [31]. More recently, Di Nardo et al. [32]

have studied the topological redundancy of a water supply network, with regard to pipe failures, applying the complex network theory [33,34]. In fact, many water supply systems consisting of up to tens of thousands nodes and hundreds of looped paths can be considered as complex networks [13]. Thus, it is possible to compute topological metrics [32,35–37] to analyze the robustness of a water distribution network.

Recently, comparisons between entropy and other indirect measures of robustness [1,26,38–40] such as resilience index [41], network resilience [42] and Surplus Power Factor [43] have been proposed, but the obtained results are contradictory. According to some authors [26,38,39], informational entropy is a good measure of network robustness. Conversely, other studies indicate that the resilience index estimates better the network hydraulic performance than entropy in the case of pipe failures [1] and for multi-objective design optimization [40].

The advantage of using informational entropy to evaluate network robustness is that only pipe flows and topology are required for its computation [39], while the main drawback is that there is not a reference value of entropy allowing for defining an acceptable level of robustness for a given WDN, nor to compare different WDN layouts. In this respect, the definition of scaling laws of flow entropy with the topological dimension of the network could be useful for WDN design and rehabilitation purposes.

This work investigates the possible relationship between topological metrics, borrowed from complex network theory, and flow entropy, through the analysis of the values that they assume for several WDNs, both real and synthetic. In particular, for each network, five of the coarsest topological characteristics of a network, the number of nodes n and links m, the average node degree k, the link density q and the average path length APL have been calculated. The results show that the flow entropy of a WDN is strongly linked to its size and topology, and that it can be expressed as a function of topological metrics. Furthermore, the maximum achievable flow entropy value has been calculated for each WDN. Scaling-laws of flow entropy with the size of the networks have been identified. Two examples of the application of the obtained results to the design of WDNs are finally provided.

2. Methods

The study of WDN using innovative topological metrics, borrowed from the theory of complex networks [13], already led to interesting results for the analysis of water network vulnerability [32,35,44], as well as for water network partitioning [45,46]. In the following sections, the topological and entropy metrics used in this paper are briefly described, and finally the deviation of actual entropy from maximum entropy is introduced as a possible measure of network robustness.

2.1. Topological Metrics

The average node degree, k, represents the mean number of links concurring in the nodes of the network, and is given by:

$$k = \frac{2 \cdot m}{n} \tag{1}$$

in which n is the number of nodes and m the number of links of the network.

The link density, q, expresses the ratio between the total number of network edges and the number of edges of a globally coupled network with the same number of nodes, thus providinga measure of network redundancy:

$$q = \frac{2 \cdot m}{n \cdot (n - 1)} \tag{2}$$

The average path length, APL [33], is the average number of steps along the shortest paths between all possible pairs of nodes in the network:

$$APL = \frac{\sum_{\forall s \neq t} \sigma(s, t)}{\frac{1}{2} n \cdot (n - 1)} \tag{3}$$

where $\sigma(s,t)$ is the number of edges along the shortest path connecting node s to node t (when there is no path between a pair of nodes, the path length is assumed to be infinite) [47]. A short average path length indicates a more interconnected network, while a long one indicates greater overall topological distances between nodes. Consequently, a network with a large *APL* value may be considered more fragmented [48].

2.2. Entropic Metrics

The Shannon's information entropy [49] is a statistical measure of the amount of uncertainty associated with the probability distribution of any discrete random variable, defined as follows:

$$E = - \sum_{k=1}^{l} p_k \ln p_k, \tag{4}$$

where E is the entropy, p_k is the probability, and l is the number of values that the variable can assume. Tanyimboh and Templeman [28], with the use of the conditional entropy formula of [50], considered all the possible flow paths from sources to demand nodes, and introduced the flow entropy S of a water distribution system by defining the probability of the water to flow along the k-th path as the ratio between the flow rate reaching the end node of the path and the total delivered flow rate [42]. The following recursive formula [24] allows the calculation of S, which is regarded as a measure of pipe flow rates uniformity:

$$S = - \sum_{i=1}^{NS} \frac{Q_i}{T} \ln\left(\frac{Q_i}{T}\right) - \frac{1}{T} \sum_{j=1}^{NN} T_j \left[\frac{Q_j}{T_j} \ln\left(\frac{Q_j}{T_j}\right) + \sum_{ji \in N_j} \frac{q_{ji}}{T_j} \ln\left(\frac{q_{ji}}{T_j}\right) \right] \tag{5}$$

On the right hand side of Equation (5), the first term is the entropy of supply nodes and the second is the entropy of demand nodes; NS is the number of supply nodes; T is the total supplied flow rate; NN is the number of demand nodes; Q_i represents the inflow at the i-th source node; T_j is the total flow rate reaching the j-th demand node; Q_j is the water demand at the j-th demand node; q_{ij} is the flow rate in the pipe connecting node j with surrounding node i; and N_j is the number of pipes carrying water from the j-th demand node towards other surrounding nodes.

The data required to assess the flow entropy are the topological layout, the water supply and the demand at all nodes, and the flow direction along each pipe. To this purpose, the hydraulic simulation of the network, carried out with the solver EPANET 2 [51], provides the flow rate and direction along each pipe.

2.3. Maximum Entropy and Network Robustness

The maximization of Equation (3) can be used to compute the maximum value of the flow entropy, *MS*, and in this case only the source flow rates, the water demands at nodes and the flow directions along the links are required. Specifically, *MS* is here computed here by means of a non-iterative procedure for multi-source networks, proposed in [52]. The entropy deficit, i.e., the deviation between the flow entropy S and the corresponding values of *MS*, given by Equation (5), is assumed to be representative of how much a network is robust, based on the idea that networks, designed to supply maximum entropy flows, would be the most robust for a given source pressure excess compared to the design pressure at nodes [23].

$$\Delta S = 1 - \frac{S}{MS} \tag{6}$$

3. Results and Discussions

Topological metrics and flow entropy metrics were computed for a set of 22 WDNs, both real and synthetic. The maximum entropy MS of each network was calculated adopting the same flow directions along the pipes as for the calculation of flow entropy S (i.e., the directions provided by the hydraulic simulation of the network for the actual set of pipe sizes). Therefore, the obtained MS cannot be considered as the maximum possible values of flow entropy, as a different choice of flow directions could lead to a higher value of MS. However, as the flow directions are mainly dictated by the position of sources and demand nodes, and by the assumed water demand at nodes, it is expected that flow directions would be only slightly (and locally) affected by changes in the size of some of the pipes. In Table 1, the computed values of the metrics are reported for all the considered networks.

Table 1. Topological metrics: number of nodes (n) and links (m), density (q), average path length (APL), flow entropy (S) and maximum flow entropy (MS), for all WDN (* denotes synthetic networks).

Network	n	m	q	APL	S	MS	ΔS
Two Loop * [53]	7	8	0.5333	1.90	2.063	2.296	0.101
Two Reservoirs * [54]	12	17	0.3778	2.59	2.829	3.008	0.059
Anytown * [55]	25	43	0.1861	2.94	4.172	5.048	0.174
GoYang * [56]	23	30	0.1299	3.75	3.113	3.658	0.149
Blacksburg * [57]	32	35	0.0805	4.37	3.358	3.473	0.033
Hanoi * [58]	32	34	0.0731	5.31	3.384	3.395	0.003
BakRyan * [59]	36	58	0.0975	4.30	3.243	3.709	0.126
Fossolo [60]	37	58	0.0921	3.67	3.677	4.441	0.172
Pescara [60]	72	99	0.0435	8.69	4.273	4.572	0.065
BWSN2008-1 * [61]	127	168	0.0213	10.15	3.939	5.567	0.292
Skiathos [62]	176	189	0.0124	11.52	5.551	6.196	0.104
Parete [1]	184	282	0.0171	8.80	6.561	9.331	0.297
Villaricca [1]	199	249	0.0130	11.29	5.206	5.497	0.053
Monteruscello [63]	206	231	0.0110	20.24	5.211	5.385	0.032
Modena [60]	272	317	0.0089	14.04	5.436	5.764	0.057
Celaya [64]	338	477	0.0086	11.81	6.8	7.734	0.121
Balerma Irrigation [65]	448	454	0.0046	23.89	6.091	6.489	0.061
Castellammare	1231	1290	0.0017	32.25	7.583	8.094	0.063
Matamoros [66]	1293	1651	0.0020	27.76	9.896	13.325	0.257
Wolf Cordera Ranch [67]	1786	1985	0.0013	25.94	7.905	9.865	0.199
Exnet * [68]	1893	2465	0.0014	20.60	10.466	12.882	0.188
San Luis Rio Colorado [66]	1908	2681	0.0015	28.86	8.097	9.443	0.143

The set of networks used as case study includes water distribution networks with very different characteristics, as indicated by the very different values assumed by the metrics:

- dimension: the smallest network has a number of nodes $n = 6$ (Two Loop), while the largest has $n = 1890$ (Exnet);
- layout: looped networks as well as branched ones are included, i.e., Balerma Irrigation can be considered a tree-network, while networks such as Parete and Sector Centro Real are very looped; compact and elongated networks are included, with low values of APL coupled with high values of density being representative of compact network layouts;
- robustness: the set of networks includes systems with very small deviation of actual entropy from maximum entropy, like Hanoi and Modena (the entropy deviation ΔS is equal to 0.0032 and 0.0616, respectively), and networks with high deviation of entropy, like Parete and BWSN2008-1 (entropy deviations of 0.297 and 0.292, respectively).

These differences indicate that the adopted set is suitable to analyze the entropy metrics from a topological point of view in a general sense.

Figure 1 shows the scatter plots of the values of S vs. various topological metrics, and the best fitting power–law equations. The diagrams show that an increasing trend exists in the relationship between flow entropy and number of nodes (Figure 1a), and between flow entropy and number of links (Figure 1b), as well as a decreasing trend for the relationship between flow entropy and link density (Figure 1c). Although in Figure 1d a positive trend of flow entropy vs. average path length is also observable, it is less clearly defined than the previous ones.

The scatter plots of *MS* vs. the same topological metrics, reported in Figure 2, confirm similar trends as in Figure 1. Specifically, a clear increasing relation of *MS* with the number of links (Figure 2b) can be noted, while Figure 2d shows a weak relation between *MS* and *APL*. The determination coefficients, R^2, of *S* (Figure 1) are slightly greater than the ones computed for the *MS* (Figure 2). Anyway, both *S* and *MS* are clearly related to network topology.

It is worth to noting that, although *APL* is considered a proxy of the topological robustness of a network [32] and the entropy has been proposed as a surrogate of network reliability, the relationships between *S* and *APL*, as well as between *MS* and *APL*, are less consistent than expected.

The best fitting relationships are the power–laws linking *S* and *MS* with the number of pipes *m*, as indicated by $R^2 = 0.94$ for the flow entropy and $R^2 = 0.90$ for the maximum flow entropy (Figures 1b and 2b, respectively). The clear dependence of *MS* and *S* on *m*, indicating that flow entropy is related to the size of the network, suggested to investigate the ratios *S/m* and *MS/m* to characterize the redundancy of a network regardless of its dimension.

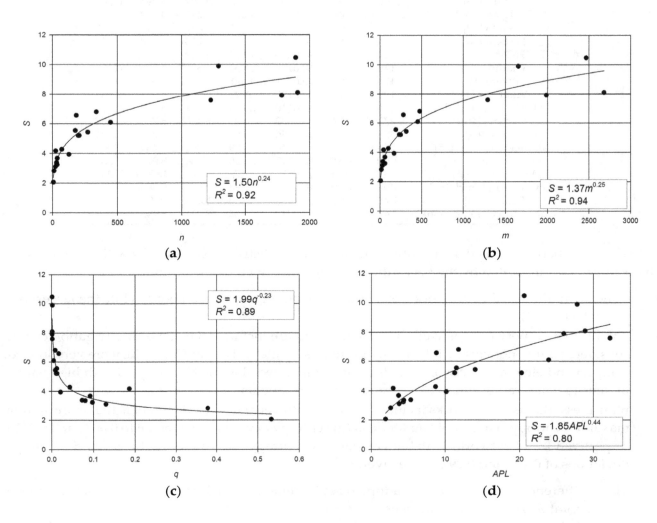

Figure 1. Scatter plots and best fitting power–laws of: (**a**) entropy vs. number of nodes; (**b**) entropy vs. number of pipes; (**c**) entropy vs. link density; (**d**) entropy vs. network average path length.

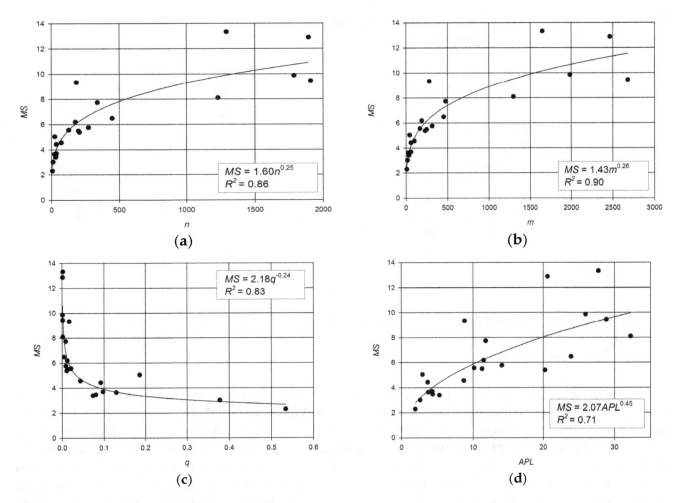

Figure 2. Scatter plots and best fitting power–laws of: (**a**) maximum entropy vs. number of nodes; (**b**) maximum entropy vs. number of pipes; (**c**) maximum entropy vs. link density; (**d**) maximum entropy vs. network average path length.

The scatter plots of S/m vs. n and MS/m vs. n, and the coefficients of determination of the relevant best fitting power–laws, are reported in Figure 3. A distinct trend is clearly visible for both the flow entropy measures, as indicated by $R^2 = 0.99$ for both the relationships. The obtained best fitting power–law equations are:

$$S = 1.05m \times n^{-0.74} \tag{7}$$

$$MS = 1.12m \times n^{-0.72} \tag{8}$$

Looking at Equation (4), and keeping in mind the adopted definition of the probability of the water flowing along a path from a source node to a demand node, it becomes clear that the maximum theoretical flow entropy (i.e., all flow paths sharing the same probability) should scale with $\ln n$. In fact, the number of possible paths in a network scales with the number of nodes (e.g., in a network with a single source, the total number of flow paths to all nodes equals the number of links $m = n + l - 1$, l being the number of loops). Therefore, it is expected that

$$\frac{MS}{m} \sim \frac{-\sum_n \frac{1}{n} \ln \frac{1}{n}}{n} = \frac{\ln n}{n} \tag{9}$$

The curve of Equation (9), also plotted in Figure 3, is not far from the scaling behavior exhibited by the maximum entropy of the considered WDNs. The observed difference can be ascribed to the fact that water flows must obey the flow balance equations at nodes, so that equal probabilities of all the flow paths are not physically possible.

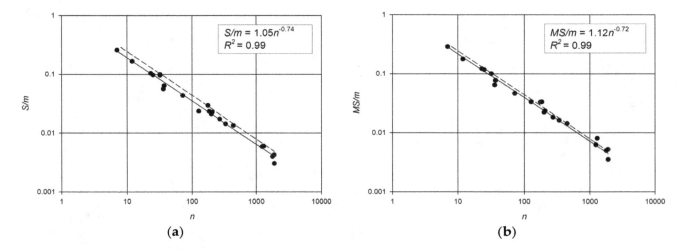

Figure 3. Scatter plots and best-fitting power law equations: (**a**) S/m vs. number of nodes; (**b**) MS/m vs. number of nodes. The dashed lines represent the expected scaling of flow entropy for a network with equiprobable flow paths.

It looks clear how both actual and maximum flow entropy strictly depend on network size and topology. The very good alignment of the values of S of WDNs designed with different criteria along a single power–law can be seen as an indirect confirmation of its suitability as a measure of network robustness. In fact, regardless of the criteria adopted for the design of pipe diameters, the smaller the hydraulic resistance of pipes (i.e., larger diameter and shorter length), the higher the flows that spontaneously tend to develop through them. The flow distribution along pipes, and so the flow entropy of the network, is thus determined by the hydraulic laws governing energy dissipation along pipes, which lead to the delivery of the demand at nodes with the minimum dissipated power [41] and, at the same time, set limits to the "disorder" of flow distribution.

The small scatter of the points from the curve of Equation (8), comparable to that of Equation (7), is likely due to the imperfect calculation of MS, as already discussed in the previous section, due to the a priori assumption of flow directions along pipes. However, as expected, the obtained trend seems not to be significantly affected by such an issue.

Equations (7) and (8) shed some light on the link between flow entropy and topology of a WDN. In fact, introducing the relationship $m = n + l - 1$, it is possible to compare the flow entropy of networks with different size and different number of loops. In example, Figure 4 shows the dependence of MS on n and l according to Equation (8). It looks clear that the more looped the network is, the higher is its entropy, thus confirming that flow entropy is a suitable measure of WDN redundancy. On the same graph, the curves representing the maximum flow entropy of WDNs with average node degree $k = 2$ and $k = 4$ are also plotted, delimiting the part of the plane to which WDNs belong. In fact, owing to the physical constraints of pipe connections at nodes, the average node degree of most WDNs falls between such values, as confirmed by the positions of the dots representing the 22 considered networks.

The obtained relationships indicate that, thanks to the high values of the coefficients of determination, it is possible to assess the maximum achievable flow entropy of a network starting from mere basic topological information such as the numbers of links and nodes.

In particular, Equation (8) provides a simple way to compute MS, without the need of a preliminary determination of flow pipe orientations, which can be easily implemented in the design of water supply networks aiming at taking into account the positive effect of redundancy on network robustness [39].

It is worth highlighting that the obtained relationships (7) and (8) have been derived for very different WDNs, both real and synthetic, from different countries, with quite different topological and hydraulic characteristics. Nonetheless, they show a clear scaling behavior in the form of power–laws, indicating that the values of the informational flow entropy are strongly related to some intrinsic and scale-invariant topological characteristic of WDNs, which likely reflects the spatial embedding of these networks, limiting their topological "disorder" (e.g., the degree connectivity of WDNs assumes a nearly constant value as the size of the network increases [32]).

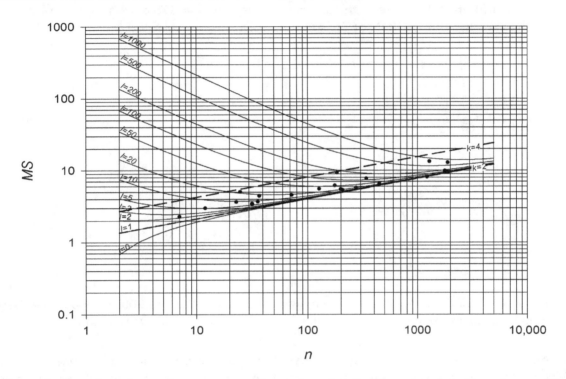

Figure 4. Scaling of maximum flow entropy with number of nodes, for networks with various numbers of loops l. The dashed lines represent maximum flow entropy of networks with fixed average node degree $k = 2$ and $k = 4$. The dots represent the considered set of 22 WDNs.

4. Examples of Application

In this section, practical examples are given of how the maximum flow entropy value MS, computed by Equation (8), can be used for WDN design or rehabilitation. Starting from the value of maximum entropy, estimated only by means of topological information, the design of the water supply network can be carried out by means of a multi-objective optimization procedure, based on the minimization of entropy deviation and pipe costs, in compliance with hydraulic constraints (i.e., the required minimum pressure at demand nodes). Specifically, the optimization problem consists of defining the optimal choice of the diameters of all pipes in the network, by minimizing the following multi-objective function (MOF):

$$\left\{ \begin{array}{l} MOF = \left\{ \Delta S; \; C = C\prime \cdot \sum_{j=1}^{m} L_j D_j^{\beta} \right\} \\ constraint : h_i > \overline{h_i} \; i = 1,, n \end{array} \right. \tag{10}$$

In Equation (10), the first component of MOF, ΔS, represents the deviation of flow entropy, calculated with Equation (5), from the maximum flow entropy, estimated by means of Equation (8) as a function of n and m; the second component C represents the total cost of the pipes of the network ($C\prime$ is the unit cost of pipes; L_j and D_j are the length and the diameter of the j-th pipe, respectively; β is a coefficient expressing the dependence of the cost of a pipe on its diameter, for which the value

$\beta = 1.5$ has been proposed [69]); h_i and $\overline{h_i}$ are, respectively, the actual and the design pressure heights at the i-th node of the network.

The application of the proposed WDN design optimization procedure, summarized by Equation (10), has been carried out for the real water supply networks of Fossolo [60], a neighborhood of the city of Bologna (Italy), and of the town of Skiathos (Greece) [62]. The first network consists of 36 nodes and 58 polyethylene pipes, and the design pressure was assumed equal to $\overline{h} = 30$ m at all nodes. The second network, made with cast iron pipes, has $n = 175$ and $m = 189$, with $\overline{h} = 22$ m. In Figure 5, the sketches of the WDN of Fossolo and Skiathos are reported.

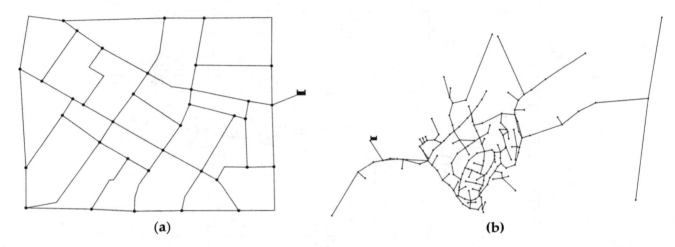

(a)   (b)

Figure 5. Layouts of the water distribution networks for which the multi-objective optimal design procedure based on maximum flow entropy has been applied: (**a**) Fossolo; (**b**) Skiathos.

The minimization of *MOF* was carried out by a heuristic optimization method based on a Genetic Algorithm (GA), a minimum search technique based on mimicking the process of natural selection in the evolution of species [70]. Such an evolutionary algorithm allows for easily introducing constraints on the unknown parameters, at the same time avoiding local minima by introducing random variations to parameter vectors. The GA parameters are the following: each individual of the population is a sequence of chromosomes corresponding to the diameters of all the pipes of the network, which can assume only the values of the existing commercial pipes reported in Table 2. The number of GA generations, the size of the population and the crossover percentage were set to 100, 100 individuals and 0.8, respectively.

The application of the proposed network design procedure led to the definition of the Pareto frontsreported in Figure 6, which represent, in the plane (C, S), the set of all the optimal solutions obtained by minimizing ΔS and the total pipe cost, in compliance with the hydraulic constraints of Equation (10). In addition, the red dots in Figure 6 represent the entropy deviation and the total pipe cost of the original network layouts. Without limiting the general validity of the obtained results, the unit cost of pipes has been assumed $C\prime = 1$. The obtained Pareto fronts show that the smallest values of ΔS correspond to the highest values of total pipe cost, as the more a network is robust, the more investment is needed for its realization (e.g., [71]). For the network of Fossolo, the minimum value of $\Delta S = 0.0004$, corresponding to a flow entropy $S = 4.66$, implies an increase of pipe cost, compared with the original layout, of about 58%. However, a flow entropy $S = 4.55$ can be obtained with an increase of cost smaller than 25%, which represents a good tradeoff between reliability improvement and cost increase. For the case of Skiathos, instead, it is worth noting that nearly the maximum flow entropy $S = 6.72$ can be achieved without any increment of overall pipe cost compared to the existing network.

Table 2. Pipe diameters of the networks of Fossolo (polyethylene pipes) and Skiathos (cast iron pipes).

FossoloDN (mm)		SkiathosDN (mm)	
16.00	73.60	40.00	125.00
20.40	90.00	50.00	140.00
26.00	102.20	63.00	150.00
32.60	147.20	75.00	160.00
40.80	184.00	80.00	225.00
51.40	204.6	90.00	
61.40	229.2	110.00	

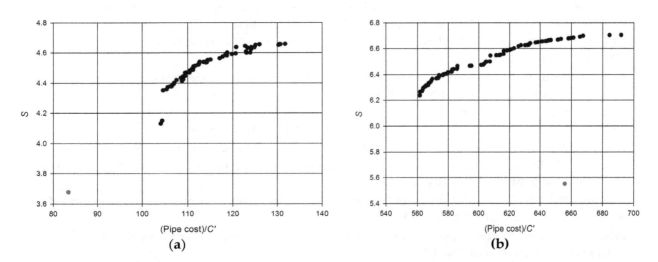

Figure 6. Pareto fronts of the proposed multi-objective optimal network design procedure (flow entropy and total cost of network pipes): (**a**) Fossolo; (**b**) Skiathos. The red dots correspond to network layouts before optimization.

5. Conclusions

The study investigates the scaling-law of informational flow entropy of water distribution networks, often assumed as a surrogate of network robustness, with their topological size. To such aim, the relationships between informational flow entropy, S, maximum informational flow entropy, MS, and some suitable topological metrics (namely, number of nodes, n; number of links, m; network link density, q; network average path length, APL) are investigated for a set of 22 networks, both real and synthetic, with different characteristics.

A clear dependence of flow entropy on topological metrics is observed, and, in particular, power–law relationships, strongly linking S/m and MS/m to the number of nodes of the network (i.e., $R^2 = 0.99$), are identified. The obtained scaling laws result in being close to the expected scaling of flow entropy in networks with equiprobable flow paths (i.e., the same flow carried to the end of any flow path connecting sources to demand nodes), although the actual flow paths cannot be equiprobable, as they must obey flow balance equations at nodes. Such a scale-invariant behavior, testified by the power–laws, probably reflects the peculiar topological feature of water distribution networks, in which each node is connected only to a few immediately surrounding nodes, thus limiting the topological "disorder" of the network, i.e., the number of possible flow paths from each node.

The obtained power–laws, providing an easy estimate of actual and maximum flow entropy of a network, allow to quantify the entropy deficit of a network, i.e., the distance of the flow entropy of a network of given topology from its maximum achievable flow entropy, which can be used in network design and rehabilitation as a measure of network robustness. In this respect, examples of application to multi-objective design of real water distribution networks show how optimal solutions in terms of pipe cost and overall network robustness are easily identified.

Acknowledgments: This research is part of the Ph.D. project "Water distribution network management optimization through Complex Network theory" within the Doctoral Course "A.D.I." granted by Università degli Studi della Campania "L. Vanvitelli".

Author Contributions: All the Authors contributed to the development of the research idea, to the conceivement of the paper, and to the discussion and comment of the results; G.F.S. and C.G. carried out the numerical experiments; G.F.S. and R.G. mostly wrote the paper, which was commented and edited by all the others.

References

1. Greco, R.; Di Nardo, A.; Santonastaso, G.F. Resilience and entropy as indices of robustness of water distribution networks. *J. Hydroinform.* **2012**, *14*, 761–771. [CrossRef]
2. Ghosh, A. Scaling Laws. In *Mechanics over Micro and Nano Scales*; Chakraborty, S., Ed.; Springer: New York, NY, USA, 2011.
3. Rodriguez-Iturbe, I.; Rinaldo, A. *Fractal River Basins: Chance and Self Organization*; Cambridge University Press: Cambridge, UK, 1996; ISBN 0521473985.
4. Hubert, P.; Tessier, Y.; Lovejoy, S.; Schertzer, D.; Schmitt, F.; Ladoy, P.; Carbonnel, J.P.; Violette, S. Multifractals and extreme rainfall events. *Geophys. Res. Lett.* **1993**, *20*, 931–934. [CrossRef]
5. Jennings, A.H. World's greatest observed point rainfalls. *Mon. Weather Rev.* **1950**, *78*, 4–5. [CrossRef]
6. Galmarini, S.; Steyn, D.G.; Ainslie, B. The scaling law relating world point-precipitation records to duration. *Int. J. Climatol.* **2004**, *24*, 533–546. [CrossRef]
7. Kolmogorov, A.N. The local structure of turbulence in incompressible viscous fluid for very large Reynolds numbers. *Proc. R. Soc. A* **1991**, *434*, 9–13. [CrossRef]
8. Frisch, U.; Sulem, P.; Nelkin, M. A simple dynamical model of intermittent fully developed turbulence. *J. Fluid Mech.* **1978**, *87*, 719–736. [CrossRef]
9. Arrault, J.; Arnéodo, A.; Davis, A.; Marshak, A. Wavelet based multifractal analysis of rough surfaces: Application to cloud models and satellite data. *Phys. Rev. Lett.* **1997**, *79*, 75–78. [CrossRef]
10. Badin, G.; Domeisen, D.I.V. Nonlinear stratospheric variability: Multifractal detrended fluctuation analysis and singularity spectra. *Proc. R. Soc. A* **2016**, *472*, 20150864. [CrossRef] [PubMed]
11. Agarwal, S.; Moon, W.; Wettlaufer, J. Trends, noise and re-entrant long-term persistence in Arctic sea ice. *Proc. R. Soc. A* **2012**, *468*, 2416–2432. [CrossRef]
12. Albert, R.; Barabási, A.-L. Statistical mechanics of complex networks. *Rev. Mod. Phys.* **2002**, *74*, 47. [CrossRef]
13. Boccaletti, S.; Latora, V.; Moreno, Y.; Chavez, M. Hwanga, D.U. Complex networks: Structure and dynamics. *Phys. Rep.* **2006**, *424*, 175–308. [CrossRef]
14. Maier, H.R.; Lence, B.J.; Tolson, B.A.; Foschi, R.O. First-order reliability method for estimating reliability, vulnerability, and resilience. *Water Resour. Res.* **2001**, *37*, 779–790. [CrossRef]
15. Yazdani, A.; Jeffrey, P. A complex network approach to robustness and vulnerability of spatially organized water distribution networks. *Phys. Soc.* **2010**, *15*, 1–18.
16. Bao, Y.; Mays, L.W. Model for water distribution system reliability. *J. Hydraul. Eng.* **1990**, *116*, 1119–1137. [CrossRef]
17. Gargano, R.; Pianese, D. Reliability as a tool for hydraulic network planning. *J. Hydraul. Eng. ASCE* **2000**, *126*, 354–364. [CrossRef]
18. Ostfeld, A. Reliability analysis of water distribution systems. *J. Hydroinform.* **2004**, *6*, 281–294. [CrossRef]
19. Wagner, J.M.; Shamir, U.; Marks, D.H. Water distribution reliability: Analytical methods. *J. Water Resour. Plan. Manag.* **1988**, *114*, 253–274. [CrossRef]
20. Shannon, C.E. A mathematical theory of communication. *Bell Syst. Tech. J.* **1948**, *27*, 379–423. [CrossRef]
21. Singh, V.P. *Entropy Theory and Its Application in Environmental and Water Engineering*, 1st ed.; John Wiley and Sons, Ltd.: Oxford, UK, 2013; ISBN 978-1-119-97656-1.
22. Awumah, K.; Goulter, I.; Bhatt, S.K. Assessment of reliability in water distribution networks using entropy based measures. *Stoch. Hydrol. Hydraul.* **1990**, *4*, 309–320. [CrossRef]
23. Tanyimboh, T.T.; Templeman, A.B. A quantified assessment of the relationship between the reliability and entropy of water distribution systems. *Eng. Optim.* **2000**, *33*, 179–199. [CrossRef]

24. Tanyimboh, T.T.; Sheahan, C. A maximum entropy based approach to the layout optimization of water distribution systems. *Civ. Eng. Environ. Syst.* **2002**, *19*, 223–253. [CrossRef]

25. Setiadi, Y.; Tanyimboh, T.T.; Templeman, A.B. Modelling errors, entropy and the hydraulic reliability of water distribution systems. *Adv. Eng. Softw.* **2005**, *36*, 780–788. [CrossRef]

26. Liu, H.; Savic, D.; Kapelan, Z.; Zhao, M.; Yuan, Y.; Zhao, H. A diameter-sensitive flow entropy method for reliability consideration in water distribution system design. *Water. Resour. Res.* **2014**, *50*, 5597–5610. [CrossRef]

27. Tanyimboh, T.T.; Templeman, A.B. Calculating maximum entropy flows in networks. *J. Oper. Res. Soc.* **1993**, *44*, 383–396. [CrossRef]

28. Tanyimboh, T.T.; Templeman, A.B. Using entropy in water distribution networks. In *Integrated Computer Applications in Water Supply*, 1st ed.; Coulbeck, B., Ed.; Research Studies Press Ltd.: Taunton, UK, 1993; Volume 1, pp. 77–90, ISBN 086380-154-4.

29. Perelman, L.; Housh, M.; Ostfeld, A. Robust optimization for water distribution systems least cost design. *Water Resour. Res.* **2013**, *49*, 6795–6809. [CrossRef]

30. Saleh, S.H.A.; Tanyimboh, T.T. Optimal design of water distribution systems based on entropy and topology. *Water Resour. Res.* **2014**, *28*, 3555–3575. [CrossRef]

31. Mays, L.W. *Water Distribution Systems Handbook*, 1st ed.; McGraw-Hill: New York, NY, USA, 2000; ISBN 780071342131.

32. Di Nardo, A.; Di Natale, M.; Giudicianni, C.; Greco, R.; Santonastaso, G.F. Complex network and fractal theory for the assessment of water distribution network resilience to pipe failures. *Water Sci. Technol. Water Supply* **2017**, *17*, ws2017124. [CrossRef]

33. Watts, D.; Strogatz, S. Collective dynamics of small world networks. *Nature* **1998**, *393*, 440–442. [CrossRef] [PubMed]

34. Barabasi, A.L.; Albert, R. Emergence of scaling in random networks. *Science* **1999**, *286*, 797–817. [CrossRef]

35. Yazdani, A.; Jeffrey, P. Robustness and Vulnerability Analysis of Water Distribution Networks Using Graph Theoretic and Complex Network Principles. In Proceedings of the 12th Annual Conference on Water Distribution Systems Analysis (WDSA), Tucson, AZ, USA, 12–15 September 2010.

36. Gutiérrez-Pérez, J.A.; Herrera, M.; Pérez-García, R.; Ramos-Martínez, E. Application of graph-spectral methods in the vulnerability assessment of water supply networks. *Math. Comput. Model.* **2013**, *57*, 1853–1859. [CrossRef]

37. Herrera, M.; Abraham, E.; Stoianov, I. A graph-theoretic framework for assessing the resilience of sectorised water distribution networks. *Water Resour. Res.* **2016**, *30*, 1685–1699. [CrossRef]

38. Gheisi, A.; Naser, G. Multistate Reliability of Water-Distribution Systems: Comparison of Surrogate Measures. *J. Water Res. Plan. Manag.* **2015**, *141*. [CrossRef]

39. Tanyimboh, T.T. Informational entropy: A failure tolerance and reliability surrogate for water distribution networks. *Water Resour. Res.* **2017**, *31*, 3189–3204. [CrossRef]

40. Creaco, E.; Fortunato, A.; Franchini, M.; Mazzola, M.R. Comparison between entropy and resilience as indirect measures of reliability in the framework of water distribution network design. *Proc. Eng.* **2014**, *70*, 379–388. [CrossRef]

41. Todini, E. Looped water distribution networks design using a resilience index based heuristic approach. *Urban Water* **2000**, *2*, 115–122. [CrossRef]

42. Prasad, T.D.; Park, N.S. Multi-Objective Genetic Algorithms for Design of Water Distribution Networks. *J. Water Res. Plan. Manag.* **2004**, *130*, 73–82. [CrossRef]

43. Vaabel, J.; Ainola, L.; Koppel, T. Hydraulic power analysis for determination of characteristics of a water. In Proceedings of the Eighth Annual Water Distribution Systems Analysis Symposium (WDSA), Cincinnati, OH, USA, 27–30 August 2006.

44. Yazdani, A.; Jeffrey, P. Complex network analysis of water distribution systems. *Chaos* **2011**, *21*, 016111. [CrossRef] [PubMed]

45. Di Nardo, A.; Di Natale, M.; Giudicianni, C.; Greco, R.; Santonastaso, G.F. Water supply network partitioning based on weighted spectral clustering. *Stud. Comp. Intell.* **2017**, *693*, 797–807. [CrossRef]

46. Herrera, M.; Izquierdo, J.; Pérez-Garcìa, R.; Montalvo, I. Multi-agent adaptive boosting on semi supervised water supply clusters. *Adv. Eng. Softw.* **2012**, *50*, 131–136. [CrossRef]

47. Guest, G.; Namey, E.E. *Public Health Research Methods*, 1st ed.; SAGE Publications, Inc.: Thousand Oaks, CA, USA, 2014.

48. Di Nardo, A.; Di Natale, M.; Giudicianni, C.; Musmarra, D.; Santonastaso, G.F.; Simone, A. Water Distribution System Clustering and Partitioning Based on Social Network Algorithms. *Proc. Eng.* **2015**, *119*, 196–205. [CrossRef]

49. Khinchin, A.I. *Mathematical Foundations of Statistical Mechanics*; Dover: New York, NY, USA, 1953.

50. Ang, W.K.; Jowitt, P.W. Path entropy method for multiple-source water distribution networks. *J. Eng. Optim.* **2005**, *37*, 705–715. [CrossRef]

51. Rossman, L.A. *EPANET2 Users Manual*; US EPA: Cincinnati, OH, USA, 2000.

52. Yassin-Kassab, A.; Templeman, A.B.; Tanyimboh, T.T. Calculating maximum entropy flows in multi-source, multi-demand networks. *Eng. Optim.* **1999**, *31*, 695–729. [CrossRef]

53. Alperovits, E.; Shamir, U. Design of optimal water distribution systems. *Water Resour. Res.* **1977**, *13*, 885–900. [CrossRef]

54. Gessler, J. Pipe network optimization by enumeration. Proceedings of Computer Applications in Water Resources, ASCE Specialty Conference, Buffalo, NY, USA, 10–12 June 1985.

55. Walski, T.; Brill, E.; Gessler, J.; Goulter, I.; Jeppson, R.; Lansey, K.; Lee, H.; Liebman, J.; Mays, L.; Morgan, D.; et al. Battle of the network models: Epilogue. *J. Water Res. Plan. Manag.* **1987**, *113*, 191–203. [CrossRef]

56. Kim, J.H.; Kim, T.G.; Kim, J.H.; Yoon, Y.N. A study on the pipe network system design using non-linear programming. *J. Korean Water Resour. Assoc.* **1994**, *27*, 59–67.

57. Sherali, H.D.; Subramanian, S.; Loganathan, G. Effective relaxations and partitioning schemes for solving water distribution network design problems to global optimality. *J. Global. Optim.* **2001**, *19*, 1–26. [CrossRef]

58. Fujiwara, O.; Khang, D.B. A two-phase decomposition method for optimal design of looped water distribution networks. *Water Resour. Res.* **1990**, *26*, 539–549. [CrossRef]

59. Lee, S.C.; Lee, S.I. Genetic algorithms for optimal augmentation of water distribution networks. *J. Korean Water Resour. Assoc.* **2001**, *34*, 567–575.

60. Bragalli, C.; D'Ambrosio, C.; Lee, J.; Lodi, A.; Toth, P. On the Optimal Design of Water Distribution Networks: A Practical MINLP Approach. *Optim. Eng.* **2012**, *13*, 219–246. [CrossRef]

61. Ostfeld, A.; Uber, J.G.; Salomons, E.; Berry, J.W.; Hart, W.E.; Phillips, C.A.; Watson, J.; Dorini, G.; Jonkergouw, P.; Kapelan, Z.; et al. The Battle of the Water Sensor Networks (BWSN): A Design Challenge for Engineers and Algorithms. *J. Water Resour. Plan. Manag.* **2008**, *134*, 556–568. [CrossRef]

62. Di Nardo, A.; Di Natale, M.; Giudicianni, C.; Laspidou, C.; Morlando, F.; Santonastaso, G.F.; Kofinas, D. Spectral analysis and topological and energy metrics for water network partitioning of Skiathos island. *Eur. Water* **2017**, *58*, 423–428.

63. Di Nardo, A.; Di Natale, M.; Gisonni, C.; Iervolino, M. A genetic algorithm for demand pattern and leakage estimation in a water distribution network. *J. Water Supply Res. Tech. Aqua* **2015**, *64*, 35–46. [CrossRef]

64. Herrera, M.; Canu, S.; Karatzoglou, A.; Perez-Garcıa, R.; Izquierdo, J. An approach to water supply clusters by semi-supervised learning. In Proceedings of theInternational Congress on Environmental Modelling and Software Modelling (iEMSs) 2010 for Environment's Sake, Fifth Biennial Meeting, Ottawa, ON, Canada, 5–8 July 2010.

65. Reca, J.; Martinez, J. Genetic algorithms for the design of looped irrigation water distribution networks. *Water Resour. Res.* **2006**, *42*, W05416. [CrossRef]

66. Tzatchkov, V.G.; Alcocer-Yamanaka, V.H.; Ortíz, V.B. Graph Theory Based Algorithms for Water Distribution Network Sectorization Projects. In Proceedings of the Eighth Annual Water Distribution Systems Analysis (WDSA) Symposium, Cincinnati, OH, USA, 27–30 August 2006.

67. Lippai, I. Colorado Springs Utilities Case Study: Water System Calibration Optimization. In Proceedings of the ASCE Pipeline Division Specialty Conference, Houston, TX, USA, 21–24 August 2005.

68. Farmani, R.; Savic, D.A.; Walters, G.A. Evolutionary multi-objective optimization in water distribution network design. *Eng. Optim.* **2005**, *37*, 167–183. [CrossRef]

69. Tanyimboh, T.T.; Saleh, S.H.A. Global maximum entropy minimum cost design of water distribution systems. In Proceedings of the ASCE/EWRI World Environmental and Water Resources Congress 2011, Palm Springs, CA, USA, 22–26 May 2011.

70. Goldberg, D.E. *Genetic Algorithms in Search, Optimization and Machine Learning*; Addison-Wesley Longman Publishing Co. Inc.: Boston, MA, USA, 1989.
71. Tricarico, C.; Gargano, R.; Kapelan, Z.; Savic, D.; de Marinis, G. Economic Level of Reliability for the Rehabilitation of Hydraulic Networks. *J. Civ. Eng. Environ. Syst.* **2006**, *23*, 191–207. [CrossRef]

10

Testing the Beta-Lognormal Model in Amazonian Rainfall Fields using the Generalized Space q-Entropy

Hernán D. Salas *, Germán Poveda and Oscar J. Mesa

Facultad de Minas, Departamento de Geociencias y Medio Ambiente, Universidad Nacional de Colombia, Sede Medellín, Carrera 80 # 65-223, Medellín 050041, Colombia; gpoveda@unal.edu.co (G.P.); ojmesa@unal.edu.co (O.J.M.)
* Correspondence: hdsalas@unal.edu.co

Abstract: We study spatial scaling and complexity properties of Amazonian radar rainfall fields using the Beta-Lognormal Model (BL-Model) with the aim to characterize and model the process at a broad range of spatial scales. The Generalized Space q-Entropy Function (GSEF), an entropic measure defined as a continuous set of power laws covering a broad range of spatial scales, $S_q(\lambda) \sim \lambda^{\Omega(q)}$, is used as a tool to check the ability of the BL-Model to represent observed 2-D radar rainfall fields. In addition, we evaluate the effect of the amount of zeros, the variability of rainfall intensity, the number of bins used to estimate the probability mass function, and the record length on the GSFE estimation. Our results show that: (i) the BL-Model adequately represents the scaling properties of the q-entropy, S_q, for Amazonian rainfall fields across a range of spatial scales λ from 2 km to 64 km; (ii) the q-entropy in rainfall fields can be characterized by a non-additivity value, q_{sat}, at which rainfall reaches a maximum scaling exponent, Ω_{sat}; (iii) the maximum scaling exponent Ω_{sat} is directly related to the amount of zeros in rainfall fields and is not sensitive to either *the number of bins* to estimate the probability mass function or *the variability of rainfall intensity*; and (iv) for small-samples, the GSEF of rainfall fields may incur in considerable bias. Finally, for synthetic 2-D rainfall fields from the BL-Model, we look for a connection between intermittency using a metric based on generalized Hurst exponents, $M(q_1, q_2)$, and the non-extensive order (q-order) of a system, Θ_q, which relates to the GSEF. Our results do not exhibit evidence of such relationship.

Keywords: hydrology; tropical rainfall; statistical scaling; Tsallis entropy; multiplicative cascades; Beta-Lognormal model

1. Introduction

1.1. Statistical Scaling and Multiplicative Random Cascades

Statistical scaling has provided a rich framework to understand and model the spatiotemporal dynamics and the complexity and intermitency of rainfall fields, including (multi-)fractal, multiscaling, and random cascade models [1–19].

The strong variability and intermittence of convective tropical rainfall constitute an adequate setting to study the scaling characteristics of rainfall in a wide range of spatio-temporal scales [19–31]. In particular, Ref. [19] found that 2-D rainfall fields over Amazonia exhibit multiscaling properties in space, which means that the relationship $M_r(\lambda) \sim \lambda^{-\tau(r)}$ exhibits a non-linear behavior, where λ is the spatial scale, r the order of the statistical moment and $\tau(r)$ is the r-th moment scaling exponent. Additionally, they show that both the diurnal cycle and the predominant atmospheric regime of Amazonian rainfall (Easterly or Westerly) exert a strong control on the scaling properties of Amazonian storms, thus shedding light towards understanding the linkages between scaling statistics and physical features of Amazonian rainfall fields.

1.2. Multiplicative Random Cascades and the Beta-LogNormal Model

The Beta-Lognormal Model (hereafter BL-Model) is a discrete 2-D random cascade non-Markovian model [13] based on the observed scaling properties of rainfall, with only two parameters: σ denoting the variability of rainfall intensity, and β representing the rainy area fraction. This model provides a framework to carry out numerical experiments controlling features of rainfall (e.g., σ and β), but also to link diverse statistical and physical characteristics across spatial scales (e.g., Refs. [19,31] and Section 1.5, respectively). In addition, the model provides a tool to investigate the robustness and sensitivity of statistical metrics to poor sampling, data sparsity and intermittency of high-resolution 2-D rainfall fields.

The construction of a spatially distributed discrete random cascade model usually begins with a given mass (or volume) of rainfall over a two-dimensional ($d = 2$) bounded region [6]. The region is successively divided into b equal parts ($b = 2^a$) at each step, and during each iteration the mass obtained at the previous step is distributed into the b subdivisions through multiplication by a set of "cascade generator" W, as shown schematically in Figure 1 (for the case of $d = 2$ and $b = 4$). If the initial area (at level 0) is assigned an average intensity R_0, this gives an initial volume $R_0 L_0^d$, where L_0 is the outer length scale of the study area. Thus, at the first level the volume is subdivided into $b = 4$ subareas denoted by Δ_1^i, $i = 1, 2, ..., 4$. At the second level, each of the previous subareas is further subdivided into $b = 4$ subareas, which are denoted by Δ_2^i, $i = 1, 2, ..., 16$, for a total of $b^2 = 16$ subareas. This subdivision is continued further down the spatial scale, leading at the nth level, to b^n subareas denoted by Δ_n^i, $i = 1, 2, ..., b^n$.

As shown in Figure 1, after the first subdivision, the four subareas (Δ_1^i, $i = 1, 2, ..., b$) are assigned volumes $R_0 L_0^d b^{-1} W_1^i$, for $i = 1, 2, ..., b$. Upon subdivision, the volumes $\mu_n \Delta_1^i$ in subareas at the nth subdivision, Δ_n^i, $i = 1, 2, ..., b^n$, are given by,

$$\mu_n \Delta_1^i = R_0 L_0^d b^{-1} \prod_{j=1}^{n} W_j^i, \tag{1}$$

where, for each cascade's level j, i represents one subarea belonging to the level. The multipliers W in Equation (1) are non-negative random cascade generators, with $E[W] = 1$ to ensure that the mass is conserved on average, from one discretization level to the next one. Over and Gupta [13,32] proposed the so-called BL-Model for the cascade generators W. The BL-Model considers W as a composite generator, $W = BY$, where B is a generator from the "Beta model" and Y is drawn from a Lognormal distribution [33]. Essentially, the Beta model partitions the region into sets with and without rain, while the Lognormal model then assigns a certain amount of rainfall to each rainy area fraction. The Beta model exhibits a discrete probability mass function with just two possible outcomes ($B = 0$ and $B = b^\beta$), given as

$$P(B = 0) = 1 - b^{-\beta} \qquad P(B = b^\beta) = b^{-\beta}, \tag{2}$$

where b is the branching number and β is a parameter. Since Y belongs to the Lognormal distribution, it can be expressed as $Y = b^{-\frac{\sigma^2 \ln(b)}{2} + \sigma X}$, where X is a standard Normal r.v. and $\sigma^2 (> 0)$ is a parameter equal to the variance of $\log_b Y$, with the condition that $E[Y] = 1$. In such case, it is easy to show that the condition $E[W] = 1$ is also satisfied. The probability distribution function of $W = BY$ can thus be expressed as

$$P(W = 0) = 1 - b^{-\beta}, \tag{3}$$

$$P(W = b^\beta Y = b^{-\frac{\sigma^2 \ln(b)}{2} + \sigma X}) = b^{-\beta}. \tag{4}$$

The parameters of the BL-Model (β and σ^2) can be estimated [13,32] through the so-called Mandelbrot–Kahane–Peyriere (MKP) function [34,35]. The MKP function characterizes the fractal or scale-invariant behavior of the multiplicative cascade process. Over and Gupta [32] theoretically derived an expression for $\chi(r)$ for the BL-Model, in terms of the cascade parameters β, σ^2, b and exponent r, such that,

$$\chi_b(r) = (\beta - 1)(r - 1) + \frac{\sigma^2 \ln(b)}{2}(r^2 - r). \tag{5}$$

Thus, provided that a rainfall field belongs to a discrete random cascade with generators satisfying the BL-Model, the expression given in Equation (5) can be matched with the empirically determined estimators, $\tau(r)/d$, to estimate β and σ^2. The first and second derivatives of $\tau(r) = d\chi(r)$ with respect to r, the latter of which is given by Equation (5), can be used to obtain [20,32],

$$\tau^{(1)}(r) = d\left[\beta - 1 + \frac{\sigma^2 \ln(b)}{2}(2r - 1)\right], \tag{6}$$

$$\tau^{(2)}(r) = d\left[\sigma^2 \ln(b)\right] \tag{7}$$

Both $\tau^{(1)}(r)$ and $\tau^{(2)}(r)$ can be computed by numerically estimating the derivatives of the empirical slopes of the scaling relation between $\tau(r)$ and r, using the log–log plotting of $M(\lambda_n, r)$ versus λ_n as,

$$M(\lambda_n, r) = [\lambda_n]^{\tau(r)}, \tag{8}$$

where $M(\lambda_n, r)$ are the sample moments $M(\lambda_n, r)$ and λ_n is the corresponding scale ratio. Equations (6) and (7) are combined together to express the cascade parameters β and σ^2 in terms of $\tau^{(1)}(r)$ and $\tau^{(2)}(r)$ as follows:

$$\beta = 1 + \frac{\tau^{(1)}(r)}{d} - \frac{\sigma^2 \ln(b)}{2}(2r - 1), \tag{9}$$

$$\sigma^2 = \tau^{(2)}(r)/d\ln(b) \tag{10}$$

Equations (9) and (10) are evaluated for a given value of r. The usual practice is to use $r = 1$, although [6] used $r = 2$ for testing a space-time model of daily rainfall in Australia.

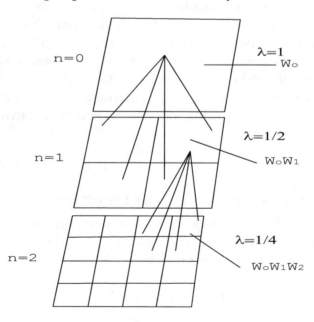

Figure 1. Schematic plot of the random cascade geometry taken from Gupta and Waymire [33].

1.3. q-Entropy

There is a wide family of generalized entropic functions with various degrees of sophistication in the literature [36–40]. In particular, Tsallis [40] proposed the concept of nonadditive entropy S_q (hereafter q-entropy), which has shown to be useful in the study of a broad range of phenomena across diverse disciplines [41,42], related to the well known Rényi entropy [43]. The q-entropy is defined as,

$$S_q = \frac{1 - \sum_{i=1}^{n} p^{\,q}(x_i)}{q - 1} \quad \left(\sum_{i=1}^{n} p(x_i) = 1; \ q \in \Re \right). \tag{11}$$

which, in the limit $q \to 1$, recovers the usual Boltzmann–Gibbs–Shannon entropy, S [44], which is additive; in other words, for a system composed of any two (probabilistically) independent subsystems, the entropy S of the sum is the sum of their entropies [45], such that, if A and B are independent,

$$S_{q=1}(A + B) = S(A + B) = S(A) + S(B). \tag{12}$$

It turns out that Tsallis entropy, S_q ($q \neq 1$), violates this property, and is therefore *nonadditive*. Thus, the additivity depends on the functional form of the entropy in terms of probabilities [45]. Therefore, if A and B are independent, then

$$S_q(A + B) = S_q(A) + S_q(B) + (1 - q)S_q(A)S_q(B). \tag{13}$$

More generally, if A and B are not probabilistically independent then,

$$S_q(A + B) = S_q(B) + S_q(A|B) + (1 - q)S_q(B)S_q(A|B). \tag{14}$$

Taking the words of Tsallis [45], the value of q is useful to characterize *the universal classes of nonadditivity*. He argues that it is determined a priori by the microscopic dynamics of the system, which means that the thermostatistical entropy is not universal but depends on the system or, more precisely, on the non-additive universality class to which the system belongs. It is worth mentioning the difference between *additivity* and *extensivity*, which is clearly explained in [45], as follows:

> *"An entropy of a system or of a subsystem is said extensive if, for a large number N of its elements (probabilistically independent or not), the entropy is (asymptotically) proportional to N. Otherwise, it is nonextensive. This means that extensivity depends on both the mathematical form of the entropic functional and the possible correlations existing between the elements of the system. Consequently, for a (sub)system whose elements are either independent or weakly correlated, the additive entropy S is extensive, whereas the nonadditive entropy S_q ($q \neq 1$) is nonextensive. In contrast, however, for a (sub)system whose elements are generically strongly correlated, the additive entropy S can be nonextensive, whereas the nonadditive entropy S_q ($q \neq 1$) can be extensive for a special value of q."*

Additionally, recent studies [28,30,31] have investigated the relationship between the q-entropy and the Generalized Pareto distribution (which is relevant in hydrological analysis). In particular, the maximization of the q-entropy under a prescribed mean leads to a Pareto probability distribution with power-law tail [28,46–48], which belongs to the family of Lévy stable distributions [49], specifically to Type II Generalized Pareto distributions. For $1 < q < 2$, the original distribution takes the form of the Zipf–Mandelbrot type [50–52], which decays as a power law for large values of x, and all moments are divergent when $3/2 < q < 2$ [53].

As such, the q-entropy constitutes a useful tool for the characterization of rainfall, and at the same time motivates interesting discussions about the physical interpretation of entropic non-linear metrics, their connection with stochastic processes (Multiplicative Cascades), and allows revisiting its applicability in geosciences (see Section 5).

1.4. Generalized Space-Time q-Entropy in Rainfall Data

Previous studies [28,30,31] introduced the Generalized Space-Time q-Entropy as a new method to study the organization degree and the scaling properties of rainfall, by considering the space-time structure of rainfall as a system conformed by correlated subsystems, which is evident in the hierarchical structure of convective rainfall. The time generalized q-entropy is defined as a function of order, q, and aggregation interval, T, as [28]:

$$S_q(T) = \frac{1 - \sum_{i=1}^{n} P^{\,q}(x_i, T)}{q - 1} \quad \left(\sum_{i=1}^{n} p(x_i) = 1; \ q \in \Re \right), \tag{15}$$

where q is the statistical order and $P(x_i, T)$ is the probability of occurrence of x_i at aggregation interval, T. Ever since the original work [28], different characteristics of the Space and Time Generalized q-entropy of rainfall were reported later on [30,31], such as:

- $S_q(T)$ decreases monotonically with q for all values of T.
- For a given value of q, estimates are inversely related to T for $q < 0$, but directly related for $q \geq 0$.
- Estimates of $S_q(T)\,|_{q=1}$ recover the standard entropy for different values of T.
- Estimates of $S_q(T)$ increase with T for values of $q \geq 0$, up to a certain saturation value (maximum q-entropy).
- The function $S(T)$ vs. T in log–log space, for different values of q, can be considered an (time) entropy analogous of the (space) *structure function* in turbulence [54].
- The scaling exponents, $\Omega(q)$, or the slope of the relation $S(T)$ vs. T in log–log space, for different values of q, exhibit a non-linear growth with q, such that $\Omega_{sat} \approx 0.5$ for $q \geq 1$. This result allowed extending the conclusions from the standard Shannon entropy to the generalized q-entropy.
- The scaling exponents of saturation, Ω_{sat}, are different in time and space for hydrological data, such as time series of rainfall, for which $\Omega_{sat}(q > 1.0) = 0.5$, for time series of streamflows, for which $\Omega_{sat}(q > 1.0) = 0.0$, and for the spatial analysis of radar rainfall fields in Amazonia, for which $\Omega_{sat}(q > 1.0) = 1.0$.

Analogous to Equation (15), the Space Generalized q-Entropy was introduced by [31], to study 2-D rainfall fields, using λ as the spatial scale, and $P(x_i, \lambda)$ the probability of occurrence of x_i associated with λ, such that

$$S_q(\lambda) = \frac{1 - \sum_{i=1}^{n} P^{\,q}(x_i, \lambda)}{q - 1} \quad \left(\sum_{i=1}^{n} p(x_i) = 1; \ q \in \Re \right). \tag{16}$$

With the aim of linking the present study with the previous ones, some important results are worth mentioning:

- Poveda [28] studied the time scaling properties of tropical rainfall in the Andes of Colombia upon temporal aggregation and introduced the Generalized Time q-Entropy Function (GTEF), as a time analogous for q-entropy of the *structure function* in turbulence [54]. He showed that the scaling exponents, $\Omega(q)$ of the relation $S_q(T)$ vs. T in log–log space, for different values of q, exhibit a non-linear growth with q up to $\Omega = 0.5$ for $q \geq 1$, putting forward the conjecture that the time dependent q-entropy, $S_q(T) \sim T^{\Omega(q)}$ with $\Omega(q) \simeq 0.5$, for $q \geq 1$.
- Salas and Poveda [30] revisited results reported in [28], and analyzed the time scaling properties of Shannon's entropy for the same data set in terms of the sensitivity to the record length, and the effect of zeros in rainfall data, and proposed the GTEF to study the scaling properties of river flows. They highlighted two important results: (i) The scaling characteristics of Shannon's entropy differ between rainfall and streamflows owing to the presence of zeros in rainfall series; and (ii) the GTEF exhibits multi-scaling for rainfall and streamflows. For rainfall, the relation $S_q(T)$ vs. T in log–log space for different values of q, exhibits a non-linear growth with q, up to $\Omega = 0.5$ for

$q \geq 1$, in contrast to the scaling properties of river flows which exhibit a non-linear growth with q, up to $\Omega = 0.0$ for $q \geq 1$.

- Poveda and Salas [31] studied diverse topics such as statistical scaling, Shannon entropy and Space-Time Generalized q-Entropy of Mesoscale Convective Systems (MCS) as seen by the Tropical Rainfall Measuring Mission (TRMM) over continental and oceanic regions of tropical South America, and in Amazonian radar rainfall fields. The main result of their study is that both the GTEF and GSEF exhibit linear growth in the range $-1.0 < q < -0.5$, and saturation of the exponent Ω_{sat} for $q \geq 1.0$, but for the spatial analysis (GSEF) the exponent tappers off at $\langle \Omega_{sat} \rangle \sim 1.0$, whereas for the temporal analysis (GTEF) the exponent saturates at $\langle \Omega_{sat} \rangle = 0.5$. In addition, results are similar for time series extracted from radar rainfall fields in Amazonia (radar S-POL) and in-situ rainfall series in the tropical Andes.

1.5. Easterly and Westerly Regimes of Amazonian Rainfall

The WETAMC/LBA campaign (January–February 1999) found that wet-season convection in Amazonia exhibits two general modes, hereinafter, the Westerly and Easterly regimes [55], which are highly correlated to changes in the 850–700 hPa zonal wind direction [56]. According to data from the NCEP-NCAR Reanalysis, as well as from the Fazenda Nossa Senhora radiosonde, the Easterly (negative values) and Westerly (positive) regimes are clearly differentiated in both data sets [19,31]. Furthermore, radar observations from southwestern Amazonia during TRMM-LBA suggest that there was relatively little difference in the daily mean rainfall totals between Easterly and Westerly regimes [57]. In addition, the TRMM-LBA and TRMM satellite observations suggested marked differences in rainfall rate distributions, with the Easterly regime (Altiplano/southern Brazil) associated with a broader rain-rate distribution and greater instantaneous rainfall rates [56]. Precipitation features during both regimes can be summarized considering that during the Easterly regime atmospheric conditions are relatively dry, with increased lightning activity and more intense and deeper convective systems. In contrast, the Westerly regime is characterized by a diminished lightning activity, less deep convection and less intense precipitation rates [56,58,59]. The regime associated with stronger vertical development, more lightning activity, and larger instantaneous rainfall rates must be associated with a more "concentrated" daily latent heat release. In addition, two mechanisms have been proposed to explain the observed changes in the overall convective structure and lightning frequency between the two regimes, these mechanisms are tied to either thermodynamics (changes in CAPE and CIN that modify the energetics of the cloud ensemble), or aerosol loading (e.g., changes in cloud condensation nuclei (CCN) concentration that modify microphysical structure of the cloud ensemble) [56].

The aforementioned results have important practical implications in the spatial and temporal scaling features of rainfall fields [28,30,31] although the connections between entropic and scaling statistics with physical characteristics remain elusive. Furthermore, the effect of the space-time structure of rainfall in the scaling of the q-Entropy as well as its sensitivity to the number of bins and to the variability of rainfall intensity and the sample-size must be investigated in depth. Therefore, in the present study, we aim to investigate how the spatial scaling and complexity of rainfall is reflected in different entropic scaling measures within the frameworks of information theory and non-extensive statistical mechanics. The rationale and objectives of this work are presented next.

1.6. Rationale and Objectives

The objectives of our study are based upon the following considerations:

- The presence of zeros in high resolution rainfall records constitute highly important information to understand, diagnose and forecast the dynamics of rainfall [28,60,61]. Salas and Poveda [30] argued that zeros (inter-storm periods) in time series of tropical convective rainfall are associated with the timescale required by nature to build up the dynamic and thermodynamic conditions of the next storm, as an atmospheric analogous of the time of energy build-up between earthquakes, avalanches and many other relaxational processes in nature [62]. Therefore, the role of zeros and their effect on scaling statistics must be investigated to further understand and model high resolution rainfall.

- The aforementioned previous works [28,30,31] are based on available rainfall data (S-POL radar, TRMM satellite and rain gauges), and it is difficult to understand differences of the q-statistics in temporal and spatial scales due to factors such as the intermittency of rainfall, record length, space-time resolution of data sets, and geographic setting.

- The *number of bins* in the probability mass function constitutes a central issue to quantify entropic measures. Previous studies have shown that the scaling exponent of Shannon entropy under aggregation in time it is not sensitive to either *the number of bins* [28] or *record length* [30]. Then, it is necessary to study the sensitivity of the q-entropic measures in order to check their robustness to characterizing 2-D tropical rainfall fields.

The objectives of this study are manifold. They involve questions based on the previous studies [28,30,31], and new ones regarding the entropic scaling measures of rainfall. The objectives of our study are thus:

- To test theBL-Model [13] for 2-D Amazonian rainfall fields considering the Easterly and Westerly climatic regimes [55,56], and using the Generalized Space q-Entropy [31].

- To examine how the spatial structure of rainfall is reflected in the q-entropic scaling measures using the BL-Model and considering the influence of zeros in the GSEF through Montecarlo experiments, aimed at understanding the saturation of the exponent Ω_{sat} reported by [28,30,31].

- To investigate the connection between parameters of the BL-Model [13] and Amazonian rainfall fields considering the identified climatic regimes [55,56].

- To quantify the sensitivity of the q-entropic scaling statistics to *the number of bins* and to the *variability of rainfall intensity*, in an attempt to check the robustness of such statistical tools in the multi-scale characterization of rainfall.

- To link two important theoretical frameworks, namely stochastic processes (Multiplicative Cascades) and Information Theory (non-extensive statistical mechanics), to advance our understanding about the scaling properties of tropical rainfall.

The paper is organized as follows. Section 2 describes the study region and data set. Section 3 discusses the methods employed. Section 4 provides an in-depth discussion of results. Section 5 provides a brief discussion about the criteria and conditions to estimate entropy in geophysical data. Finally, Section 6 contains the conclusions.

2. Study Region and Data Sets

General Information

We use a set of 2-D radar rainfall fields gathered in Amazonia during the January–February 1999 Wet Season Atmospheric Meso-scale Campaign/LBA (WETAMC/LBA), which was designed to study the dynamical, microphysical, electrical, and diabatic heating characteristics of tropical convection over southwestern Amazonia [19,59,63,64]. The WETAMC campaign was developed in the state of Rondônia (Brazil). The data set used in the present study consists on radar scans of storm intensities recorded by the S-POL radar (S-band, dual polarimetric) located at 61.9982° W, 11.2213° S. Data consist of 2 km resolution microwave band reflectivity which is directly related to rainfall intensity, over a circle of ∼31,000 km². Scans produced by the Colorado State University Radar Meteorology Group were available every 7–10 min at the URL http://radarmet.atmos.colostate.edu/trmm-lba/rainlba.html.

Additionally, information about zonal wind velocity at 700 hPa during the study period was obtained from the NCEP/NCAR Reanalysis [65], over the region inside 61° W to 62.8° W and 10.4° S to 12.1° S, corresponding to the area covered by the S-POL radar. Data were obtained 4 times per day at 0000, 0600, 1200 and 1800 LST. In addition, radiosonde data (62.37° W, 10.75° S) were used from the WETAMC campaign in Ouro Preto d'Oeste at the Fazenda Nossa Senhora site, located inside the S-POL radar coverage region. Radiosonde data were obtained in the URL http://www.master.iag. usp.br/lba/.

3. Methods

A set of experiments were developed to study the sensitivity of the GSEF of 2-D Amazonian rainfall fields attempting to link the spatial structure of rainfall and the emerging scaling exponents of the q-entropic analysis [28,30]. To that aim we use the BL-Model proposed by [13] to generate 2-D rainfall fields as a multiplicative random cascade, by varying the model parameters β and σ. A detailed description of each experiment is presented below.

3.1. Parameters of the BL-Model and Amazonian Precipitation Features

The first experiment is carried out by controlling the percentage of wet (rainy) and dray (non-rainy) areas, and the second one by controlling the average intensity of the rainfall field. A set of 1000 simulations for each parameter were carried out. The mass of rainfall over a two-dimensional ($d = 2$) region was considered the unit, the branching number $b = 4$ for 2-D cascades, and the level of subdivision $n = 6$ (see Figure 1) during all experiments, consistently with the observed scans from S-POL radar which are 64 rows × 64 columns matrices ($b^n = 4^6 = 4096$ values). On the other hand, with the aim to link the numerical results from the BL-Model with the S-POL observations, we compared the samples of the estimated cascade's parameters (Equations (9) and (10)) considering $\beta_{Easterly}$ vs. $\beta_{Westerly}$ and, $\sigma_{Easterly}$ vs. $\sigma_{Westerly}$ using the k-sample tests based on the likelihood ratio [66], which are more robust than traditional methods (e.g., Kolmogorov–Smirnov, Cramer–von Mises, and Anderson–Darling), but also because the climatic regimes prevailing in the study region are statistically different in terms of the continental-scale flow and lightning activity, as well as in the vertical structure of convection and other precipitation features [56].

3.2. Bin-Counting Methods and Entropic Estimators

The correct estimation of diverse informational entropy statistical parameters require to take into account diverse practical considerations. Gong and others [67] argue that there are four practical problems in the estimation of entropy using hydrologic data: (i) the zero effect; (ii) the widely used bin-counting method for estimation of PDFs; (iii) the measure effect; and (iv) the skewness effect. We focus our attention on the second practical issue within the framework of scaling theory using Shannon theoretical entropy inequality [68],

$$S(T) \leq \ln \sqrt{2\pi e V(T)}, \tag{17}$$

where $V(T)$ is the variance of the process at aggregation interval T, with the equality holding just for the for the Gaussian distribution. We use a set of parametric and non-parametric bin-counting methods, such as those introduced by Sturges [69], Dixon and Kronmal [70], Scott [71], Freedman and Diaconis [72], Knuth [73,74], Shimazaki and Shinomoto [75,76], and a recent method for estimating entropy in hydrologic data proposed by Gong et al. [67]. In work, we use common techniques reported in the literature, although there are other methods [77]. Finally, we discuss the effect of *the number of bins* on the scaling of q-entropy in rainfall fields [28,30,31].

3.3. Sample-Size and Entropy Estimators

Information-theory statistics require an adequate sample size for a proper estimation and interpretation. In spite of the existence of a large body of literature dealing with the problem of estimation of distributions for data sparsity and poor sampling [78,79], the problem constitutes a challenge in geosciences. A recent study on the estimation of Shannon entropy in hydrological records under small-samples [80], employed three different estimators: (i) maximum likelihood (ML); (ii) Chao–Shen (CS); and (iii) James–Stein-type shrinkage (JSS). Their results exhibited that the ML estimator had the worst performance of the three methods, with the largest *Mean Squared Errors* (MSE) for all sample sizes. In particular, when sample sizes are small (less than 200 data points), the entropy estimator was dramatically underestimated, although errors turned out to decrease quickly when sample sizes increased. Furthermore, when the sample size was larger than 100 data points, the accuracy of the CS and JSS estimators were basically the same, with MSE nearly equal to zero. It is worth mentioning that the said study [80] did not deal either with the effect of *the number of bins* or the role of zeros in the estimation of entropy.

At the root of the problem of the numerical estimation of entropy, is the sample-size necessary for adequately estimating the underlying probability mass function (pmf) of the process. For example, it is well known that, for the ML estimators, the larger the sample size, the better the estimates will be. In addition, high temporal resolution precipitation data sets (e.g., minutes or hours) are mostly (approximately 90%) constituted by zeros [30], which it is not the case for low-resolution data (e.g., months or years), so an adequate characterization of the tails of the Probability Distribution Function (PDF) requires longer data sets for high-resolution rainfall data than for low-resolution.

With the aim of studying the influence of sample-size in the GSEF, we will compare the q-entropy of the S-POL radar fields (matrices with 64 rows and 64 columns) and synthetic 2-D fields of the BL-Model at different cascade levels, n, to obtain fields with different sizes, ζ, e.g., the cascade level $n = 1$ means a 2×2 field, (2 rows and 2 columns) and, consequently, a cascade level $n = 7$ means a 128×128 field. In other words, a field of the BL-Model with $n = 1$ has a sample-size $\zeta_i = 4$ and a field of the BL-Model with $n = 7$ has a sample-size $\zeta_i = 16,384$. Using the synthetic fields, we quantify the q-entropy: (i) estimating the pmf for all the values in the synthetic rainfall field (including zeros); (ii) separing values in two subsets $P(x) = \{P(x = 0), P(x > 0)\}$; and (iii) the minimal quantity of non-zero values in the fields, to ensure robustness in the estimation of S_q. Our results are discussed in Section 4.5.

3.4. Intermittency and q-Order

High space-time resolution tropical rainfall is a highly complex and intermittent process. To analyze its intermittent behavior, several techniques have been proposed in the literature such as: spectral scale invariance analysis [31,81–84]; moment-scaling analysis [1,3,18,19,31,32,84], and intermittency exponents [84,85]. In this sense, the Generalized Space q-Entropy Function (GSEF) (Section 1.4) can be thought as a measure of (multi-) fractal behavior of a process [30,31]. In addition, the BL-Model (Section 1.2) is directly related to the Mandelbrot-Kahane-Peyriere (MKP) function [34,35] characterizing the fractal (or scale-invariant) behavior of a multiplicative cascade process. Therefore, an interesting question arises: is there any relationship between the GSEF and intermittency estimators? In order to shed light about such question, the BL-Model is used to estimate the q-order, Θ_q, and the multifractality measure, $M(q_1, q_2)$, defined by Bickel [86], Equation (22).

First, we carry out an experiment for high-resolution-spatial rainfall based on Bickel [86], who showed the relationship between intermittency (multifractality) and the *non-extensive order* (hereafter q-order) for point processes of dimension D ranging from 0.1 to 0.9. The order in a system is defined in terms of its distance from equilibrium, so the higher disordered, the closer to the equilibrium state. The q-order can be defined as,

$$\Theta(q) \equiv 1 - \frac{S_q}{S_q^{max}} \tag{18}$$

where S_q is the q-entropy (Equation (11)) and S_q^{max} is the maximum possible value of S_q for the equilibrium condition, which probabilistically can be denoted as $p_j = 1/N$ for all $j = 1, 2, ..., N$, whose S_q^{max} can be written as,

$$S_q^{max} = \frac{1 - N^{(1-q)}}{q - 1} \tag{19}$$

At this point, it is necessary to emphasize that Equation (19) is a generalization for the *extensive order* defined as a measure of complexity using the Boltzman-Gibbs-Shannon entropy [87]. In addition, the non-extensive order satisfies $0 \leq \Theta(q) \leq 1$ with $\Theta(q) = 0$ if $P_j = 1/N \; \forall j$ and $\Theta(q) = 1$ if $P_j = \delta_{jl} \; \forall j$, given any integer l between 1 and N [86,88].

Second, a generalization for $\Theta(q) \equiv \Theta_q$, at multiple spatial or temporal scales, is possible using a similar procedure to Equation (15). Therefore, if the spatial scale is denoted as λ, and the maximum possible value of q-entropy at each scale λ as $S_q^{max}(\lambda)$, the q-order can be rewritten as,

$$\Theta_q(\lambda) \equiv 1 - \frac{S_q(\lambda)}{S_q^{max}(\lambda)}. \tag{20}$$

From previous studies [30,31], it is easy to note that $\Theta_q(\lambda) \neq S_q(\lambda)$ but for the scaling laws $\Theta_q(\lambda) \sim \lambda^{\Omega(q)}$ and $S_q(\lambda) \sim \lambda^{\Omega(q)}$, the scaling exponents $\Omega(q)$ are exactly the same, which means that GSEF, $\Omega(q)$, does not change under such transformation.

Third, considering that a process $x(t)$ exhibiting a multifractal spectrum whose spectrum of generalized Hurst exponents, $H(q)$, is defined as [86],

$$\psi(q, T) = \langle |x(t+T) - x(t)|^q \rangle^{1/q} \propto T^{H(q)}, \tag{21}$$

which holds for some range T with non-extensive parameter q. For $\psi(q = 1, T)$ is the mean (first moment) of the absolute displacement and for $\psi(q = 2, T)$ is the standard deviation of this displacement. Therefore, the intermittency of a nonstationary process $x(t)$ can be quantified by its *multifractality*, $M(q_1, q_2)$, as the difference between two generalized Hurst exponents,

$$M(q_1, q_2) = \begin{cases} -q_1 q_2 \frac{H(q_2) - H(q_1)}{q_2 - q_1} & q_1 \neq q_2 \\ \lim_{q_1 \to q_2} M(q_1, q_2) = -q_2^2 [\partial H(q_2)/\partial q_2] & q_1 = q_2 \end{cases} \tag{22}$$

normalized such that $0 \leq M(q_1, q_2) \leq 1$ for nondegenerate processes, with $M(q_1, q_2) = 0$ for monofractals [89].

From previously mentioned considerations, the generalized Hurst exponents for 2-D rainfall fields can be computed as suggested by Carbone [90] in combination with the Equation (21) as follows,

$$\hat{\psi}_k(q) = \left(\frac{1}{(n_x - k)(n_y - k)} \sum_{i=1}^{n_x - k} \sum_{j=1}^{n_y - k} |x(i+k, j+k) - x(i, j)|^q \right)^{1/q}, \tag{23}$$

where $x \in \Re^2$, n_x and n_y are the number of rows and columns, respectively. The estimates $\hat{\psi}_k(q_1)$ and $\hat{\psi}_k(q_2)$ for $k = k_{min}, 2k_{min}, 4k_{min}, \cdots, k_{max}$, with k_{min} and k_{max} such that $\log \hat{\psi}_k(q_1)$ and $\log \hat{\psi}_k(q_2)$ exhibit linear relationship with $\log k$. Then, $\hat{H}(q_1)$ and $\hat{H}(q_2)$ are the slopes of the least-square regressions of $\log \hat{\psi}_k(q_1)$ vs. $\log k$ and $\log \hat{\psi}_k(q_2)$ vs. $\log k$, respectively. Finally, the *multifractality* is estimated as,

$$\hat{M}(q_1, q_2) = \begin{cases} -q_1 q_2 \frac{\hat{H}(q_2) - \hat{H}(q_1)}{q_2 - q_1} & q_1 \neq q_2 \\ \lim_{q_1 \to q_2} \hat{M}(q_1, q_2) & q_1 = q_2 \end{cases} \tag{24}$$

In this work, we explore the relationship between intermittency (multifractality), $\hat{M}(q_1, q_2)$, and q-order, Θ_q, for synthetic rainfall fields from the BL-Model. The results will be discussed in Section 4.6.

4. Results

4.1. Linking Parameters of the BL-Model with Precipitation Features

Following previous studies [19,31], we classify the available information from the S-POL radar considering the Easterly (negative values) and Westerly (positive) climatic regimes. Then, for both climate regimes, we estimate the parameters of the BL-Model using Equations (9) and (10). Subsequently, we estimate the pmf (histograms) and the Cumulative Distribution Functions (CDFs) of the parameters β and σ^2 (Figures 2 and 3). Results show that the histograms of β and σ^2 are statistically different for both climatic regimes, and, additionally, that the CDFs of the parameter β for the Easterly and Westerly regimes are significantly different according to a k-sample test, based on the likelihood ratio [66] at 95% confidence level, but no so for the parameter σ^2.

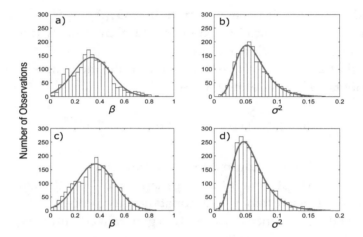

Figure 2. Histograms for Beta-Lognormal Model parameters in rainfall scans of the S-POL radar: (**a,b**) Westerly events; and (**c,d**) Easterly events. (red) the Gaussian function for β and the Generalized Extreme Value function for σ^2.

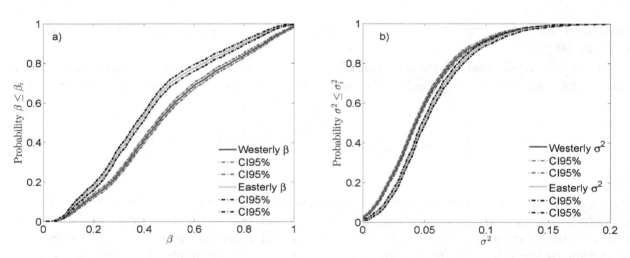

Figure 3. Empirical Cumulative Distribution Functions for cascade parameters: β (**a**); and σ^2 (**b**) using 4227 scans of the S-POL radar considering climatic regimes of Amazonia (1867 scans for Westerly and 2360 for Easterly). The figure shows the 95% confidence bounds using Greenwood's formula.

Secondly, we estimated the GSEF using a set of 1000 synthetic rainfall fields generated with the average values of β and σ^2 from the S-POL scans (see Figure 4). Figure 5b shows the GSEF of rainfall fields generated using the BL-Model, and the average of the observed S-POL rainfall fields. In addition, Figure 6a shows that the scaling exponents of the GSEF, $\Omega(q)$-observed vs. $\Omega(q)$-simulated, exhibit a very good fit. Furthermore, Figure 6b shows that the BL-Model represents adequately the relationship $S_q(\lambda) \sim \lambda^{\Omega}$, with saturation for $q \geq 2.5$ and $\Omega_{sat} \sim 0.5$. However, observed and simulated rainfall fields exhibit significant differences in the interval $0.5 \leq q \leq 2.5$; the S-POL scans do not exhibit power-laws for $1.0 \leq q \leq 1.5$, albeit in this interval the model shows power-laws with $R^2 \geq 0.7$.

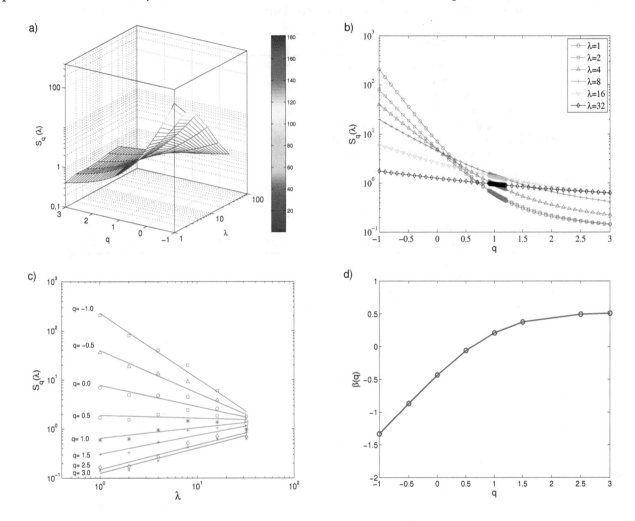

Figure 4. Space Generalized q-Entropy for the S-POL radar scan 01/10/1999 18:23:15 LST: (**a**) 3D plot of the Tsallis' entropy, S_q, for different scale factors, λ, and q-values from -1.0 to 3.0; (**b**) projection of $S(q,\lambda)$ vs. q for different values of λ; (**c**) projection of $S(q,\lambda)$ vs. λ, for different values of q, or spatial structure function for entropy; and (**d**) values of the regression slopes of the spatial structure function for entropy, Ω, as function of q, exhibiting a non-linear growth up to $\langle \Omega \rangle \sim 0.50$ for $q > 2.5$.

Finally, considering that the BL-Model has two parameters, β and σ, it is necessary to link them with precipitation features associated with both climatic regimes in Amazonian rainfall, as follows:

- The cascade parameter, β, (Table 1), for the Easterly events is greater than the Westerly events, indicating more spatially concentrated rainfall fields (more zeros in the Easterly scans). This result is related to diverse precipitation features observed during the Easterly regime, given that the atmospheric conditions are relatively dry, with increased lightning activity and more intense and deeper convective systems [56,58,59].

- The cascade parameter, σ, (Table 1), exhibits smaller (larger) values during the Easterly (Westerly) regime, indicating that the *variability of rainfall intensity* for the Westerly events is higher than for Easterly events. This result is coherent with diverse features observed during the Westerly regime, which is characterized by less lightning activity, less deep convection and less intense precipitation rates [56,58,59].

Table 1. Scans of S-POL radar according to the two identified Amazonian climate regimes.

Description	Westerly	Easterly
Total number of scans	2607	3884
Average β for all scans	0.421	0.491
Average σ for all scans	0.235	0.221
q-value where the SGEF saturates for all scans	1.50	1.50
Average scaling exponent of saturation, Ω_{sat}, for all scans	1.0	1.0
Scans with more than 200 values non-zero (Denoted as $*$)	1867	2360
Scans with all values zeros	86	21
Percentage of scans $*$	71.6%	60.8%
Percentage of scans with less than 200 values non-zero	28.4%	39.2%
Average β for scans $*$	0.336	0.365
Average σ for scans $*$	0.248	0.242
q-value where the SGEF saturates for scans $*$	2.5	2.5
Average scaling exponent of saturation, Ω_{sat}, for scans $*$	0.38 ± 0.15	0.4 ± 0.15

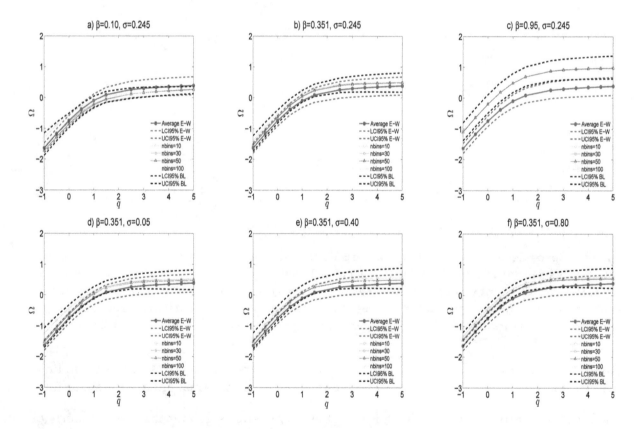

Figure 5. Space Generalized q-Entropy Function for spatially distribute rainfall as a random cascade varying cascade's parameters σ and β for 1000 simulated independently fields including zeros in the histogram. $LCI95\%$ and $UCI95\%$ are the lower and upper confidence intervals for Easterly and Westerly events (E-W) and the BL-Model model (BL): (**a**–**c**) varying the cascade's parameter β; and (**d**–**f**) varying the cascade's parameter σ. In all cases, varying the number of bins (nbins = 10, 30, 50 and 100).

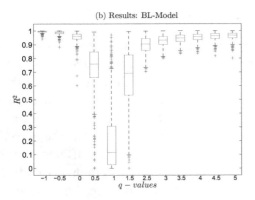
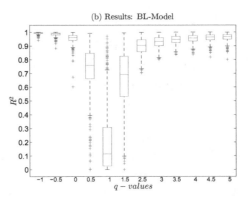

Figure 6. Validation of the BL-Model using Generalized Space q-Entropy: (**a**) comparison $\Omega(q)$-Observed vs. $\Omega(q)$-Simulated, (Solid line) relation 1:1; and (**b**) box plots of the coefficients of determination, R^2, for the power fits $S_q(\lambda) \sim \lambda^{\Omega}$ from 1000 synthetic fields of BL-Model with $\beta = 0.351$ and $\sigma = 0.245$. $R^2 \geq 0.85$ in the intervals $-1.0 \leq q \leq 0.0$ and $q \geq 2.5$. The histogram for estimate S_q includes zeros.

4.2. The Role of Zeros in the Generalized Space q-Entropy

With the aim of studying the influence of zeros in the estimation of the GSEF, we simulated rainfall fields using the BL-Model with cascade parameters $\beta = 0.10, 0.35$ and 0.95. In addition, 1000 simulations were carried out for each value of β. Modeling a 2-D rainfall field with $\beta = 0$ corresponds to the situation in which the cascade assigns a uniformly distributed unit rainfall intensity over the whole area. In contrast, β close to 1.0 indicates that rainfall is concentrated in a very small area. Our experiments estimate the GSEF for varying values of β, while keeping σ constant and equal to the average value for the Amazonian scans, $\sigma = 0.245$ (see Table 1). Results show that the saturation scaling exponent Ω_{sat} in the GSEF is significantly affected by the fraction of non-rainy cells (Figure 5a–c). A similar result is found for the minimum value of the scaling exponent Ω_{min} in the GSEF, which is significantly increased with the amount of zeros present in the rainfall fields. Figure 7 shows that the scaling exponents Ω_{sat} and Ω_{min} increase with the value of β. This behavior is explained by the loss rate of zeros during the change in spatial resolution, as is explained below.

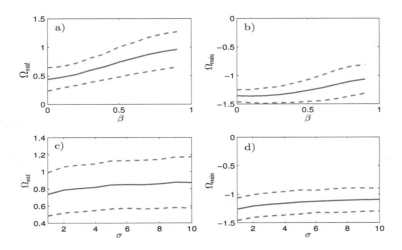

Figure 7. Sensitivity analysis of the saturation Ω_{sat}, and minimum Ω_{min} scaling exponents in the SGEF for 1000 independent rainfall fields generated by random cascade model [13]. Confidence intervals for 95% in dash line and mean value in solid line. (**a**) Varying cascade's parameter β and considering $\sigma = 0.25$ constant; (**b**) varying cascade's parameter β and considering $\sigma = 0.25$ constant; (**c**) varying cascade's parameter σ and considering $\beta = 0.5$ constant; and (**d**) varying cascade's parameter σ and considering $\beta = 0.5$ constant.

4.3. The Role of Rainfall Intensity Variability in Generalized Space q-Entropy

The influence of *the variability in rainfall intensity* on the GSFE was examined with a similar strategy. We generated rainfall fields with cascade parameters $\sigma = 0.05$, 0.40 and 0.80 (1000 simulations for each σ value), with a constant parameter $\beta = 0.351$ (see Table 1). Results show that the scaling exponent of saturation, Ω_{sat}, in the GSEF increases slowly in comparison with the case where the cascade parameter β is considered variable. This result implies that the increase or decrease in uncertainty across spatial scales is dominated by the dry areas and not by *the variability in rainfall intensity*. In general, the scaling exponents Ω_{sat} and Ω_{min} were not affected by changes in σ (Figure 7c,d). A possible explanation for this result is that the scaling exponents Ω_{sat} and Ω_{min} are directly related to the loss rate of zeros in rainfall fields when data are averaged going from higher resolution (2 km pixel size) to lower resolution scales (32 km pixel size) [30]. Then, if the amount of zeros is constant and the variability of rainfall intensity increases, the loss rate of zeros remains the same regardless of the spatial resolution, which means the scaling exponent Ω_{sat} is not affected by the cascade parameter σ.

4.4. Bin-Counting Methods and the Generalized Space q-Entropy

First, we discuss the numerical estimation of Shannon entropy for an i.i.d. Gaussian random variable under increasing aggregation intervals, T, using the analytic inequality given by Equation (17), and the multiple bin-counting methods mentioned in Section 3.2. From that equation, it is easy to see why entropy increases under aggregation of T. The theorem [91] proves that if $X_1, X_2, X_3, ..., X_n$ are i.i.d. random variables, then the expected value $E(X_j) = \mu$, with finite variance $V(X_j) = \sigma^2$. Defining the sum $S_n = X_1 + X_2 + X_3 + ... + X_n$, then the average is $A_n = \frac{S_n}{n}$, $E(S_n) = n\mu$ and $V(S_n) = n\sigma^2$.

On the other hand, we revisit the classical problem [92] of *how* and *how well* diverse information-theoretic quantities, can be estimated given a finite set of i.i.d. r.v., which lies at the heart of the majority of applications of entropy in data analysis. Paninski's paper focuses on the non-parametric estimation of entropy, and compares different estimation methods without delving into the role of the number of bins. For our proposes, we study the sensitivity of Shannon entropy (Equation (25)) to the *number of bins*. According to Shannon [44], discrete data entropy can be estimated as,

$$S(X) = - \sum_{i=1}^{n} p(x_i) \log_a p(x_i) \tag{25}$$

where $p(x_1), p(x_2), \ldots, p(x_k)$ represents the probability mass function, such that $\sum_{i=1}^{n} p(x_i) = 1$, and $p(x_i) \geq 0$, $\forall i$. Figure 8 shows that the main differences among the different estimation methods are the following:

- The bin-counting method proposed by Dixon and Kronmal [70] is the nearest to the method presented by Gong et al. (2014) for Gaussian r.v. under aggregation, for a number of aggregation intervals greater than 70.
- The theoretical inequality given by Equation (17) is better captured by Scott's method, although this method shows lower values than the theoretical expression, for aggregation intervals $T \geq 100$.
- The difference between the theoretical inequality (Equation (17)) and Gong et al.'s [67] method is explained because the "Discrete Entropy" and the "Continuous Entropy" (also referred to as "Differential Entropy") are related as:

$$\lim_{\Delta \to 0} [H_\Delta(X_d) + \log(\Delta)] = h(Xc), \tag{26}$$

where $H_\Delta(X_d)$ is the discrete entropy with the bin-width Δ and $h(Xc)$ is the corresponding continuous entropy. Thus, the continuous entropy of a r.v. requires to add $\log(\Delta)$ in the numerical estimation. In our numerical estimation, Δ was selected as the average of the bin-width estimated for the six methods (see the Section 3.2) for each aggregation interval, T, so the behavior of

Gong et al.'s method is approximately the average of the set of methods. At this point, it is worth noting that although Gong et al.'s method is designed for hydrological records, it is not free from sensitivity to *the number of bins*.

- To check the sensitivity of the scaling exponents of the GSEF to *the number of bins*, we developed a numerical experiment using the BL-Model and 1000 independent simulations for each number of bins $n = 10, 30, 50$ and 100, with parameters $\beta = 0.10, 0.351$, and 0.950 and $\sigma = 0.05, 0.40$ and 0.80. Figures 5 and 7 show that the GSEF is not statistically affected either by *the number of bins* or by the value of σ when $nbins > 30$ and the sample-size is bigger than 200 data. Consequently, the GSEF is not affect by the bin-counting method.

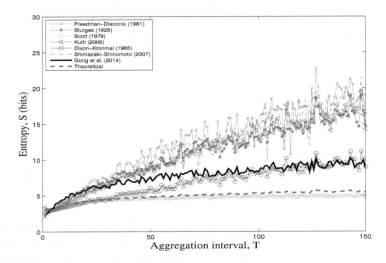

Figure 8. Numerical estimation of Shannon's Entropy using multiple bin-counting methods and the theoretical inequality (Equation (17), for a Gaussian r.v. for different levels of aggregation, T).

4.5. Sample-Size and the Generalized Space q-Entropy

We consider 6491 scans of the S-POL radar, each one with 4096 data (64 rows and 64 columns; pixel-size 2 km), of which on average 82% are zeros. Furthermore, approximately 34% of all scans have less than 200 non-zero values. Thus, using the complete data set, the probability mass function (pmf) of some scans could be concentrated in the first bins, with small informational content to the entropic estimator whereas the highest rain values could appear with a very low probability, contributing to augment the entropy. To quantify the effect of sample-size in estimating q-entropy we performed the following experiments:

First, we generated rainfall fields using the BL-Model with constant values of $\beta = 0.351, \sigma = 0.245$, $q = 2.5$ in S_q, and *the number of bins* of the pmf, $nbins = 50$. Then, we calculated by S_q changing the amount of cascade levels $n = 1, 2, ..., 8$ to obtain synthetic rainfall fields with different number of values in the scan (or sample-size) $\xi = 4, 16, ..., 65,537$, and finally we estimated S_q in the following two manners:

- The pmf to calculate S_q was built considering all values in the synthetic rainfall field including zeros.
- The pmf to calculate S_q was built considering all values separated in two subsets: (i) values greater than zero (rain), i.e., $P(x > 0)$; and (ii) values equal to zero (dry), $P(x = 0)$. For the subset (i), the pmf was built and then corrected by the probability of rainfall $(1 - P(x = 0))$ thus, the probability of occurrence of rainfall can be written as $P(x) = \{P(x = 0), P(x > 0)\}$.

Figure 9 shows that, in both cases, S_q decreases with sample size, ξ, but the variance of S_q is slightly greater when the pmf was built using all values including zeros (Figure 9a) than when the pmf was built using two separated subsets (Figure 9b). Figure 10 shows that values of q-entropy S_q differ when zeros are included in the pmf and when $P(x = 0)$ is calculated separately. Figure 10a shows that values of S_q with zeros in the pmf versus S_q without zeros in the pmf are not significantly different when the sample size $\xi \leq 4096$ (i.e., 64×64 matrices or cascade level 6). In contrast, Figure 10b shows that values of S_q with the zeros in pmf versus S_q without the zeros in the pmf are significantly different in fields of sample size $\xi > 4096$ (i.e., 256×256 matrices or cascade level 8).

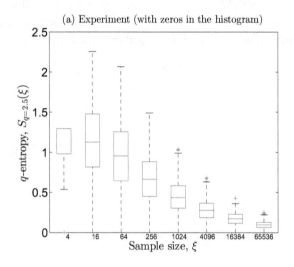
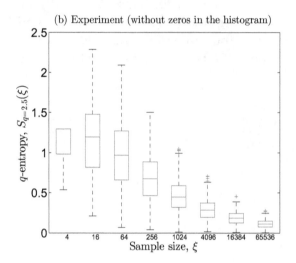

Figure 9. Boxplots for q-Entropy $S_{q=2.5}$ vs. sample sizes, ξ, in 1000 independent random cascade simulations of the Beta-Lognormal model with parameters $\beta = 0.351$ and $\sigma = 0.245$: (**a**) q-Entropy including zeros in the histogram; and (**b**) q-Entropy without including zeros in the histogram. In both cases, (*nbins* = 50).

Secondly, we performed a detailed examination of previous results [31] obtained using all the S-POL radar scans, and re-calculated the GSEF varying the sample-size. Results show considerable differences between sample-sizes with *less* and *more* than 200 non-zero values. Figure 11 shows differences between the GSEF for all S-POL scans, and the GSEF considering only scans with more than 200 non-zero values, but including zeros in the pmf. Furthermore, Figure 12 shows that the power laws $S_q(\lambda) \sim \lambda^\Omega$, considering the two climate regimes in Amazonian rainfall, exhibit $R^2 \geq 0.85$ in the intervals $-1.0 \leq q \leq 0.5$ and $q \geq 2.5$. For scans with more than 200 non-zero values, the scaling exponents $\Omega(q)$ of the relation $S_q(\lambda)$ vs. λ exhibit a non-linear growth with q, up to $\Omega \sim 0.5$ for $q \geq 2.5$, while in our previous study [31], the GSEF exhibited a non-linear growth with q, up to $\Omega \sim 1.0$ for $q \geq 1.0$.

These results turned out to be even more interesting with respect to those presented by [31] for the Generalized Time q-Entropy Function (GTEF) in Amazonian rainfall, whose scaling exponents $\Omega(q)$ of the relation $S_q(T)$ vs. T in log–log space, exhibit a non-linear growth with q, up to $\Omega \sim 0.5$ for $q \geq 1.0$. A thorough analysis showed that for the 400 time-series of the S-POL radar used by [31], only the 5% had less than 800 non-zero values, and that the 99% of the time-series had more than 200 non-zero values. Additionally, the scaling exponent of saturation, Ω_{sat}, for the GTEF remains the same, as well as the q-value for saturation. Therefore, our results suggest that the scaling exponent of S_q across a range of scales in space and time reaches the same maximum value $\Omega_{sat} = \Omega \sim 0.5$, but the non-additive q value of saturation differs between space scaling ($q \sim 2.5$) and time scaling ($q \sim 1.0$). According to Tsallis [45], these results reflect the differences between the space and time dynamics of the system, although their connection with the physics of rainfall is an open problem.

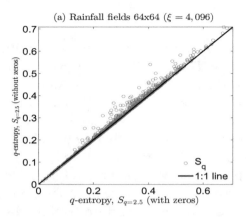
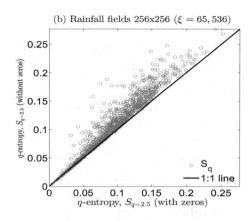

Figure 10. Comparison of q-Entropy $S_{q=2.5}$ in 1000 independent random cascade fields of the Beta-Lognormal model with parameters $\beta = 0.351$ and $\sigma = 0.245$: (a) q-Entropy S_q including zeros in the histogram vs. q-Entropy S_q without zeros in the histogram, cascade's level = 6, i.e., $\xi = 4096$; and (b) q-Entropy S_q including zeros in the histogram vs. q-Entropy S_q without zeros in the histogram, cascade's level = 8, i.e., $\xi = 65,536$. In both cases, (*nbins* = 50).

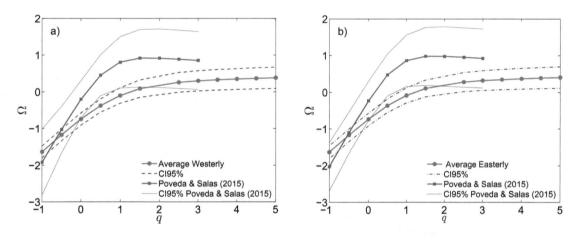

Figure 11. Space Generalized q-Entropy Functions (SGEFs) for the climate regimes of Amazonian rainfall from the S-POL radar: (a) Westerly events; and (b) Easterly events. (circles) Average SGEF for scans with more than 200 values greater than zero, (dashed lines) 95% confidence intervals (CI); (squares) average SGEF for all the scans available of each climate regime; (solid lines) 95% confidence intervals (CI).

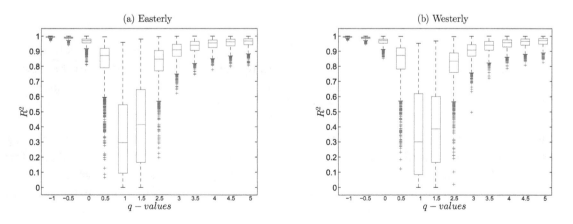

Figure 12. Coefficient of Determination, R^2, for the power fits $S_q(\lambda) \sim \lambda^{\Omega}$ from scans radars of the climate regimes in Amazonian rainfall. $R^2 \geq 0.85$ in the intervals $-1.0 \leq q \leq 0.5$ and $q \geq 2.5$. The histogram for estimate S_q includes zeros: (a) Easterly events; and (b) Westerly events.

4.6. Rainfall Intermittency and q-Order

As mentioned in Section 3.4, Bickel [86] showed a positive correlation between multifractality, $\hat{M}(q_1, q_2)$, and q-order, Θ_q, for point process with dimension D ranging from 0.1 to 0.9. In this study, we looked for an analogous relationship for synthetic 2-D rainfall fields from the BL-Model. Our results show that for $\hat{\psi}_k(q_1)$ vs. k and $\hat{\psi}_k(q_2)$ vs. k, both cases exhibit linear relationship in the log–log graph to estimate the generalized Hurst exponents $H(q_1)$ and $H(q_2)$ and subsequently $\hat{M}(q_1, q_2)$ as we explained in Section 3.4. Figure 13 shows a typical regression for a synthetic rainfall field created with the BL-Model with $\beta = 0.351$ and $\sigma = 0.245$, with average intermittency $\langle \hat{M}(q_1, q_2) \rangle = 0.519$. However, there is no evidence of a clear-cut relationship between $\hat{M}(q_1, q_2)$ and Θ_q in our numerical experiments (figures not shown here). Those results can be explained because the point-process model used by Bickel [86] is a Markovian model whose stochastic properties and probability distribution function (PDF) differ from point-process models for rainfall [93,94], which do not explicitly consider statistical scaling properties [95]. In addition, the BL-Model is a non-Markovian rainfall model based on the spatial statistical (multi) scaling properties, whose PDF is well known across spatial scales emerging as power laws, $S_q \sim \lambda^{\Omega(q)}$. The study of the linkages between intermittency and q-entropic statistics are outside the scope of this work that deserves to be explored in detail in future works.

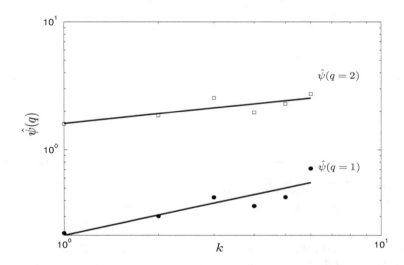

Figure 13. Typical least-square regressions $\hat{\psi}_k(q_1) \sim k^{H(q_1)}$ and $\hat{\psi}_k(q_2) \sim k^{H(q_2)}$ with $H(q = 1) = 0.097$ and $H(q = 2) = 0.412$, for a synthetic 2-D rainfall field from the BL-Model with $\beta = 0.351$, $\sigma = 0.245$ and cascade level $n = 8.0$.

5. Discussion

In spite of the increasing interest in entropic techniques in geosciences, few studies have discussed diverse underlying assumptions regarding the data to guarantee their applicability. In the case studied, high-resolution rainfall records are neither i.i.d. nor with continuous pdf, so the appropriate estimation of entropy needs clarity on the implicit assumptions in data analysis.

First, the i.i.d. condition for rainfall is not satisfied because: (i) the spatial dynamics of mesoscale rainfall has strong spatial correlations [33] (e.g., for Amazonian rainfall see [31]); and (ii) the temporal dynamics of tropical rainfall reflects long-term correlations [28] (see Figure 14). However, by definition, q-entropy, S_q, for $q \neq 1$, considers probabilistically dependent subsystems, with non-negligible global correlations, whereas Shannon entropy (S_q, for $q \to 1$) considers probabilistically independent subsystems [45,96]. Hence, the non-i.i.d. nature of data is not a restriction in the framework of non-extensive entropy, whereas in the framework of extensive entropy such non-i.i.d nature must be used under specific assumptions (e.g., for weakly correlated sub-systems).

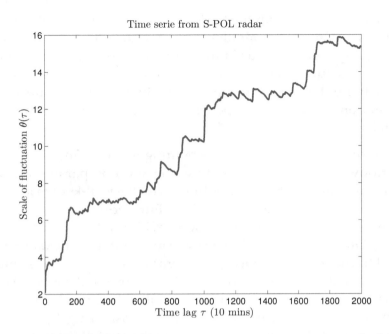

Figure 14. Scale of fluctuation, $\theta(\tau)$, for a time serie of Amazonian rainfall from S-POL radar.

Second, high-resolution Amazonian rainfall records do not satisfy the condition of continuous pdf because zeros constitute more than 80% of data in spatiotemporal scales. Therefore, although continuity in pdf is a fundamental requirement to estimate the additive (Shannon) entropy using the most common estimators [77,97], the condition of pdf's continuity for q-entropy is not clear in the literature. This is a relevant topic for further research.

Finally, an alternative option to deal with the conditions behind entropic estimators consists in finding a transformation that generates i.i.d data exhibiting a continuous pdf. However, that transformation constitutes a great challenge in geosciences, more so having in mind that such a transformation include multi-scale statistical properties.

6. Conclusions

Using 2-D radar rainfall fields from Amazonia, we investigate the spatial scaling and complexity properties of Amazonian rainfall using the Generalized Space q-Entropy Function (GSEF), defined as a set of continuous power laws covering a broad range of spatial scales, $S_q(\lambda) \sim \lambda^{\Omega(q)}$, to test for the validity of the random multiplicative cascade BL-Model in representing 2-D properties of observed rainfall fields. The spatial scaling analysis considered the Westerly and Easterly weather regimes in the Amazon basin. Our results show that for both climate regimes the GSEFs are not statistically different whereas the BL-Model parameters σ and β are statistically different.

We tested the skill of the BL-Model in reproducing the space scaling properties of q-entropy reported in previous works. Our results evidence that the BL-Model appropriately reproduces the relationship $S_q(\lambda) \sim \lambda^{\Omega}$, with saturation for $q \geq 2.5$ and $\Omega_{sat} \sim 0.5$. Furthermore, the power laws, $S_q(\lambda) \sim \lambda^{\Omega}(q)$, observed in S-POL rainfall scans exhibit $R^2 \geq 0.85$ in the intervals $-1.0 \leq q \leq 0.5$ and $q \geq 2.5$, whereas synthetic rainfall fields generated with the BL-Model exhibit power laws in the intervals $-1.0 \leq q \leq 0.5$ and $q \geq 1.5$. This result evidences that the q-entropy allows to successfully characterizing the spatial scaling properties of high resolution Amazonian rainfall, thus confirming the validity of this tool in the study of systems conformed by strongly correlated subsystems, for which the Shannon entropy ($S_{q \to 1}$) is no longer valid. In particular, the spatial scaling structure of Amazonian rainfall can be characterized by a non-additivity value, $q_{sat} \sim 2.5$, at which rainfall reaches its the maximum scaling exponent Ω_{sat}.

Using Montecarlo experiments with the BL-Model, we studied the influence of zeros and rainfall intensity on the estimation of the GSEF, aiming to explain the differences of the saturation exponent Ω_{sat} found between multiple data sets used in previous works [30,31]. Our results evidence that: (i) the scaling exponent of saturation Ω_{sat} is related to the non-rainy area fraction, represented by β; and (ii) the variability in rainfall intensity, represented by σ, does not affect significantly the GSEF. Then, changes in saturation of the scaling exponent Ω_{sat} are related to the intermittence properties of high-resolution rainfall.

In addition, we studied the influence of bin-counting methods and sample-size in the estimation of entropy and q-entropy. We used a set of parametric and non-parametric bin-counting methods showing the difficulties in estimating Shannon entropy with the well-known inequality linking variance and entropy for Gaussian i.i.d. random variables. Furthermore, we explored the sensitivity of the GSEF to *the number of bins* (*nbins*). Our results evidenced that the GSEF is a robust measure provided *nbins* $\geq$ 30. On the other hand, we performed a detailed examination of the results by Poveda and Salas [31] to check the influence of the sample-size, ζ_i, in the estimation of the q-entropy. We studied synthetic 2-D fields of the BL-Model from 2×2 (rows and columns) to 128×128 (rows and columns) quantifying the q-entropy with respect to: (i) all values inside the rainfall fields (including zeros) in the probability mass function (pmf); (ii) pmf considering $P(x) = \{P(x = 0), P(x > 0)\}$; and (iii) the minimum amount of non-zero values inside the rainfall fields. Our results evidenced that for small-samples the generalized space q-entropy function may incur in considerable bias, and our experiments showed that a rainfall field requires at least 200 non-zero values so that the estimation of q-entropy be robust.

Finally, we explored a possible relationship between a measure of multifractality $M(q_1, q_2)$ and the q-order Θ_q. Our results suggest that the relationship found by Bickel [86] could be related to the point process used therein. In our case, for the BL-Model based on multiplicative cascades there is not evidence of such links between $\hat{M}(q_1, q_2)$ and Θ_q.

Acknowledgments: We thank the Colorado State University Radar Meteorology Group for radar data and the LBA project for providing access to the Fazenda Nossa Senhora radiosonde. The work of H.D.S. is supported by COLCIENCIAS—Call for National Doctorates 617 (2014). The work of G.P. and O.J.M. is supported by Universidad Nacional de Colombia at Medellin.

Author Contributions: H.D.S., G.P. and O.J.M. conceived and designed the study. H.D.S. prepared the data and performed computations and simulations. H.D.S., G.P. and O.J.M. discussed and analyzed the data and results. H.D.S. and G.P. wrote the manuscript.

References

1.   Deidda, R.; Benzi, R.; Siccardi, F. Multifractal modeling of anomalous scaling laws in rainfall. *Water Resour. Res.* **1999**, *35*, 1853–1867.

2.   Devineni, N.; Lall, U.; Xi, C.; Ward, P. Scaling of extreme rainfall areas at a planetary scale. *Chaos* **2015**, *25*, 075407.

3.   Foufoula-Georgiou, E. On scaling theories of space-time rainfall: Some recent results and open problems. In *Stochastic Methods in Hydrology: Rainfall, Land Forms and Floods*; Barndor-Neilsen, O.E., Gupta, V.K., Pérez-Abreu, V., Waymire, E., Eds.; World Science: Hackensack, NJ, USA, 1998; pp. 25–72.

4.   Gupta, V.K.; Waymire, E. Multiscaling properties of spatial rainfall and river flow distributions. *J. Geophys. Res.* **1990**, *95*, 1999–2009.

5.   Harris, D.; Foufoula-Georgiou, E.; Droegemeier, K.K.; Levit, J.J. Multiscale statistical properties of a high-resolution precipitation forecast. *J. Hydrometeorol.* **2001**, *2*, 406–418.

6.   Jothityangkoon, C.; Sivapalan, M.; Viney, N. Test of a space-time model of daily rainfall in soutwestern Australia based on nonhomogeneous random cascades. *Water Resour. Res.* **2000**, *36*, 267–284.

7.   Gentine, P.; Troy, T.J.; Lintner, B.R.; Findell, K.L. Scaling in Surface Hydrology: Progress and Challenges. *J. Contemp. Water Res. Educ.* **2012**, *147*, 28–40.

8. Lovejoy, S. Area-perimeter relation for rain and cloud areas. *Science* **1982**, *216*, 185–187.

9. Lovejoy, S.; Schertzer, D. Multifractal analysis techniques and the rain and cloud fields from 10^3 to 10^6 m. In *Non-Linear Variability in Geophysics: Scaling and Fractals*; Kluwer Academic Publishers: Norwell, MA, USA, 1991; pp. 111–144.

10. Lovejoy, S.; Schertzer, D. Scale invariance and multifractals in the atmosphere. In *Encyclopedia of the Environment*; Pergamon Press: New York, NY, USA, 1993; pp. 527–532.

11. Nordstrom, K.M.; Gupta, V.K. Scaling statistics in a critical, nonlinear physical model of tropical oceanic rainfall. *Nonlinear Process. Geophys.* **2003**, *10*, 1–13.

12. Nykanen, D.K. Linkages between orographic forcing and the scaling properties of convective rainfall in mountainous regions. *J. Hydrometeorol.* **2008**, *9*, 327–347.

13. Over, T.M.; Gupta, V.K. Statistical analysis of mesoscale rainfall: Dependence of a random cascade generator on large-scale forcing. *J. Appl. Meteorol.* **1994**, *33*, 1526–1542.

14. Perica, S.; Foufoula-Georgiou, E. Linkage of scaling and thermodynamic parameters of rainfall: Results from midlatitude mesoscale convective systems. *J. Geophys. Res.* **1996**, *101*, 7431–7448.

15. Perica, S.; Foufoula-Georgiou, E. Model for multiscale disagreggation of spatial rainfall based on coupling meteorological and scaling descriptions. *J. Geophys. Res.* **1996**, *101*, 26347–26361.

16. Singleton, A.; Toumi, R. Super-Clausius-Clapeyron scaling of rainfall in a model squall line. *Q. J. R. Meteorol. Soc.* **2013**, *139*, 334–339.

17. Yano, J.-I.; Fraedrich, K.; Blender, R. Tropical convective variability as $1/f$ noise. *J. Clim.* **2001**, *14*, 3608–3616.

18. Barker, H.W.; Qu, Z.; Bélair, S.; Leroyer, S.; Milbrandt, J.A.; Vaillancourt, P.A. Scaling properties of observed and simulated satellite visible radiances. *J. Geophys. Res. Atmos.* **2017**, *122*, doi:10.1002/2017JD027146.

19. Morales, J.; Poveda, G. Diurnally driven scaling properties of Amazonian rainfall fields: Fourier spectra and order-q statistical moments. *J. Geophys. Res.* **2009**, *114*, D11104, doi:10.1029/2008JD011281.

20. Over, T. M. Modeling Space-Time Rainfall at the Mesoscale Using Random Cascades. Ph.D. Thesis, University of Colorado, Boulder, CO, USA, 1995; 249p.

21. Gorenburg, I.P.; McLaughlin, D.; Entekhabi, D. Scale-recursive assimilation of precipitation data. *Adv. Water Resour.* **2001**, *24*, 941–953.

22. Bocchiola, D. Use of Scale Recursive Estimation for assimilation of precipitation data from TRMM (PR and TMI) and NEXRAD. *Adv. Water Resour.* **2007**, *30*, 2354–2372.

23. Lovejoy, S.; Schertzer, D.; Allaire, V.C. The remarkable wide range spatial scaling of TRMM precipitation. *Atmos. Res.* **2008**, *90*, 10–32.

24. Gebremichael, M.; Over, T.M.; Krajewski, W.F. Comparison of the Scaling Characteristics of Rainfall Derived from Space-Based and Ground-Based Radar Observations. *J. Hydrometeorol.* **2006**, *7*, 1277–1294.

25. Gebremichael, M.; Krajewski, W.F.; Over, T.M.; Takayabu, Y.N.; Arkin, P.; Katayama, M. Scaling of tropical rainfall as observed by TRMM precipitation radar. *Atmos. Res.* **2008**, *88*, 337–354.

26. Hurtado, A.F.; Poveda, G. Linear and global space-time dependence and Taylor hypotheses for rainfall in the tropical Andes. *J. Geophys. Res.* **2009**, *114*, D10105, doi:10.1029/2008JD0110.

27. Varikoden, H.; Samah, A.A.; Babu, C.A. Spatial and temporal characteristics of rain intensity in the peninsular Malaysia using TRMM rain rate. *J. Hydrol.* **2010**, *387*, 312–319.

28. Poveda, G. Mixed memory, (non) Hurst effect, and maximum entropy of rainfall in the Tropical Andes. *Adv. Water Resour.* **2011**, *34*, 243–256.

29. Venugopal, V.; Sukhatme, J.; Madhyastha, K. Scaling Characteristics of Global Tropical Rainfall. In Proceedings of the European Geosciences Union General Assembly, Vienna, Austria, 27 April–2 May 2014.

30. Salas H.D.; Poveda, G. Scaling of entropy and multi-scaling of the time generalized q-entropy in rainfall and streamflows. *Physica A* **2015**, *423*, 11–26.

31. Poveda, G.; Salas, H.D. Statistical scaling, Shannon entropy and generalized space-time q-entropy of rainfall fields in Tropical South America. *Chaos* **2015**, *25*, 075409.

32. Over, T.M.; Gupta, V.K. A space-time theory of mesoscale rainfall using random cascades. *J. Geophys. Res.* **1996**, *101*, 26319–26331.

33. Gupta, V.K.; Waymire, E.C. A statistical analysis of mesoscale rainfall as a random cascade. *J. Appl. Meteorol.* **1993**, *32*, 251–267.

34. Mandelbrot, B.B. Intermittent turbulence in self-similar cascades: Divergence of high moments and dimension of the carrier. *J. Fluid Mech.* **1974**, *62*, 331–358.

35. Kahane, J.P.; Peyriere, J. Sur certains martingales de Benoit Mandelbrot. *Adv. Math.* **1976**, *22*, 131–145.

36. Mittal, D.P. On continuous solutions of a functional equation. *Metrika* **1976**, *22*, 31–40.

37. Rényi, A. On measures of information and entropy. In Proceedings of the Fourth Berkeley Symposium on Mathematics, Statistics and Probability, Berkeley, CA, USA, 20 June–30 July 1960; pp. 547–561.

38. Havrda, J.H.; Charvát, F. Quantification method of classification processes: Concept of structural α-entropy. *Kybernetika* **1967**, *3*, 30–35.

39. Sharma, B.D.; Taneja, I.J. Entropy of type (α,β) and other generalized measures of Information Theory. *Metrika* **1975**, *22*, 205–215.

40. Tsallis, C. Possible Generalization of Boltzmann-Gibbs Statistics. *J. Stat. Phys.* **1988**, *52*, 479–487.

41. Gell-Mann, M.; Tsallis, C. *Nonextensive Entropy—Interdisciplinary Applications*; Oxford University Press: New York, NY, USA, 2004.

42. Singh, V.P. *Introduction to Tsallis Entropy Theory in Water Engineering*; CRC Press: Boca Raton, FL, USA, 2016.

43. Furuichi, S. Information theoretical properties of Tsallis entropies. *J. Math. Phys.* **2006**, *47*, 023302.

44. Shannon, C.E. A mathematical theory of communication. *Bell Syst. Tech. J.* **1948**, *27*, 379–423.

45. Tsallis, C. *Introduction to Nonextensive Statistical Mechanics: Approaching a Complex World*; Springer: New York, NY, USA, 2009; doi:10.1007/978-0-387-85359-8.

46. Bercher, J.-F. Tsallis distribution as a standard maximum entropy solution with 'tail' constraint. *Phys. Lett. A* **2008**, *59*, 5657.

47. Tsallis, C.; Mendes, R.S.; Plastino, A.R. The role of constraints within generalized nonextensive statistics. *Physica A* **1998**, *261*, 534–554.

48. Plastino, A. Why Tsallis' statistics? *Physica A* **2004**, *344*, 608–613.

49. Rathie, P.N.; Da Silva, S. Shannon, Lévy, and Tsallis: A note. *Appl. Math. Sci.* **2008**, *2*, 1359–1363.

50. Zipf, G.K. *Selective Studies and the Principle of Relative Frequency*; Addison Wesley: Cambridge, UK, 1932.

51. Zipf, G.K. *Human Behavior and the Principle of Least Effort*; Addison Wesley: Cambridge, UK, 1949.

52. Mandelbrot, B.B. Structure formelle des textes et communication. *J. Word* **1953**, *10*, 1–27.

53. Abe, S. Geometry of escort distributions. *Phys. Rev. E* **2003**, *68*, 031101.

54. Frisch, U. *Turbulence: The Legacy of A. N. Kolmogorov*; Cambridge University Press: Cambridge, UK, 1995; 296p.

55. Cifelli, R.; Petersen, W.A.; Carey, L.D.; Rutledge, S.A.; da Silva-Dias, M.A.F. Radar observations of the kinematics, microphysical, and precipitation characteristics of two MCSs in TRMM LBA. *J. Geophys. Res.* **2002**, *107*, 8077, doi:10.1029/2000JD000264.

56. Petersen, W.; Nesbitt, S.W.; Blakeslee, R.J.; Cifelli, R.; Hein, P.; Rutledge, S.A. TRMM observations of intraseasonal variability in convective regimes over the Amazon. *J. Clim.* **2002**, *14*, 1278–1294.

57. Carey, L.D.; Cifelli, R.; Petersen, W.A.; Rutledge, S.A. Characteristics of Amazonian rain measured during TRMMLBA. In Proceedings of the 30th International Conference on Radar Meteorology, Munich, Germany, 18–24 July 2001; pp. 682–684.

58. Laurent, H.; Machado, L.A.; Morales, C.A.; Durieux, L. Characteristics of the Amazonian mesoscale convective systems observed from satellite and radar during the WETAM/LBA experiment. *J. Geophys. Res.* **2002**, *107*, 8054, doi:10.1029/2001JD000337.

59. Anagnostou, E.N.; Morales, C.A. Rainfall estimation from TOGA radar observations during LBA field campaign. *J. Geophys. Res.* **2002**, *107*, 8068, doi:10.1029/2001JD000377.

60. Sivakumar, B. Nonlinear dynamics and chaos in hydrologic systems: Latest developments and a look forward. *Stoch. Environ. Res. Risk Assess.* **2009**, *23*, 1027–1036, doi:10.1007/s00477-008-0265-z.

61. Gires, A.; Tchiguirinskaia, I.; Schertzer, D.; Lovejoy, S. Development and analysis of a simple model to represent the zero rainfall in a universal multifractal framework. *Nonlinear Process. Geophys.* **2013**, *20*, 343–356.

62. Peters, O.; Christensen, K. Rain: Relaxations in the sky. *Phys. Rev. E* **2002**, *66*, 036120.

63. Machado, L.A.T.; Laurent, H.; Lima, A.A. Diurnal march of the convection observed during TRMM-WETAMC/LBA. *J. Geophys. Res.* **2002**, *107*, 8064, doi:10.1029/2001JD000338.

64. Silva Dias, M.A.F.; Rutledge, S.; Kabat, P.; Silva Dias, P.L.; Nobre, C.; Fisch, G.; Dolman, A.J.; Zipser, E.; Garstang, M.; Manzi, A.O.; et al. Cloud and rain processes in a biosphere-atmosphere interaction context in the Amazon region. *J. Geophys. Res.* **2002**, *107*, 8072, doi:10.1029/2001JD000335.

65. Kalnay, E.; Kanamitsu, M.; Kistler, R.; Collins, W.; Deaven, D.; Gandin, L.; Iredell, M.; Saha, S.; White, G.; Woollen, J.; et al. The NCEP/NCAR 40-year Reanalysis Project. *Bull. Am. Meteorol. Soc.* **1996**, *77*, 437–471.

66. Zhang, J.; Wu, Y. *k*-Sample tests based on the likelihood ratio. *Comput. Stat. Data Anal.* **2007**, *51*, 4682–4691.

67. Gong, W.; Yang, D.; Gupta, H.V.; Nearing, G. Estimating information entropy for hydrological data: One-dimensional case. *Water Resour. Res.* **2014**, *50*, doi:10.1002/2014WR015874.

68. Cover, T.M.; Thomas, J.A. *Elements of Information Theory*; John Wiley & Sons, Inc.: Hoboken, NJ, USA, 2006.

69. Sturges, H. The choice of a class-interval. *J. Am. Stat. Assoc.* **1926**, *21*, 65–66.

70. Dixon, W.J.; Kronmal, R.A. The choice of origin and scale of graphs. *J. Assoc. Comput. Mach.* **1965**, *12*, 259–261.

71. Scott, D.W. On optimal and data-based histograms. *Biometrika* **1979**, *66*, 605–610.

72. Freedman, D.; Diaconis, P. On the Histogram as a Density Estimator: L_2 Theory, Z. Wahrscheinlichkeitstheorie verw. *Gebiete* **1981**, *57*, 453–476.

73. Knuth, K.H. Optimal data-based binning for histograms. *arXiv* **2006**, arXiv:physics/0605197.

74. Gencaga, D.; Knuth, K.H.; Rossow, W.B. A recipe for the estimation of information flow in a dynamical system. *Entropy* **2015**, *17*, 438–470, doi:10.3390/e17010438.

75. Shimazaki, H.; Shinomoto, S. A method for selecting the bin size of a time histogram. *Neural Comput.* **2007**, *19*, 1503–1527.

76. Shimazaki, H.; Shinomoto, S. Kernel bandwidth optimization in spike rate estimation. *J. Comput. Neurosci.* **2010**, *29*, 171–182.

77. Beirlant, J.; Dudewicz, E.J.; Györfi, L.; Van der Meulen, E.C. Nonparametric entropy estimation: An overview. *Int. J. Math. Stat. Sci.* **1997**, *6*, 17–39.

78. Pires, C.A.L.; Perdigão, R.A.P. Minimum Mutual Information and Non-Gaussianity through the Maximum Entropy Method: Estimation from Finite Samples. *Entropy* **2013**, *15*, 721–752.

79. Trendafilov, N.; Kleinsteuber, M.; Zou, H. Sparse matrices in data analysis. *Comput. Stat.* **2014**, *29*, 403–405.

80. Liu, D.; Wang, D.; Wang, Y.; Wu, J.; Singh, V.P.; Zeng, X.; Wang, L.; Chen, Y.; Chen, X.; Zhang, L.; et al. Entropy of hydrological systems under small samples: Uncertainty and variability. *J. Hydrol.* **2016**, *532*, 163–176, doi:10.1016/j.jhydrol.2015.11.019.

81. Mandelbrot, B.B. *Multifractals and 1/f Noise. Wild Self-Affinity in Physics (1963–1976)*; Springer: New York, NY, USA, 1998; p. 442.

82. Fraedrich, K.; Larnder, C. Scaling regimes of composite rainfall time series. *Tellus* **1993**, *45*, 289–298.

83. Verrier, S.; Mallet, C.; Barthés, L. Multiscaling properties of rain in the time domain, taking into account rain support biases. *J. Geophys. Res.* **2011**, *116*, doi:10.1029/2011JD015719.

84. Mascaro, G.; Deidda, R.; Hellies, M. On the nature of rainfall intermittency as revealed by different metrics and sampling approaches. *Hydrol. Earth Syst. Sci.* **2013**, *17*, 355.

85. Molini, A.; Katul, G.G.; Porporato, A. Revisiting rainfall clustrering and intermittency across different climatic regimes. *Water Resour. Res.* **2009**, *45*, doi:10.1029/2008WR007352.

86. Bickel, D.R. Generalized entropy and multifractality of time-series: Relationship between order and intermittency. *Chaos Solitons Fractals* **2002**, *13*, 491–497.

87. Shiner, J.S.; Davison, M. Simple measure of complexity. *Phys. Rev. E* **1999**, *59*, 1459–1464.

88. Badin, G.; Domeisen, D.I.V. Nonlinear stratospheric variability: Multifractal detrended fluctuation analysis and singularity spectra. *Proc. R. Soc. A Math. Phys. Eng. Sci.* **2016**, *472*, 20150864, doi:10.1098/rspa.2015.0864.

89. Bickel, D.R. Simple estimation of intermittency in multifractal stochastic processes: Biomedical applications. *Phys. Lett. A* **1999**, *262*, 251–256.

90. Carbone, A. Algorithm to estimate the Hurst exponent of high-dimensional fractals. *Phys. Rev. E* **2007**, *76*, 056703.

91. Grinstead, C.M.; Snell, J.L. *Introduction to Probability*, 2nd ed.; American Mathematical Society: Providence, RI, USA, 1997.

92. Paninski, L. Estimation of Entropy and Mutual Information. *Neural Comput.* **2003**, *15*, 1191–1253.

93. Rodríguez-Iturbe, I.; Power, B.F.; Valdes, J.B. Rectangular pulses point process models for rainfall: Analysis of empirical data. *J. Geophys. Res. Atmos.* **1987**, *92*, 9645–9656.

94. Cowpertwait, P.; Isham, V.; Onof, C. Point process models of rainfall: Developments for fine-scale structure. *Proc. R. Soc. Lond. A Math. Phys. Eng. Sci.* **2007**, *463*, 2569–2587.

95. Olsson, J.; Burlando, P. Reproduction of temporal scaling by a rectangular pulses rainfall model. *Hydrol. Process.* **2002**, *16*, 611–630.

96. Boon, J.P.; Tsallis, C. Special issue overview nonextensive statistical mechanics: New trends, new perspectives. *Europhys. News* **2005**, *36*, 185–186.

97. Robinson, P.M. Consisten nonparametric entropy-based testing. *Rev. Econ. Stud.* **1991**, *58*, 437–453.

11

Entropy-Based Parameter Estimation for the Four-Parameter Exponential Gamma Distribution

Songbai Song *, Xiaoyan Song * and Yan Kang *

College of Water Resources and Architectural Engineering, Northwest A & F University, Yangling 712100, China
* Correspondence: ssb6533@nwsuaf.edu.cn (S.S.); songxiaoyan107@mails.ucas.ac.cn (X.S.); kangyan@nwsuaf.edu.cn (Y.K.)

Academic Editors: Huijuan Cui, Bellie Sivakumar, Vijay P. Singh and Kevin H. Knuth

Abstract: Two methods based on the principle of maximum entropy (POME), the ordinary entropy method (ENT) and the parameter space expansion method (PSEM), are developed for estimating the parameters of a four-parameter exponential gamma distribution. Using six data sets for annual precipitation at the Weihe River basin in China, the PSEM was applied for estimating parameters for the four-parameter exponential gamma distribution and was compared to the methods of moments (MOM) and of maximum likelihood estimation (MLE). It is shown that PSEM enables the four-parameter exponential distribution to fit the data well, and can further improve the estimation.

Keywords: four-parameter exponential gamma distribution; principle of maximum entropy; precipitation frequency analysis; methods of moments; maximum likelihood estimation

1. Introduction

Hydrological frequency analysis is a statistical prediction method that consists of studying past events that are characteristic of a particular hydrological process in order to determine the probabilities of the occurrence of these events in the future [1,2]. It is widely used for planning, design, and management of water resource systems. The probability distributions containing four or more parameters may exhibit some useful properties [3]: (1) versatility and (2) ability to represent data from mixed populations. Among these distributions, some popular distributions are Wakeby, two-component lognormal, two-component extreme value distributions, and the four-parameter kappa distribution. Since the pioneering stream flow records frequency analysis of Herschel and Freeman during the period from 1880 to 1890, hydrological frequency analysis has undergone extensive further development. There are a multitude of methods for estimating parameters of hydrologic frequency distributions. Some of the popular methods include [3,4]: (1) the method of moments; (2) the method of probability weighted moments; (3) the method of mixed moments; (4) L-moments; (5) the maximum likelihood estimation; (6) the least square method; and (7) the entropy-based parameter estimation method.

Among the above parameter estimation methods, entropy, which is a measure of uncertainty of random variables, has attracted much attention and has been used for a variety of applications in hydrology [5–23]. For example, an entropy-based derivation of daily rainfall probability distribution [24], the Burrr XII-Singh-Maddala (BSM) distribution function derived from the maximum entropy principle using the Boltzmann-Shannon entropy with some constraints [25]. *"Entropy-Based Parameter Estimation in Hydrology"* is the first book focusing on parameter estimation using entropy for a number of distributions frequently used in hydrology [3], including the uniform distribution, exponential distribution, normal distribution, two-parameter lognormal distribution, three-parameter lognormal distribution, extreme value type I distribution, log-extreme value type I distribution, extreme value type III distribution, generalized extreme value distribution, Weibull distribution, gamma distribution, Pearson type III distribution, log-Pearson type III distribution, beta distribution, two-parameter log-logistic distribution, three-parameter log-logistic distribution, two-parameter Pareto distribution, two-parameter generalized Pareto distribution, three-parameter generalized Pareto distribution and two-component extreme value distribution. Recently, two entropy-based methods, called the ordinary entropy method (ENT) and the parameter space expansion method (PSEM), that are both based on the principle of maximum entropy (POME) have been applied for estimating the parameters of the extended Burr XII distribution and the four-parameter kappa distribution [5,16]. The results of the estimation show that the entropy method enables these two distributions to fit the data better than the other estimation methods. In the above method of entropy-based parameter estimation of a distribution, the distribution parameters are expressed in terms of the given constraints, and then the method can provide a way to derive the distribution from the specified constraints. The general procedure for the ENT for a hydrologic frequency distribution involves the following steps [3]: (1) define the given information in terms of the constraints; (2) maximize the entropy subject to the given information; and (3) relate the parameters to the given information. The PSEM employs an enlarged parameter space and maximizes the entropy subject to the parameters and the Lagrange multipliers [3]. The parameters of the distribution can be estimated by the maximization of the entropy function.

The Pearson III distribution is recommended as a standard distribution to fit hydrological data in China. In addition, generalized Pareto distribution (GPD), generalized extreme value (GEV) and three-parameter Burr type XII distribution also have been applied flood frequency analysis [26].

Inspired in large part by the two-parameter gamma distribution, a four-parameter exponential gamma distribution has been developed to apply in many areas, such as wind and flood frequency in Yellow River basin, Yangtse River basin, Aumer Basin and Liaohe River basin of China [4]. Depending on the parameter values, the four-parameter exponential gamma distribution can be turned into a Pearson type III distribution, Weibull distribution, Maxwell distribution, Kritsky and Menkel distribution, Chi-square distribution, Poisson distribution, half-normal distribution and half-Laplace distribution. The properties of the four-parameter exponential gamma and relations between this distribution and other distributions have been investigated [4]. These investigations suggest that the four-parameter exponential gamma distribution may have a potential in hydrology. Despite the advances mentioned above, the entropy-based parameter estimation for the four-parameter exponential gamma distribution has received comparatively little attention from the hydrologic community.

The objective of this paper is to apply two entropy-based methods that both use the POME for the estimation of the parameters of the four-parameter exponential gamma distribution; compute the annual precipitation quantiles using this distribution for different return periods; and compare these parameters with those estimated when the methods of moments (MOM) and maximum likelihood estimation (MLE) were employed for parameter estimation.

2. Four-Parameter Exponential Gamma Distribution

2.1. Probability Density Function and Cumulative Distribution Function

The probability density function (PDF) of the four-parameter exponential gamma distribution can be expressed as [4]:

$$f(x) = \frac{\beta^\alpha}{b\Gamma(\alpha)}(x-\delta)^{\frac{\alpha}{b}-1}e^{-\beta(x-\delta)^{\frac{1}{b}}}; \delta \leq x < \infty \tag{1}$$

where α β, δ and b are, respectively, the shape, scale, location and transformation parameter.

Depending on the values of the four parameters α β, δ and b, Equation (1) turns into the following special cases:

(1) If $b = 1$, then Equation (1) becomes the Pearson type III distribution:

$$f(x) = \frac{\beta^\alpha}{\Gamma(\alpha)}(x - \delta)^{\alpha-1}e^{-\beta(x-\delta)} \tag{2}$$

If $\delta = 0$, Equation (2) becomes a gamma distribution:

$$f(x) = \frac{\beta^\alpha}{\Gamma(\alpha)}x^{\alpha-1}e^{-\beta x} \tag{3}$$

(2) If $\delta = 0$, $b = \frac{1}{m}$, $\alpha = \frac{\alpha}{m}$ and $\beta = \frac{1}{d}$, then Equation (1) reduces to the Weibull distribution:

$$f(x) = \frac{m}{d^\alpha}(x - \delta)^{m-1}e^{-\frac{1}{d}(x-\delta)^m} \tag{4}$$

(3) If $\delta = 0$, $b = \frac{1}{m}$, $\alpha = \frac{\alpha}{m}$ and $\beta = \frac{1}{d}$, then Equation (1) becomes the three-parameter Weibull distribution:

$$f(x) = \frac{m}{d^{\frac{\alpha}{m}}\Gamma(\frac{\alpha}{m})}x^{\alpha-1}e^{-\frac{x^m}{d}} \tag{5}$$

(4) If $\delta = 0$ and $\beta = \frac{\alpha}{\alpha^{\frac{1}{b}}}$, then Equation (1) reduces to the Kritsky and Menkel distribution:

$$f(x) = \frac{\alpha^\alpha}{\alpha^{\frac{\alpha}{b}}b\Gamma(\alpha)}x^{\frac{\alpha}{b}-1}e^{-\alpha(\frac{x}{\alpha})^{\frac{1}{b}}} \tag{6}$$

(5) If $\delta = 0$, $b = 1$, $\alpha = \frac{n}{2}$ and $\beta = \frac{1}{2}$, then Equation (1) becomes the Chi-square distribution:

$$f(x) = \frac{1}{2^{\frac{n}{2}}\Gamma(\frac{n}{2})}x^{\frac{n}{2}-1}e^{-\frac{x}{2}} \tag{7}$$

(6) If $\delta = 0$, $b = 1$, $\alpha = k + 1$ and $\beta = 1$, then Equation (1) reduces to the Poisson distribution

$$f(x) = \frac{x^k}{k!}e^{-x} \tag{8}$$

(7) If $\delta = a$, $b = \alpha = \frac{1}{2}$ and $\beta = \frac{1}{2\sigma^2}$, then Equation (1) becomes the half-normal distribution:

$$f(x) = \frac{2}{\sigma\sqrt{2\pi}}e^{-\frac{1}{2\sigma^2}(x-a)^2} \tag{9}$$

(8) If $\delta = 0$, $b = \frac{1}{2}$, $\alpha = \frac{3}{2}$ and $\beta = \frac{1}{a^2}$, then Equation (1) becomes the half-normal distribution:

$$f(x) = \frac{4}{a^3\sqrt{\pi}}x^2e^{-\frac{x^2}{a^2}} \tag{10}$$

(9) If $b = 1$, $\alpha = 1$ and $\beta = \frac{1}{\lambda}$, then Equation (1) reduces to the half-Laplace distribution:

$$f(x) = \frac{1}{2\lambda}e^{-\frac{x-\delta}{\lambda}} \tag{11}$$

The cumulative distribution function (CDF) of the four-parameter exponential gamma distribution can be expressed as:

$$P = F(X \leq x_p) = \int_\delta^{x_p} f(x)dx = \int_\delta^{x_p} \frac{\beta^\alpha}{b\Gamma(\alpha)}(x - \delta)^{\frac{\alpha}{b}-1}e^{-\beta(x-\delta)^{\frac{1}{b}}}dx \tag{12}$$

Let $t = \beta(x - \delta)^{\frac{1}{b}}$, then $x = \delta + \frac{1}{\beta^b} t^b$, $x - \delta = \frac{1}{\beta^b} t^b$ and $dx = \frac{b}{\beta^b} t^{b-1} dt$. Substitution of the above quantities in Equation (12) yields [4]:

$$P = \int_0^{t_p} \frac{1}{\Gamma(\alpha)} t^{\alpha-1} e^{-t} dt \tag{13}$$

where $t_p = \beta(x_p - \delta)^{\frac{1}{b}}$ and can be determined by the incomplete gamma function.

2.2. Quantile Corresponding to the Probability of Exceedance

The quantile corresponding to the probability of exceedance p, x_p, is obtained by Equation (14) or Equation (15):

$$x_p = \delta + \frac{1}{\beta^b} t_p^b \tag{14}$$

$$x_p = \bar{x}(1 + \Phi_p C_v) \tag{15}$$

here, $\bar{x}$ and C_v are the mean and coefficient of the variation of a sample, respectively, and Φ_p is the frequency factor corresponding to x_p. Given the expectation and variance of the population, the frequency factorr Φ_p is given by [4]:

$$\Phi_p = \frac{x_p - \mu'_1}{\sqrt{\mu_2}} \Phi_p = \frac{t_p^b \Gamma(\alpha) - \Gamma(\alpha+b)}{\sqrt{\Gamma(\alpha)\Gamma(\alpha+2b) - \Gamma^2(\alpha+b)}} \tag{16}$$

where μ'_1 and μ_2 are the expectation and variance of the population.

If $b = 10$, the frequency factors of the four-parameter exponential gamma distribution, Φ_p, are very close to that of the log-normal distribution (Table 1). If $C_s = 1.1395$, the Φ_p values are very close to that of the Gumbel distribution (Table 2).

Table 1. Frequency factors of the four-parameter exponential gamma distribution and log-normal distribution ($b = 10$).

C_s	Distribution	P (%)							
		0.01	0.1	1	5	20	50	90	99
0.2	Four-parameter exponential gamma	4.17	3.39	2.47	1.70	0.830	−0.033	−1.26	−2.18
	log-normal	4.17	3.39	2.47	1.70	0.830	−0.033	−1.26	−2.18
2.0	Four-parameter exponential gamma	9.44	6.23	3.53	1.90	0.614	−0.240	−0.963	−1.26
	log-normal	9.51	6.24	3.52	1.89	0.614	−0.240	−0.967	−1.28

Table 2. Frequency factors of the four-parameter exponential gamma distribution and Gumbel distribution under $C_s = 1.1395$.

Distribution	P (%)							
	0.01	0.1	1	5	20	50	90	99
Four-parameter exponential gamma	6.80	4.92	3.12	1.87	0.728	−0.166	−1.10	−1.61
log-normal	6.80	4.94	3.14	1.87	0.728	−0.164	−1.10	−1.64

2.3. Cumulants and Moments

The first three cumulants of the four-parameter exponential gamma distribution are expressed as [4]:

$$k_1 = \delta + \frac{\Gamma(b+\alpha)}{\beta^b \Gamma(\alpha)} \tag{17}$$

$$k_2 = \frac{\Gamma(\alpha)\Gamma(\alpha + 2b) - \Gamma^2(\alpha + b)}{\beta^{2b}\Gamma^2(\alpha)} \tag{18}$$

$$k_3 = \frac{\Gamma^2(\alpha)\Gamma(\alpha + 3b) - 3\Gamma(\alpha)\Gamma(\alpha + b)\Gamma(\alpha + 2b) + 2\Gamma^3(\alpha + b)}{\beta^{3b}\Gamma^3(\alpha)} \tag{19}$$

Using the relations between moments and cumulants and Equations (17)–(19), the expression for the first four moments of the four-parameter exponential gamma distribution are given below:

$$v_1 = E(X) = \delta + \frac{\Gamma(\alpha + b)}{\beta^b\Gamma(\alpha)} \tag{20}$$

$$v_2 = \frac{\Gamma(\alpha + 2b)}{\beta^{2b}\Gamma(\alpha)} + 2\delta\frac{\Gamma(\alpha + b)}{\beta^b\Gamma(\alpha)} + \delta^2 \tag{21}$$

$$v_3 = \frac{\Gamma(\alpha + 3b)}{\beta^{3b}\Gamma(\alpha)} + 3\delta\frac{\Gamma(2b + \alpha)}{\beta^{2b}\Gamma(\alpha)} + 3\delta^2\frac{\Gamma(b + \alpha)}{\beta^b\Gamma(\alpha)} + \delta^3 \tag{22}$$

$$v_4 = \frac{\Gamma(\alpha + 4b)}{\beta^{4b}\Gamma(\alpha)} + 4\delta\frac{\Gamma(\alpha + 3b)}{\beta^{3b}\Gamma(\alpha)} + 6\delta^2\frac{\Gamma(\alpha + 2b)}{\beta^{2b}\Gamma(\alpha)} + 4\delta^3\frac{\Gamma(\alpha + b)}{\beta^b\Gamma(\alpha)} + \delta^4 \tag{23}$$

In next sections, we use two methods of parameter estimation, ENT and PSEM, to derive the parameters estimation expression of the four-parameter exponential gamma distribution.

3. Ordinary Entropy Method

For ENT, three steps are involved in the estimation of the parameters of a probability distribution: (1) specification of appropriate constraints, (2) derivation of the entropy function of the distribution, and (3) derivation of the relations between parameters and constraints [3,16].

3.1. Specification of Constraints

Taking the natural logarithm of Equation (1), we obtain:

$$\ln f(x) = \alpha \ln \beta - \ln b - \ln \Gamma(\alpha) + \left(\frac{\alpha}{b} - 1\right)\ln(x - \delta) - \beta(x - \delta)^{\frac{1}{b}} \tag{24}$$

Multiplying Equation (24) by $[-f(x)]$ and integrating from δ to ∞, we obtain the entropy function:

$$S = -\int_\delta^\infty f(x)\ln f(x)dx = -\alpha \ln \beta + \ln b + \ln \Gamma(\alpha) - \left(\frac{\alpha}{b} - 1\right)E[\ln(x - \delta)] + \beta \cdot E\left[(x - \delta)^{\frac{1}{b}}\right] \tag{25}$$

To maximize S in Equation (25), the following constraints for Equation (25) should be satisfied

$$\int_\delta^\infty f(x)dx = 1 \tag{26}$$

$$\int_\delta^\infty \ln(x - \delta)f(x)dx = E[\ln(x - \delta)] \tag{27}$$

$$\int_\delta^\infty \ln(x - \delta)^{\frac{1}{b}}f(x)dx = E\left[\ln(x - \delta)^{\frac{1}{b}}\right] \tag{28}$$

$$\int_\delta^\infty (x - \delta)^{\frac{1}{b}}f(x)dx = E\left[(x - \delta)^{\frac{1}{b}}\right] \tag{29}$$

3.2. Construction of Partition Function and Zeroth Lagrange Multiplier

The least-biased pdf, $f(x)$, consistent with Equations (26) to (29) and corresponding to the POME takes the form:

$$f(x) = \exp\left[-\lambda_0 - \lambda_1 \ln(x - \delta) - \lambda_2 \ln(x - \delta)^{\frac{1}{b}} - \lambda_3 (x - \delta)^{\frac{1}{b}}\right] \tag{30}$$

where λ_0, λ_1, λ_2 and λ_3 are Lagrange multipliers. Substitution of Equation (30) in Equation (26) yields:

$$\int_\delta^\infty \exp\left[-\lambda_0 - \lambda_1 \ln(x - \delta) - \lambda_2 \ln(x - \delta)^{\frac{1}{b}} - \lambda_3 (x - \delta)^{\frac{1}{b}}\right] dx = 1 \tag{31}$$

The argument of the exponential function on the left side of Equation (31) has two parts: zeroth Lagrange multiplier without the random variable and four Lagrange multipliers with the random variable. The zeroth Lagrange multiplier part is separated out and is expressed as:

$$\exp(\lambda_0) = \int_\delta^\infty \exp\left[-\lambda_1 \ln(x - \delta) - \lambda_2 \ln(x - \delta)^{\frac{1}{b}} - \lambda_3 (x - \delta)^{\frac{1}{b}}\right] dx \tag{32}$$

To calculate the above integral, let $y = \lambda_3 (x - \delta)^{\frac{1}{b}}$, then $x = \left(\frac{y}{\lambda_3}\right)^b + \delta$, $dx = \frac{b}{\lambda_3^b} y^{b-1} dy$. Substituting the above quantities in Equation (32), we obtain:

$$\begin{aligned}
\exp(\lambda_0) &= \int_0^\infty \left(\frac{y}{\lambda_3}\right)^{-(\lambda_1 + \frac{\lambda_2}{b})} \cdot \exp(-y) \frac{b}{\lambda_3^b} y^{b-1} dy \\
&= \frac{b}{\lambda_3^{b-(b\lambda_1 + \lambda_2)}} \int_0^\infty y^{b-(b\lambda_1 + \lambda_2)-1} \cdot e^{-y} dy = \frac{b}{\lambda_3^{b-(b\lambda_1 + \lambda_2)}} \Gamma[b - (b\lambda_1 + \lambda_2)]
\end{aligned} \tag{33}$$

Taking the logarithm of Equation (33) results in the zeroth Lagrange λ_0 multiplier as a function of Lagrange multipliers λ_1, λ_2 and λ_3, with the expression given as:

$$\lambda_0 = \ln b - [b - (b\lambda_1 + \lambda_2)] \ln \lambda_3 + \ln \Gamma[b - (b\lambda_1 + \lambda_2)] \tag{34}$$

$$\lambda_0 = \ln \int_\delta^\infty \exp\left[-\lambda_1 \ln(x - \delta) - \lambda_2 \ln(x - \delta)^{\frac{1}{b}} - \lambda_3 (x - \delta)^{\frac{1}{b}}\right] dx \tag{35}$$

3.3. Relation between Lagrange Multiplier and Constraints

Differentiating Equation (35) with λ_1, λ_2 and λ_3, we obtain the derivatives of λ_0 with respect to λ_1, λ_2 and λ_3, the detailed derivations are given in Appendix B:

$$\frac{\partial \lambda_0}{\partial \lambda_1} = -E[\ln(x - \delta)] \tag{36}$$

$$\frac{\partial \lambda_0}{\partial \lambda_2} = -E\left[\ln(x - \delta)^{\frac{1}{b}}\right] \tag{37}$$

$$\frac{\partial \lambda_0}{\partial \lambda_3} = -E\left[(x - \delta)^{\frac{1}{b}}\right] \tag{38}$$

Furthermore, we can write:

$$\frac{\partial^2 \lambda_0}{\partial \lambda_3^2} = E\left[(x - \delta)^{\frac{1}{b}}\right]^2 - \left\{E\left[(x - \delta)^{\frac{1}{b}}\right]\right\}^2 \tag{39}$$

Additionally, differentiating Equation (34) with λ_1, λ_2 and λ_3, we obtain:

$$\frac{\partial \lambda_0}{\partial \lambda_1} = b \ln \lambda_3 - b\psi[b - (b\lambda_1 + \lambda_2)] \tag{40}$$

$$\frac{\partial \lambda_0}{\partial \lambda_2} = \ln \lambda_3 - \psi[b - (b\lambda_1 + \lambda_2)] \tag{41}$$

$$\frac{\partial \lambda_0}{\partial \lambda_3} = -\frac{b - (b\lambda_1 + \lambda_2)}{\lambda_3} \tag{42}$$

$$\frac{\partial^2 \lambda_0}{\partial \lambda_3^2} = \frac{b - (b\lambda_1 + \lambda_2)}{\lambda_3^2} \tag{43}$$

Equating Equations (36) and (40), we obtain:

$$- E[\ln(x - \delta)] = b \ln \lambda_3 - b\psi[b - (b\lambda_1 + \lambda_2)] \tag{44}$$

Equating Equations (37) and (41), we obtain:

$$- E\left[\ln(x - \delta)^{\frac{1}{b}}\right] = \ln \lambda_3 - \psi[b - (b\lambda_1 + \lambda_2)] \tag{45}$$

Equating Equations (38) and (42), we obtain:

$$- E\left[(x - \delta)^{\frac{1}{b}}\right] = -\frac{b - (b\lambda_1 + \lambda_2)}{\lambda_3} \tag{46}$$

Equating Equations (39) and (43), we obtain:

$$E\left[(x - \delta)^{\frac{1}{b}}\right]^2 - \left\{E\left[(x - \delta)^{\frac{1}{b}}\right]\right\}^2 = \frac{b - (b\lambda_1 + \lambda_2)}{\lambda_3^2} \tag{47}$$

3.4. Relation between Lagrange Multiplier and Parameters

Introduction of Equation (34) in Equation (30) produces:

$$
\begin{aligned}
f(x) &= \exp\left\{-\ln b + [b - (b\lambda_1 + \lambda_2)]\ln \lambda_3 - \ln \Gamma[b - (b\lambda_1 + \lambda_2)] - \lambda_1 \ln(x - \delta) - \lambda_2 \ln(x - \delta)^{\frac{1}{b}} - \lambda_3(x - \delta)^{\frac{1}{b}}\right\} \\
&= \exp(\ln b^{-1}) \cdot \exp\left[\ln \lambda_3^{b - (b\lambda_1 + \lambda_2)}\right] \cdot \exp\left\{\ln \frac{1}{\Gamma[b - (b\lambda_1 + \lambda_2)]}\right\} \\
&\quad \cdot \exp\left[\ln(x - \delta)^{-\lambda_1}\right] \cdot \exp\left[\ln(x - \delta)^{-\frac{\lambda_2}{b}}\right] \cdot \exp\left[-\lambda_3(x - \delta)^{\frac{1}{b}}\right] \\
&= \frac{1}{b} \cdot \lambda_3^{b - (b\lambda_1 + \lambda_2)} \cdot \frac{1}{\Gamma[b - (b\lambda_1 + \lambda_2)]} \cdot (x - \delta)^{-\lambda_1} \cdot (x - \delta)^{-\frac{\lambda_2}{b}} \cdot \exp\left[-\lambda_3(x - \delta)^{\frac{1}{b}}\right] \\
&= \frac{\lambda_3^{b - (b\lambda_1 + \lambda_2)}}{b \cdot \Gamma[b - (b\lambda_1 + \lambda_2)]} \cdot (x - \delta)^{-(\lambda_1 + \frac{\lambda_2}{b})} \cdot e^{-\lambda_3(x - \delta)^{\frac{1}{b}}}
\end{aligned}
\tag{48}
$$

a comparison of Equation (48) with Equation (1) shows that:

$$\lambda_3 = \beta \tag{49}$$

$$b\lambda_1 + \lambda_2 = b - \alpha \tag{50}$$

3.5. Relation between Parameters and Constraints

The four-parameter exponential gamma distribution has four parameters α, β, δ and b that are related to the Lagrange multipliers by Equations (49) and (50). In turn, these parameters are related to the known constrains by Equations (44)–(47). Eliminating the Lagrange multipliers among these four sets of Equations, we can obtain the following Equations:

$$
\begin{cases}
b \ln \beta - b\psi(\alpha) + E[\ln(x - \delta)] = 0 \\
\ln \beta - \psi(\alpha) + E\left[\ln(x - \delta)^{\frac{1}{b}}\right] = 0 \\
E\left[(x - \delta)^{\frac{1}{b}}\right] - \frac{\alpha}{\beta} = 0 \\
E\left[(x - \delta)^{\frac{1}{b}}\right]^2 - \left\{E\left[(x - \delta)^{\frac{1}{b}}\right]\right\}^2 - \frac{\alpha}{\beta^2} = 0
\end{cases}
\tag{51}
$$

4. Parameter Space Expansion Method

4.1. Specification of Constraints

Following reference [3], the constraints consistent with the POME method and appropriate for the four-parameter exponential gamma distribution are specified by Equations (26), (27) and (29).

4.2. Construction of Zeroth Lagrange Multiplier

The least-biased pdf corresponding to POME and consistent with Equations (26), (27) and (29) takes the form:

$$f(x) = \exp\left[-\lambda_0 - \lambda_1 \ln(x - \delta) - \lambda_2 (x - \delta)^{\frac{1}{b}}\right] \tag{52}$$

where λ_0, λ_1 and λ_2 are Lagrange multipliers. Substitution of Equation (52) into Equation (26) yields

$$\int_\delta^\infty \exp\left[-\lambda_0 - \lambda_1 \ln(x - \delta) - \lambda_2 (x - \delta)^{\frac{1}{b}}\right] dx = 1 \tag{53}$$

$$\exp(\lambda_0) = \int_\delta^\infty (x - \delta)^{-\lambda_1} \exp\left[-\lambda_2 (x - \delta)^{\frac{1}{b}}\right] dx \tag{54}$$

Let $y = \lambda_2 (x - \delta)^{\frac{1}{b}}$, then $x = \left(\frac{y}{\lambda_2}\right)^b + \delta$, $dx = \frac{b}{\lambda_2^b} y^{b-1} dy$. Substituting the above quantities in Equation (54) and changing the limits of integration, we obtain:

$$\begin{aligned}
\exp(\lambda_0) &= \int_0^\infty \left(\frac{y}{\lambda_2}\right)^{-b\lambda_1} \cdot \exp(-y) \frac{b}{\lambda_2^b} y^{b-1} dy \\
&= \frac{b}{\lambda_2^{b-b\lambda_1}} \int_0^\infty y^{b-b\lambda_1-1} \cdot e^{-y} dy = \frac{b}{\lambda_2^{b-b\lambda_1}} \Gamma(b - b\lambda_1)
\end{aligned} \tag{55}$$

This yields the zeroth Lagrange multiplier:

$$\lambda_0 = \ln b - (b - b\lambda_1) \ln \lambda_2 + \ln \Gamma(b - b\lambda_1) \tag{56}$$

4.3. Derivation of Entropy Function

Introduction of Equation (56) into Equation (52) yields:

$$\begin{aligned}
f(x) &= \exp\left[-\ln b + (b - b\lambda_1) \ln \lambda_2 - \ln \Gamma(b - b\lambda_1) - \lambda_1 \ln(x - \delta) - \lambda_2 (x - \delta)^{\frac{1}{b}}\right] \\
&= \frac{\lambda_2^{b-b\lambda_1}}{b\Gamma(b-b\lambda_1)} (x - \delta)^{-\lambda_1} \exp\left[-\lambda_2 (x - \delta)^{\frac{1}{b}}\right]
\end{aligned} \tag{57}$$

a comparison of Equation (57) with Equation (1) shows that:

$$\lambda_1 = 1 - \frac{\alpha}{b} \tag{58}$$

$$\lambda_2 = \beta \tag{59}$$

taking the logarithm of Equation (57) yields:

$$\ln f(x) = (b - b\lambda_1) \ln \lambda_2 - \ln b - \ln \Gamma(b - b\lambda_1) - \lambda_1 \ln(x - \delta) - \lambda_2 (x - \delta)^{\frac{1}{b}} \tag{60}$$

then, making use of Equation (60), the entropy function can be written as:

$$S = -\int_\delta^\infty f(x) \ln f(x) dx$$

$$= \int_\delta^\infty \left[-(b - b\lambda_1) \ln \lambda_2 + \ln b + \ln \Gamma(b - b\lambda_1) + \lambda_1 \ln(x - \delta) + \lambda_2 (x - \delta)^{\frac{1}{b}} \right] f(x) dx \qquad (61)$$

$$= -(b - b\lambda_1) \ln \lambda_2 + \ln b + \ln \Gamma(b - b\lambda_1) + \lambda_1 E[\ln(x - \delta)] + \lambda_2 E\left[(x - \delta)^{\frac{1}{b}} \right]$$

4.4. Relation between Parameters and Constraints

Taking partial derivatives of (61) with respect to b, δ, λ_1, and λ_2, and equating each derivative to zero yields:

$$\frac{\partial S}{\partial \lambda_1} = 0 = b \ln \lambda_2 - b\psi(b - b\lambda_1) + E[\ln(x - \delta)] \qquad (62)$$

$$\frac{\partial S}{\partial \lambda_2} = 0 = -\frac{b - b\lambda_1}{\lambda_2} + E\left[(x - \delta)^{\frac{1}{b}} \right] \qquad (63)$$

$$\frac{\partial S}{\partial b} = 0 = -(1 - \lambda_1) \ln \lambda_2 + \frac{1}{b} + (1 - \lambda_1) \cdot \psi(b - b\lambda_1) - \frac{\lambda_2}{b^2} E\left[(x - \delta)^{\frac{1}{b}} \ln(x - \delta) \right] \qquad (64)$$

$$\frac{\partial S}{\partial \delta} = 0 = -\lambda_1 E\left(\frac{1}{x - \delta} \right) - \frac{\lambda_2}{b} E\left[(x - \delta)^{\frac{1}{b} - 1} \right] \qquad (65)$$

Introduction of Equations (58)–(59) into Equations (62)–(65) and recalling Equations (62)–(65) yields, respectively:

$$\begin{cases} b \ln \beta - b\psi(\alpha) + E[\ln(x - \delta)] = 0 \\ \frac{\alpha}{\beta} - E\left[(x - \delta)^{\frac{1}{b}} \right] = 0 \\ -\frac{\alpha}{b} \ln \beta + \frac{1}{b} + \frac{\alpha}{b} \cdot \psi(\alpha) - \frac{\beta}{b^2} E\left[(x - \delta)^{\frac{1}{b}} \ln(x - \delta) \right] = 0 \\ \left(\frac{\alpha}{b} - 1 \right) E\left(\frac{1}{x - \delta} \right) - \frac{\beta}{b} E\left[(x - \delta)^{\frac{1}{b} - 1} \right] = 0 \end{cases} \qquad (66)$$

The expectations of Equation (66) are replaced by their sample estimates, and the simplification of Equation (66) leads to:

$$\begin{cases} -\frac{\alpha}{b} \sum_{i=1}^n \left(\frac{1}{x_i - \delta} \right) + \sum_{i=1}^n \left(\frac{1}{x_i - \delta} \right) + \frac{\beta}{b} \sum_{i=1}^n (x_i - \delta)^{\frac{1}{b} - 1} = 0 \\ n \ln \beta - n\psi(\alpha) + \frac{1}{b} \sum_{i=1}^n \ln(x_i - \delta) = 0 \\ \frac{n\alpha}{\beta} - \sum_{i=1}^n (x_i - \delta)^{\frac{1}{b}} = 0 \\ -\frac{n}{b} - \frac{\alpha}{b^2} \sum_{i=1}^n \ln(x_i - \delta) + \frac{\beta}{b^2} \sum_{i=1}^n \left[(x_i - \delta)^{\frac{1}{b}} \ln(x_i - \delta) \right] = 0 \end{cases} \qquad (67)$$

Equations (51) has the second moments and results in some biases. Therefore, Equation (67) should be used for the estimation of the parameters.

5. Two Other Parameter Estimation Methods

Two other methods of parameter estimation frequently used in hydrology are the method of moments (MOM) and the MLE method.

5.1. Method of Moments

The four-parameter exponential gamma distribution has four parameters α, β, δ and b. Therefore, four moments are needed for the parameters estimation. The detailed derivation of the four moments is presented in Appendix A:

$$
\begin{cases}
\mu_2 = \frac{\Gamma(\alpha)\Gamma(\alpha+2b)-\Gamma^2(\alpha+b)}{\beta^{2b}\Gamma^2(\alpha)} \\
E\left[\frac{1}{x-\delta}\right] = \frac{\beta^b\Gamma(\alpha-b)}{\Gamma(\alpha)} \\
\mu'_1 = \frac{\Gamma(\alpha+b)}{\beta^b\Gamma(\alpha)} + \delta \\
E\left[(x-\delta)^{\frac{1}{b}}\right] = \frac{\alpha}{\beta}
\end{cases}
\tag{68}
$$

For a sample, $x = \{x_1, x_2, \cdots, x_n\}$, the estimation equations become:

$$
\begin{cases}
\frac{1}{n}\sum_{i=1}^{n}(x_i-\overline{x})^2 = \frac{\Gamma(\alpha)\Gamma(\alpha+2b)-\Gamma^2(\alpha+b)}{\beta^{2b}\Gamma^2(\alpha)} \\
\frac{1}{n}\sum_{i=1}^{n}\left(\frac{1}{x_i-\delta}\right) = \frac{\beta^b\Gamma(\alpha-b)}{\Gamma(\alpha)} \\
\overline{x} = \frac{\Gamma(\alpha+b)}{\beta^b\Gamma(\alpha)} + \delta \\
\frac{1}{n}\sum_{i=1}^{n}(x_i-\delta)^{\frac{1}{b}} = \frac{\alpha}{\beta}
\end{cases}
\tag{69}
$$

where n is the sample size; $\overline{x} = \frac{1}{n}\sum_{i=1}^{n}x_i$.

5.2. Method of Maximum Likelihood Estimation

For the MLE method, the log-likelihood function L for a sample $x = \{x_1, x_2, \cdots, x_n\}$ is given by:

$$
\ln L = n\alpha\ln\beta - n\ln b - n\ln\Gamma(\alpha) + \frac{\alpha}{b}\sum_{i=1}^{n}\ln(x_i-\delta) - \sum_{i=1}^{n}\ln(x_i-\delta) - \beta\sum_{i=1}^{n}(x_i-\delta)^{\frac{1}{b}}
\tag{70}
$$

The MLE's of parameters $\hat{\alpha}$, $\hat{\beta}$, $\hat{\delta}$ and $\hat{b}$ are taken to be the values that yield the maximum of $\ln L$. Differentiating Equation (70) partially with respect to each parameter and equating each partial derivative to zero produces:

$$
\begin{cases}
-\frac{\alpha}{b}\sum_{i=1}^{n}\frac{1}{x_i-\delta} + \sum_{i=1}^{n}\frac{1}{x_i-\delta} + \frac{\beta}{b}\sum_{i=1}^{n}(x_i-\delta)^{\frac{1}{b}-1} = 0 \\
n\ln\beta - n\psi(\alpha) + \frac{1}{b}\sum_{i=1}^{n}\ln(x_i-\delta) = 0 \\
\frac{n\alpha}{\beta} - \sum_{i=1}^{n}(x_i-\delta)^{\frac{1}{b}} = 0 \\
-\frac{n}{b} - \frac{\alpha}{b^2}\sum_{i=1}^{n}\ln(x_i-\delta) + \frac{\beta}{b^2}\sum_{i=1}^{n}\left[(x_i-\delta)^{\frac{1}{b}}\ln(x_i-\delta)\right] = 0
\end{cases}
\tag{71}
$$

These are the parameter estimation Equations, and the obtained results are the same as those of the PSEM method.

6. Evaluation and Comparison of Parameter Estimation Methods

The PSEM as presented in this paper is used for six annual precipitation data sets observed from 1959 to 2008 without any missing records at the Weihe River basin of China. All data are obtained from the National Climate of China Meteorological Administration and are complete. The characteristics of these data are summarized in Table 3. Obviously, all annual precipitation records have very low first-order serial correlation coefficients, ρ. Using Anderson's test of independence, the results have shown that these gauge data have an independent structure at 90% confidence levels. Hence, they are suitable for the application of meteorological frequency analysis.

Table 3. Characteristics of data used for parameter estimation.

Site Name	Mean	Standard Deviation	Coefficient of Variation	Skewness	Kurtosis	First-Order Serial Correlation Coefficient
Xi'an	571.9	126.9575	0.2220	0.2938	3.1935	−0.11399
Zhouzhi	635.2	158.7627	0.2499	0.6613	3.6873	0.16198
Lantian	713.7	150.0908	0.2103	0.2787	3.1394	0.02126
Huxian	633.8	147.5611	0.2328	0.3582	3.1987	0.05029
Lintong	579.5	129.2021	0.2230	0.6108	3.5747	0.04745
Wugong	606.7	158.2829	0.2609	0.5710	3.0826	0.09791

None of the above-discussed three methods yielded explicit solutions for the estimation of parameters of the four-parameter exponential gamma distribution. The parameter estimation Equations were therefore solved for α, β, δ and b by the four-dimensional Levenberg–Marquardt method.

Equations (67)–(68) and (71) can be simplified as the form of $\begin{cases} F_1 = 0 \\ F_2 = 0 \\ F_3 = 0 \\ F_4 = 0 \end{cases}$. Then, according to the

above procedures the Matlab (Version R2007b) computer codes were developed and used to calculate the parameters. To verify the validities of parameters, the left side functions F_1, F_2, F_3 and F_4 in Equations (67)–(68) and (71) are listed Table 4. It is seen that these compute quantities are close to zero, indicating satisfactory performance of the four dimensional Levenberg–Marquardt algorithm.

Table 4. The left side functions F_1, F_2, F_3 and F_4 in Equations (67)–(68) and (71).

Site Name	Methods	F_1	F_2	F_3	F_4
Xi'an	PSEM	0.00581	0.00350	0.00000	−0.08445
	MLE	0.00581	0.00350	0.00000	−0.08445
	MOM	0.00842	−0.00000	0.00000	0.07950
Zhouzhi	PSEM	0.00144	0.00174	0.00000	−0.01937
	MLE	0.00144	0.00174	0.00000	−0.01937
	MOM	0.01241	−0.00007	0.00000	0.02139
Lantian	PSEM	0.00244	0.00206	0.00000	−0.07525
	MLE	0.00244	0.00206	0.00000	−0.07525
	MOM	0.00493	−0.00000	0.00000	0.01965
Huxian	PSEM	0.00242	0.00199	0.00000	−0.05926
	MLE	0.00242	0.00199	0.00000	−0.05926
	MOM	0.00533	−0.00006	0.00000	0.08050
Lintong	PSEM	0.00086	0.00107	0.00000	−0.08474
	MLE	0.00086	0.00107	0.00000	−0.08474
	MOM	0.03094	−0.00002	0.00000	0.02216
Wugong	PSEM	−0.00047	−0.00132	0.00000	0.00726
	MLE	−0.00047	−0.00132	0.00000	0.00726
	MOM	0.00285	−0.00001	0.00000	0.00850

The values of the distribution parameters are given in Table 5. The results of PSEM and MLE are the same. To evaluate and compare the performance of the three methods, the relative error (RERR) was employed that can be defined as:

$$REER = \frac{1}{n}\sum_{i=1}^{n}\left(\frac{x_{0i} - x_{pi}}{x_{0i}}\right)^2 \tag{72}$$

where x_{0i} and x_{pi} are the observed and predicted values of a given (i-th) quantile, respectively, and n is the sample size. The RERR values are summarized in Table 5.

Table 5. Parameter values estimated by the three methods.

Site Name	Methods	$\hat{\delta}$	$\hat{\alpha}$	$\hat{\beta}$	$\hat{b}$	RERR
Xi'an	PSEM	0.01000	88.34381	4.42388	2.11605	0.00127
	MLE	0.01000	88.34381	4.42388	2.11605	0.00127
	MOM	7.20138	62.88730	1.79612	1.77889	0.00131
Zhouzhi	PSEM	0.01000	79.70604	4.26549	2.19876	0.00189
	MLE	0.01000	79.70604	4.26549	2.19876	0.00189
	MOM	210.63187	27.69157	1.27634	1.95521	0.00274
Lantian	PSEM	0.01000	94.39458	3.88381	2.05573	0.00128
	MLE	0.01000	94.39458	3.88381	2.05573	0.00128
	MOM	68.28209	60.56026	1.69273	1.80523	0.00136
Huxian	PSEM	0.01000	84.61866	4.25523	2.15294	0.00119
	MLE	0.01000	84.61866	4.25523	2.15294	0.00119
	MOM	212.47184	30.39572	1.33086	1.92191	0.00213
Lintong	PSEM	0.01000	90.81538	4.35225	2.08997	0.00129
	MLE	0.01000	90.81538	4.35225	2.08997	0.00129
	MOM	162.17872	36.92153	1.48868	1.87230	0.00161
Wugong	PSEM	0.01000	75.75303	4.39321	2.24406	0.00158
	MLE	0.01000	75.75303	4.39321	2.24406	0.00158
	MOM	114.20629	35.23074	1.35950	1.89735	0.00183

Examination of the data in Table 5 shows that the parameters estimated using PSEM and MLE are comparable to MOM in terms of RERR and it is thus difficult to distinguish them from one another. However, PSEM and MLE yield the best parameter estimates. Thus, the parameters estimated by PSEM should be employed as the ones of four-parameter exponential gamma distribution in case study sites.

To measure the agreement between a theoretical probability distribution and an empirical distribution for the samples, Kolmogorov–Smirnov (K–S) test D_n was used to assess the goodness-of-fit.

Let $x_1 < x_2 < \cdots < x_n$ be order statistics for a sample size n whose population is defined by a continuous cumulative distribution function $F(x)$ and $F_0(x_i)$ be a specified distribution that contains a set of parameters θ ($\hat{\theta}$ is estimated value from a sample size n). For an annual precipitation series, the null hypothesis H_0 that the true distribution was F_0 with parameters θ was tested. K–S test D_n can be expressed as:

$$D_n = \max_{1 \leq i \leq n} \left(\hat{\delta}_i \right) \tag{73}$$

$$\hat{\delta}_i = \max \left[\frac{i}{n} - F_0\left(x_i; \hat{\theta}\right), F_0\left(x_i; \hat{\theta}\right) - \frac{i-1}{n} \right] \tag{74}$$

The sample values of K–S test statistic D_n, are shown in Table 6. The critical value D_n^* of the four-parameter exponential gamma distribution (at the significance level a = 0.05, for sample size n) is 0.18654. From Table 6 it can be seen that the statistics of observed annual precipitation are all less than their corresponding critical values, respectively. Therefore, it is concluded that annual precipitation series are all accepted by the K–S test.

Table 6. Sample values of K–S test statistic D_n of case study sites.

Site Name	D_n	Site Name	D_n
Xi'an	0.07764	Huxian	0.07268
Zhouzhi	0.07290	Lintong	0.07755
Lantian	0.04485	Wugong	0.10037

7. Conclusions

Hydrologic frequency analysis, in spite of having developed a great number of distribution models and parameter estimation methods for reliable parameters and quantiles estimates, comes up against practical difficulties imposed by the short sample ranges. The Pearson Type III distribution is recommended as a standard distribution in hydrological frequency analysis in China. A large number of studies have shown that fitting small and large return period segments of Pearson Type III distribution is affected by its skewness value. Different studies employing the same parameter estimation methods may obtain different results. The use of four-parameter exponential gamma distribution has emerged as an attempt to reduce the estimate errors of small and large return period segments. The advantage of the proposed entropy method is that the first moments are made about the calculation of the distribution parameters, instead of variance, skewness and kurtosis. The results of the case estimates show that the entropy method enables the four-parameter exponential gamma distribution to fit the data well. The entropy-based parameter estimation also provides a new way to estimate parameters of the four-parameter exponential gamma distribution. The disadvantage of the method is that it will be computationally cumbersome because four parameters are involved. However, this should not be an insurmountable difficultly, given the currently available numerical tools and computer progress. Also, there are significant differences between among the MOM, PSEM and MLE estimates. Such large differences may be caused by the system of non-linear equations of parameter estimation involving the second central moment of the variable for the MOM, first moments for PSEM and MLE. In addition, the confidence intervals of quantiles for the four-parameter exponential gamma distribution deserve thorough investigation.

In summary, the following conclusions can be drawn from the present study: (1) for parameter estimation, PSEM yields the same results as MLE, whereas MOM performs with the highest bias; (2) PSEM is comparable to the MOM; (3) the four-parameter exponential gamma distribution fits the observed annual precipitation data well; (4) the quantile discharge values estimated by the three methods are close to each other; (5) the four-parameter exponential gamma distribution is a versatile distribution and results in nine different distributions, depending on its parameter values.

Acknowledgments: The authors acknowledge the financial support of National Natural Science Foundation of China (Grant Nos. 51479171, 41501022 and 51409222). The authors also wish to express their cordial gratitude to the editors, and anonymous reviewers for their illuminating comments which have greatly helped to improve the quality of this manuscript.

Author Contributions: Songbai Song and Xiaoyan Song designed the computations; Xiaoyan Song and Yan Kang made Anderson's test of the data; Songbai Song and Xiaoyan Song wrote the paper. All authors have read and approved the final manuscript.

Appendix A. First Four Original Moments and Central Moments

Consider that the first original moments of the four-parameter exponential gamma distribution are μ'_1, μ'_2, μ'_3 and μ'_4, respectively; μ_2, μ_3, μ_4 are the second, third and fourth central moments, respectively;C_v is coefficient of variation; C_s is coefficient of skewness; C_e is coefficient of kurtosis. A detailed derivation of the above moments is given below:

$$\mu'_1 = \int_\delta^\infty xf(x)dx = \int_\delta^\infty \frac{\beta^\alpha}{b\Gamma(\alpha)}x(x-\delta)^{\frac{\alpha}{b}-1}e^{-\beta(x-\delta)^{\frac{1}{b}}}dx = \int_0^\infty \frac{\beta^\alpha}{b\Gamma(\alpha)}\left(\delta+\frac{1}{\beta^b}t^b\right)\left(\frac{1}{\beta^b}t^b\right)^{\frac{\alpha}{b}-1}e^{-t}\frac{b}{\beta^b}t^{b-1}dt$$

$$= \int_0^\infty \frac{1}{\Gamma(\alpha)}\left(\delta+\frac{1}{\beta^b}t^b\right)t^{\alpha-1}e^{-t}dt = \frac{1}{\Gamma(\alpha)}\left[\delta\int_0^\infty t^{\alpha-1}e^{-t}dt + \frac{1}{\beta^b}\int_0^\infty t^{\alpha+b-1}e^{-t}dt\right] \quad (A1)$$

$$= \frac{1}{\Gamma(\alpha)}\left[\delta\cdot\Gamma(\alpha)+\frac{1}{\beta^b}\Gamma(\alpha+b)\right] = \frac{\Gamma(\alpha+b)}{\beta^b\Gamma(\alpha)}+\delta$$

$$\mu'_2 = \int_\delta^\infty x^2f(x)dx = \int_\delta^\infty \frac{\beta^\alpha}{b\Gamma(\alpha)}x^2(x-\delta)^{\frac{\alpha}{b}-1}e^{-\beta(x-\delta)^{\frac{1}{b}}}dx$$

$$= \int_0^\infty \frac{\beta^\alpha}{b\Gamma(\alpha)}\left(\delta+\frac{1}{\beta^b}t^b\right)^2\left(\frac{1}{\beta^b}t^b\right)^{\frac{\alpha}{b}-1}e^{-t}\frac{b}{\beta^b}t^{b-1}dt = \int_0^\infty \frac{1}{\Gamma(\alpha)}\left(\delta+\frac{1}{\beta^b}t^b\right)^2 t^{\alpha-1}e^{-t}dt$$

$$= \frac{1}{\Gamma(\alpha)}\left[\delta^2\int_0^\infty t^{\alpha-1}e^{-t}dt + \frac{2\delta}{\beta^b}\int_0^\infty t^{\alpha+b-1}e^{-t}dt + \frac{1}{\beta^{2b}}\int_0^\infty t^{\alpha+2b-1}e^{-t}dt\right] \quad (A2)$$

$$= \frac{1}{\Gamma(\alpha)}\left[\delta^2\Gamma(\alpha)+\frac{2\delta}{\beta^b}\Gamma(\alpha+b)+\frac{1}{\beta^{2b}}\Gamma(\alpha+2b)\right]$$

$$= \delta^2 + \frac{2\delta}{\beta^b\Gamma(\alpha)}\Gamma(\alpha+b)+\frac{1}{\beta^{2b}\Gamma(\alpha)}\Gamma(\alpha+2b)$$

$$\mu'_3 = \int_\delta^\infty x^3f(x)dx = \int_\delta^\infty \frac{\beta^\alpha}{b\Gamma(\alpha)}x^3(x-\delta)^{\frac{\alpha}{b}-1}e^{-\beta(x-\delta)^{\frac{1}{b}}}dx$$

$$= \int_0^\infty \frac{\beta^\alpha}{b\Gamma(\alpha)}\left(\delta+\frac{1}{\beta^b}t^b\right)^3\left(\frac{1}{\beta^b}t^b\right)^{\frac{\alpha}{b}-1}e^{-t}\frac{b}{\beta^b}t^{b-1}dt = \int_0^\infty \frac{1}{\Gamma(\alpha)}\left(\delta+\frac{1}{\beta^b}t^b\right)^3 t^{\alpha-1}e^{-t}dt$$

$$= \frac{1}{\Gamma(\alpha)}\left[\delta^3\int_0^\infty t^{\alpha-1}e^{-t}dt + \frac{3\delta^2}{\beta^b}\int_0^\infty t^{\alpha+b-1}e^{-t}dt + \frac{3\delta}{\beta^{2b}}\int_0^\infty t^{\alpha+2b-1}e^{-t}dt + \frac{1}{\beta^{3b}}\int_0^\infty t^{\alpha+3b-1}e^{-t}dt\right] \quad (A3)$$

$$= \frac{1}{\Gamma(\alpha)}\left[\delta^3\Gamma(\alpha)+\frac{3\delta^2}{\beta^b}\Gamma(\alpha+b)+\frac{3\delta}{\beta^{2b}}\Gamma(\alpha+2b)+\frac{1}{\beta^{3b}}\Gamma(\alpha+3b)\right]$$

$$\mu'_4 = \int_\delta^\infty x^4f(x)dx = \int_\delta^\infty \frac{\beta^\alpha}{b\Gamma(\alpha)}x^4(x-\delta)^{\frac{\alpha}{b}-1}e^{-\beta(x-\delta)^{\frac{1}{b}}}dx$$

$$= \int_0^\infty \frac{\beta^\alpha}{b\Gamma(\alpha)}\left(\delta+\frac{1}{\beta^b}t^b\right)^4\left(\frac{1}{\beta^b}t^b\right)^{\frac{\alpha}{b}-1}e^{-t}\frac{b}{\beta^b}t^{b-1}dt = \int_0^\infty \frac{1}{\Gamma(\alpha)}\left(\delta+\frac{1}{\beta^b}t^b\right)^4 t^{\alpha-1}e^{-t}dt \quad (A4)$$

$$= \frac{1}{\Gamma(\alpha)}\left[\delta^4\int_0^\infty t^{\alpha-1}e^{-t}dt + \frac{4\delta^3}{\beta^b}\int_0^\infty t^{\alpha+b-1}e^{-t}dt + \frac{6\delta^2}{\beta^{2b}}\int_0^\infty t^{\alpha+2b-1}e^{-t}dt + \frac{4\delta}{\beta^{3b}}\int_0^\infty t^{\alpha+3b-1}e^{-t}dt + \frac{1}{\beta^{4b}}\int_0^\infty t^{\alpha+4b-1}e^{-t}dt\right]$$

$$= \delta^4 + \frac{4\delta^3\Gamma(\alpha+b)}{\beta^b\Gamma(\alpha)}+\frac{6\delta^2\Gamma(\alpha+2b)}{\beta^{2b}\Gamma(\alpha)}+\frac{4\delta\cdot\Gamma(\alpha+3b)}{\beta^{3b}\Gamma(\alpha)}+\frac{\Gamma(\alpha+4b)}{\beta^{4b}\Gamma(\alpha)}$$

$$\mu_2 = Var(x) = \mu'_2 - (\mu'_1)^1 = \delta^2 + \frac{2\delta}{\beta^b\Gamma(\alpha)}\Gamma(\alpha+b)+\frac{1}{\beta^{2b}\Gamma(\alpha)}\Gamma(\alpha+2b)-\left[\frac{\Gamma(\alpha+b)}{\beta^b\Gamma(\alpha)}+\delta\right]^2$$

$$= \delta^2 + \frac{2\delta}{\beta^b\Gamma(\alpha)}\Gamma(\alpha+b)+\frac{1}{\beta^{2b}\Gamma(\alpha)}\Gamma(\alpha+2b)-\frac{\Gamma^2(\alpha+b)}{\beta^{2b}\Gamma^2(\alpha)}-\frac{2\delta}{\beta^b\Gamma(\alpha)}\Gamma(\alpha+b)-\delta^2 \quad (A5)$$

$$= \frac{1}{\beta^{2b}\Gamma(\alpha)}\Gamma(\alpha+2b)-\frac{\Gamma^2(\alpha+b)}{\beta^{2b}\Gamma^2(\alpha)} = \frac{\Gamma(\alpha+2b)\Gamma(\alpha)-\Gamma^2(\alpha+b)}{\beta^{2b}\Gamma^2(\alpha)}$$

$$\mu_3 = E[X - E(X)]^3 = E[X^3 - 3X^2 E(X) + 3X E^2(X) - E^3(X)]$$

$$= E(X^3) - 3E(X^2)E(X) + 2E^3(X) = \mu'_3 - 3\mu'_2 \mu'_1 + 2(\mu'_1)^3$$

$$= \frac{1}{\Gamma(\alpha)}\left[\delta^3 \Gamma(\alpha) + \frac{3\delta^2}{\beta^b}\Gamma(\alpha+b) + \frac{3\delta}{\beta^{2b}}\Gamma(\alpha+2b) + \frac{1}{\beta^{3b}}\Gamma(\alpha+3b)\right]$$

$$-3\left[\delta^2 + \frac{2\delta \cdot \Gamma(\alpha+b)}{\beta^b \Gamma(\alpha)} + \frac{\Gamma(\alpha+2b)}{\beta^{2b}\Gamma(\alpha)}\right] \cdot \left[\frac{\Gamma(\alpha+b)}{\beta^b \Gamma(\alpha)} + \delta\right] + 2\left[\frac{\Gamma(\alpha+b)}{\beta^b \Gamma(\alpha)} + \delta\right]^3 \quad \text{(A6)}$$

$$= \left(\delta^3 + 2\delta^3 - 3\delta^3\right) + \left[\frac{3\delta^2 \Gamma(\alpha+b)}{\beta^b \Gamma(\alpha)} - \frac{3\delta^2 \Gamma(\alpha+b)}{\beta^b \Gamma(\alpha)} - \frac{6\delta^2 \cdot \Gamma(\alpha+b)}{\beta^b \Gamma(\alpha)} + \frac{6\delta^2 \Gamma(\alpha+b)}{\beta^b \Gamma(\alpha)}\right]$$

$$+\left[\frac{3\delta \cdot \Gamma(\alpha+2b)}{\beta^{2b}\Gamma(\alpha)} - \frac{3\delta \cdot \Gamma(\alpha+2b)}{\beta^{2b}\Gamma(\alpha)}\right] + \left[\frac{6\delta \cdot \Gamma^2(\alpha+b)}{\beta^{2b}\Gamma^2(\alpha)} - \frac{6\delta \cdot \Gamma^2(\alpha+b)}{\beta^{2b}\Gamma^2(\alpha)}\right] + \frac{\Gamma(\alpha+3b)}{\beta^{3b}\Gamma(\alpha)} + \frac{2\Gamma^3(\alpha+b)}{\beta^{3b}\Gamma^3(\alpha)} - \frac{3\Gamma(\alpha+b)\Gamma(\alpha+2b)}{\beta^{3b}\Gamma^2(\alpha)}$$

$$= \frac{\Gamma^2(\alpha)\Gamma(\alpha+3b) - 3\Gamma(\alpha)\Gamma(\alpha+b)\Gamma(\alpha+2b) + 2\Gamma^3(\alpha+b)}{\beta^{3b}\Gamma^3(\alpha)}$$

$$\mu_4 = E[X - E(X)]^4 = E[X^4 - 4X^3 E(X) + 6X^2 E^2(X) - 4X E^3(X) + E^4(X)]$$

$$= E(X^4) - 4E(X^3)E(X) + 6E(X^2)E^2(X) - 3E^4(X)$$

$$= \mu'_4 - 4\mu'_3 \mu'_1 + 6\mu'_2 (\mu'_1)^2 - 3(\mu'_1)^4$$

$$= \delta^4 + \frac{4\delta^3 \Gamma(\alpha+b)}{\beta^b \Gamma(\alpha)} + \frac{6\delta^2 \Gamma(\alpha+2b)}{\beta^{2b}\Gamma(\alpha)} + \frac{4\delta \cdot \Gamma(\alpha+3b)}{\beta^{3b}\Gamma(\alpha)} + \frac{\Gamma(\alpha+4b)}{\beta^{4b}\Gamma(\alpha)}$$

$$-4\frac{1}{\Gamma(\alpha)}\left[\delta^3 \Gamma(\alpha) + \frac{3\delta^2}{\beta^b}\Gamma(\alpha+b) + \frac{3\delta}{\beta^{2b}}\Gamma(\alpha+2b) + \frac{1}{\beta^{3b}}\Gamma(\alpha+3b)\right] \cdot \left[\frac{\Gamma(\alpha+b)}{\beta^b \Gamma(\alpha)} + \delta\right]$$

$$+6\left[\delta^2 + \frac{2\delta \cdot \Gamma(\alpha+b)}{\beta^b \Gamma(\alpha)} + \frac{\Gamma(\alpha+2b)}{\beta^{2b}\Gamma(\alpha)}\right] \cdot \left[\frac{\Gamma(\alpha+b)}{\beta^b \Gamma(\alpha)} + \delta\right]^2 - 3\left[\frac{\Gamma(\alpha+b)}{\beta^b \Gamma(\alpha)} + \delta\right]^4$$

$$= \left(\delta^4 - 4\delta^4 + 6\delta^4 - 3\delta^4\right) + \left[\frac{4\delta^3 \Gamma(\alpha+b)}{\beta^b \Gamma(\alpha)} - \frac{4\delta^3 \Gamma(\alpha+b)}{\beta^b \Gamma(\alpha)} - \frac{12\delta^3 \Gamma(\alpha+b)}{\beta^b \Gamma(\alpha)} + \frac{12\delta^3 \cdot \Gamma(\alpha+b)}{\beta^b \Gamma(\alpha)}\right] \quad \text{(A7)}$$

$$+\left[\frac{12\delta^3 \cdot \Gamma(\alpha+b)}{\beta^b \Gamma(\alpha)} - \frac{6\delta^3 \cdot \Gamma(\alpha+b)}{\beta^b \Gamma(\alpha)} - \frac{6\delta^3 \cdot \Gamma(\alpha+b)}{\beta^b \Gamma(\alpha)}\right] + \left[\frac{6\delta^2 \Gamma(\alpha+2b)}{\beta^{2b}\Gamma(\alpha)} + \frac{6\delta^2 \Gamma(\alpha+2b)}{\beta^{2b}\Gamma(\alpha)} - \frac{12\delta^2 \Gamma(\alpha+2b)}{\beta^{2b}\Gamma(\alpha)}\right]$$

$$+\left[\frac{24\delta^2 \cdot \Gamma^2(\alpha+b)}{\beta^{2b}\Gamma^2(\alpha)} - \frac{12\delta^2 \Gamma^2(\alpha+b)}{\beta^{2b}\Gamma^2(\alpha)} - \frac{12\delta^2 \cdot \Gamma^2(\alpha+b)}{\beta^{2b}\Gamma^2(\alpha)}\right] + \left[\frac{6\delta^2 \Gamma^2(\alpha+b)}{\beta^{2b}\Gamma^2(\alpha)} - \frac{3\delta^2 \Gamma^2(\alpha+b)}{\beta^{2b}\Gamma^2(\alpha)} - \frac{3\delta^2 \Gamma^2(\alpha+b)}{\beta^{2b}\Gamma^2(\alpha)}\right]$$

$$+\left[\frac{4\delta \cdot \Gamma(\alpha+3b)}{\beta^{3b}\Gamma(\alpha)} - \frac{4\delta \cdot \Gamma(\alpha+3b)}{\beta^{3b}\Gamma(\alpha)}\right] + \left[\frac{12\delta \cdot \Gamma(\alpha+b)\Gamma(\alpha+2b)}{\beta^{3b}\Gamma^2(\alpha)} - \frac{12\delta \cdot \Gamma(\alpha+b)\Gamma(\alpha+2b)}{\beta^{3b}\Gamma^2(\alpha)}\right]$$

$$+\left[\frac{12\delta \cdot \Gamma^3(\alpha+b)}{\beta^{3b}\Gamma^3(\alpha)} - \frac{6\delta \cdot \Gamma^3(\alpha+b)}{\beta^{3b}\Gamma^3(\alpha)} - \frac{6\delta \cdot \Gamma^3(\alpha+b)}{\beta^{3b}\Gamma^3(\alpha)}\right] + \frac{\Gamma(\alpha+4b)}{\beta^{4b}\Gamma(\alpha)} - \frac{4\Gamma(\alpha+b)\Gamma(\alpha+3b)}{\beta^{4b}\Gamma^2(\alpha)} + \frac{6\Gamma^2(\alpha+b)\Gamma(\alpha+2b)}{\beta^{4b}\Gamma^3(\alpha)} - \frac{3\Gamma^4(\alpha+b)}{\beta^{4b}\Gamma^4(\alpha)}$$

$$= \frac{\Gamma(\alpha+4b)}{\beta^{4b}\Gamma(\alpha)} - \frac{4\Gamma(\alpha+b)\Gamma(\alpha+3b)}{\beta^{4b}\Gamma^2(\alpha)} + \frac{6\Gamma^2(\alpha+b)\Gamma(\alpha+2b)}{\beta^{4b}\Gamma^3(\alpha)} - \frac{3\Gamma^4(\alpha+b)}{\beta^{4b}\Gamma^4(\alpha)}$$

$$= \frac{\Gamma^3(\alpha)\Gamma(\alpha+4b) - 4\Gamma^2(\alpha)\Gamma(\alpha+b)\Gamma(\alpha+3b) + 6\Gamma(\alpha)\Gamma^2(\alpha+b)\Gamma(\alpha+2b) - 3\Gamma^4(\alpha+b)}{\beta^{4b}\Gamma^4(\alpha)}$$

$$C_v = \frac{(\mu_2)^{1/2}}{\mu'_1} = \frac{\left[\frac{\Gamma(\alpha+2b)\Gamma(\alpha) - \Gamma^2(\alpha+b)}{\beta^{2b}\Gamma^2(\alpha)}\right]^{1/2}}{\frac{\Gamma(\alpha+b)}{\beta^b \Gamma(\alpha)} + \delta} = \frac{\sqrt{\Gamma(\alpha+2b)\Gamma(\alpha) - \Gamma^2(\alpha+b)}}{\Gamma(\alpha+b) + \delta \cdot \beta^b \Gamma(\alpha)} \quad \text{(A8)}$$

$$C_s = \frac{\mu_3}{(\mu_2)^{3/2}} = \frac{\frac{\Gamma^2(\alpha)\Gamma(\alpha+3b) - 3\Gamma(\alpha)\Gamma(\alpha+b)\Gamma(\alpha+2b) + 2\Gamma^3(\alpha+b)}{\beta^{3b}\Gamma^3(\alpha)}}{\left[\frac{\Gamma(\alpha+2b)\Gamma(\alpha) - \Gamma^2(\alpha+b)}{\beta^{2b}\Gamma^2(\alpha)}\right]^{3/2}}$$

$$= \frac{\Gamma^2(\alpha)\Gamma(\alpha+3b) - 3\Gamma(\alpha)\Gamma(\alpha+b)\Gamma(\alpha+2b) + 2\Gamma^3(\alpha+b)}{\beta^{3b}\Gamma^3(\alpha)} \frac{\beta^{3b}\Gamma^3(\alpha)}{\left[\Gamma(\alpha)\Gamma(\alpha+2b) - \Gamma^2(\alpha+b)\right]^{3/2}} \quad \text{(A9)}$$

$$= \frac{\Gamma^2(\alpha)\Gamma(\alpha+3b) - 3\Gamma(\alpha)\Gamma(\alpha+b)\Gamma(\alpha+2b) + 2\Gamma^3(\alpha+b)}{\left[\Gamma(\alpha)\Gamma(\alpha+2b) - \Gamma^2(\alpha+b)\right]^{3/2}}$$

$$C_e = \frac{\mu_4}{(\mu_2)^2} - 3 = \frac{\Gamma^3(\alpha)\Gamma(\alpha+4b)-4\Gamma^2(\alpha)\Gamma(\alpha+b)\Gamma(\alpha+3b)+6\Gamma(\alpha)\Gamma^2(\alpha+b)\Gamma(\alpha+2b)-3\Gamma^4(\alpha+b)}{\beta^{4b}\Gamma^4(\alpha)}$$

$$\cdot \frac{\beta^{4b}\Gamma^4(\alpha)}{\left[\Gamma(\alpha)\Gamma(\alpha+2b)-\Gamma^2(\alpha+b)\right]^2} - 3$$

$$= \frac{\Gamma^3(\alpha)\Gamma(\alpha+4b)-4\Gamma^2(\alpha)\Gamma(\alpha+b)\Gamma(\alpha+3b)+6\Gamma(\alpha)\Gamma^2(\alpha+b)\Gamma(\alpha+2b)-3\Gamma^4(\alpha+b)}{\left[\Gamma(\alpha)\Gamma(\alpha+2b)-\Gamma^2(\alpha+b)\right]^2}$$

$$= \frac{\Gamma^3(\alpha)\Gamma(\alpha+4b)-4\Gamma^2(\alpha)\Gamma(\alpha+b)\Gamma(\alpha+3b)+6\Gamma(\alpha)\Gamma^2(\alpha+b)\Gamma(\alpha+2b)-3\Gamma^4(\alpha+b)}{\left[\Gamma(\alpha)\Gamma(\alpha+2b)-\Gamma^2(\alpha+b)\right]^2}$$

$$+ \frac{-3\Gamma^2(\alpha)\Gamma^2(\alpha+2b)+6\Gamma(\alpha)\Gamma(\alpha+2b)\Gamma^2(\alpha+b)-3\Gamma^4(\alpha+b)}{\left[\Gamma(\alpha)\Gamma(\alpha+2b)-\Gamma^2(\alpha+b)\right]^2}$$

$$= \frac{\Gamma^3(\alpha)\Gamma(\alpha+4b)-4\Gamma^2(\alpha)\Gamma(\alpha+b)\Gamma(\alpha+3b)+12\Gamma(\alpha)\Gamma^2(\alpha+b)\Gamma(\alpha+2b)-3\Gamma^2(\alpha)\Gamma^2(\alpha+2b)-6\Gamma^4(\alpha+b)}{\left[\Gamma(\alpha)\Gamma(\alpha+2b)-\Gamma^2(\alpha+b)\right]^2}$$

(A10)

Appendix B. Derivatives of λ_0 with Respect to λ_1, λ_2 and λ_3

$$\frac{\partial \lambda_0}{\partial \lambda_1} = \frac{-\int_\delta^\infty \ln(x-\delta)\cdot\exp\left[-\lambda_1\ln(x-\delta)-\lambda_2\ln(x-\delta)^{\frac{1}{b}}-\lambda_3(x-\delta)^{\frac{1}{b}}\right]dx}{\int_\delta^\infty \exp\left[-\lambda_1\ln(x-\delta)-\lambda_2\ln(x-\delta)^{\frac{1}{b}}-\lambda_3(x-\delta)^{\frac{1}{b}}\right]dx}$$

$$= \frac{-\int_\delta^\infty \ln(x-\delta)\cdot\exp\left[-\lambda_0-\lambda_1\ln(x-\delta)-\lambda_2\ln(x-\delta)^{\frac{1}{b}}-\lambda_3(x-\delta)^{\frac{1}{b}}\right]dx}{\int_\delta^\infty \exp\left[-\lambda_0-\lambda_1\ln(x-\delta)-\lambda_2\ln(x-\delta)^{\frac{1}{b}}-\lambda_3(x-\delta)^{\frac{1}{b}}\right]dx}$$

(A11)

$$= \frac{-\int_\delta^\infty \ln(x-\delta)\cdot f(x)dx}{\int_\delta^\infty \cdot f(x)dx} = -E[\ln(x-\delta)]$$

$$\frac{\partial \lambda_0}{\partial \lambda_2} = \frac{-\int_\delta^\infty \ln(x-\delta)^{\frac{1}{b}}\cdot\exp\left[-\lambda_1\ln(x-\delta)-\lambda_2\ln(x-\delta)^{\frac{1}{b}}-\lambda_3(x-\delta)^{\frac{1}{b}}\right]dx}{\int_\delta^\infty \exp\left[-\lambda_1\ln(x-\delta)-\lambda_2\ln(x-\delta)^{\frac{1}{b}}-\lambda_3(x-\delta)^{\frac{1}{b}}\right]dx}$$

$$= \frac{-\int_\delta^\infty \ln(x-\delta)^{\frac{1}{b}}\cdot\exp\left[-\lambda_0-\lambda_1\ln(x-\delta)-\lambda_2\ln(x-\delta)^{\frac{1}{b}}-\lambda_3(x-\delta)^{\frac{1}{b}}\right]dx}{\int_\delta^\infty \exp\left[-\lambda_0-\lambda_1\ln(x-\delta)-\lambda_2\ln(x-\delta)^{\frac{1}{b}}-\lambda_3(x-\delta)^{\frac{1}{b}}\right]dx}$$

(A12)

$$= \frac{-\int_\delta^\infty \ln(x-\delta)^{\frac{1}{b}}\cdot f(x)dx}{\int_\delta^\infty f(x)dx} = -E\left[\ln(x-\delta)^{\frac{1}{b}}\right]$$

$$\frac{\partial \lambda_0}{\partial \lambda_3} = \frac{-\int_\delta^\infty (x-\delta)^{\frac{1}{b}}\cdot\exp\left[-\lambda_1\ln(x-\delta)-\lambda_2\ln(x-\delta)^{\frac{1}{b}}-\lambda_3(x-\delta)^{\frac{1}{b}}\right]dx}{\int_\delta^\infty \exp\left[-\lambda_1\ln(x-\delta)-\lambda_2\ln(x-\delta)^{\frac{1}{b}}-\lambda_3(x-\delta)^{\frac{1}{b}}\right]dx}$$

$$= \frac{-\int_\delta^\infty (x-\delta)^{\frac{1}{b}}\cdot\exp\left[-\lambda_0-\lambda_1\ln(x-\delta)-\lambda_2\ln(x-\delta)^{\frac{1}{b}}-\lambda_3(x-\delta)^{\frac{1}{b}}\right]dx}{\int_\delta^\infty \exp\left[-\lambda_0-\lambda_1\ln(x-\delta)-\lambda_2\ln(x-\delta)^{\frac{1}{b}}-\lambda_3(x-\delta)^{\frac{1}{b}}\right]dx}$$

(A13)

$$= \frac{-\int_\delta^\infty (x-\delta)^{\frac{1}{b}}\cdot f(x)dx}{\int_\delta^\infty f(x)dx} = -E\left[(x-\delta)^{\frac{1}{b}}\right]$$

Furthermore, we can write:

$$
\frac{\partial^2 \lambda_0}{\partial \lambda_3^2} = \frac{\int_\delta^\infty \left[(x-\delta)^{\frac{1}{b}}\right]^2 \cdot \exp\left[-\lambda_1 \ln(x-\delta) - \lambda_2 \ln(x-\delta)^{\frac{1}{b}} - \lambda_3(x-\delta)^{\frac{1}{b}}\right] dx}{\left\{\int_\delta^\infty \exp\left[-\lambda_1 \ln(x-\delta) - \lambda_2 \ln(x-\delta)^{\frac{1}{b}} - \lambda_3(x-\delta)^{\frac{1}{b}}\right] dx\right\}^2}
$$
$$
\cdot \int_\delta^\infty \exp\left[-\lambda_1 \ln(x-\delta) - \lambda_2 \ln(x-\delta)^{\frac{1}{b}} - \lambda_3(x-\delta)^{\frac{1}{b}}\right] dx
$$
$$
+ \frac{-\int_\delta^\infty (x-\delta)^{\frac{1}{b}} \cdot \exp\left[-\lambda_1 \ln(x-\delta) - \lambda_2 \ln(x-\delta)^{\frac{1}{b}} - \lambda_3(x-\delta)^{\frac{1}{b}}\right] dx}{\left\{\int_\delta^\infty \exp\left[-\lambda_1 \ln(x-\delta) - \lambda_2 \ln(x-\delta)^{\frac{1}{b}} - \lambda_3(x-\delta)^{\frac{1}{b}}\right] dx\right\}^2}
$$
$$
\cdot \int_\delta^\infty (x-\delta)^{\frac{1}{b}} \exp\left[-\lambda_1 \ln(x-\delta) - \lambda_2 \ln(x-\delta)^{\frac{1}{b}} - \lambda_3(x-\delta)^{\frac{1}{b}}\right] dx
$$
$$
= \frac{\int_\delta^\infty \left[(x-\delta)^{\frac{1}{b}}\right]^2 \cdot \exp\left[-\lambda_0 - \lambda_1 \ln(x-\delta) - \lambda_2 \ln(x-\delta)^{\frac{1}{b}} - \lambda_3(x-\delta)^{\frac{1}{b}}\right] dx}{\left\{\int_\delta^\infty \exp\left[-\lambda_0 - \lambda_1 \ln(x-\delta) - \lambda_2 \ln(x-\delta)^{\frac{1}{b}} - \lambda_3(x-\delta)^{\frac{1}{b}}\right] dx\right\}^2} \quad \frac{\partial \lambda_0}{\partial \lambda_3} = -\frac{b - (b\lambda_1 + \lambda_2)}{\lambda_3} \tag{A14}
$$
$$
+ \frac{-\int_\delta^\infty (x-\delta)^{\frac{1}{b}} \cdot \exp\left[-\lambda_0 - \lambda_1 \ln(x-\delta) - \lambda_2 \ln(x-\delta)^{\frac{1}{b}} - \lambda_3(x-\delta)^{\frac{1}{b}}\right] dx}{\left\{\int_\delta^\infty \exp\left[-\lambda_0 - \lambda_1 \ln(x-\delta) - \lambda_2 \ln(x-\delta)^{\frac{1}{b}} - \lambda_3(x-\delta)^{\frac{1}{b}}\right] dx\right\}^2}
$$
$$
\cdot \int_\delta^\infty (x-\delta)^{\frac{1}{b}} \exp\left[-\lambda_0 - \lambda_1 \ln(x-\delta) - \lambda_2 \ln(x-\delta)^{\frac{1}{b}} - \lambda_3(x-\delta)^{\frac{1}{b}}\right] dx
$$
$$
= \frac{\int_\delta^\infty \left[(x-\delta)^{\frac{1}{b}}\right]^2 \cdot f(x) dx}{\left\{\int_\delta^\infty f(x) dx\right\}^2} \cdot \int_\delta^\infty f(x) dx - \frac{\int_\delta^\infty (x-\delta)^{\frac{1}{b}} \cdot f(x) dx}{\left\{\int_\delta^\infty f(x) dx\right\}^2} \cdot \int_\delta^\infty (x-\delta)^{\frac{1}{b}} e f(x) dx
$$
$$
= \frac{\int_\delta^\infty \left[(x-\delta)^{\frac{1}{b}}\right]^2 \cdot f(x) dx}{\left\{\int_\delta^\infty f(x) dx\right\}^2} \cdot \int_\delta^\infty f(x) dx - \frac{\int_\delta^\infty (x-\delta)^{\frac{1}{b}} \cdot f(x) dx}{\left\{\int_\delta^\infty f(x) dx\right\}^2} \cdot \int_\delta^\infty (x-\delta)^{\frac{1}{b}} e f(x) dx
$$
$$
= E\left[(x-\delta)^{\frac{1}{b}}\right]^2 - \left\{E\left[(x-\delta)^{\frac{1}{b}}\right]\right\}^2
$$

References

1. Rao, A.R.; Hamed, K.H. *Flood Frequency Analysis*; CRC Press: Boca Raton, FL, USA, 1999.
2. Meylan, P.; Favre, A.C.; Musy, A. *Predictive Hydrology: A Frequency analysis Approach*; CRC Press; Taylor & Francis Group, Sciences Publishers: Enfield, NH, USA, 2012.
3. Singh, V.P. *Entropy-Based Parameter Estimation in Hydrology*; Kluwer Academic Publishers: Boston, MA, USA, 1998.
4. Sun, J.; Qin, D.; Sun, H. *Generalized Probability Distribution in Hydrometeorology*; China Water and Power Press: Beijing, China, 2001.
5. Singh, V.P.; Deng, Z.Q. Entropy-Based Parameter Estimation for Kappa Distribution. *J. Hydrol. Eng.* **2003**, *8*, 81–92. [CrossRef]
6. Singh, V.P. *Entropy Theory and Its Application in Environmental and Water Engineering*; John Wiley & Sons: New York, NY, USA, 2013.
7. Singh, V.P. The use of entropy in hydrology and water resources. *Hydrol. Process.* **1997**, *11*, 587–626. [CrossRef]
8. Singh, V.P. *Entropy Theory in Hydrologic Science and Engineering*; McGraw-Hill Education: New York, NY, USA, 2015.
9. Singh, V.P. *Entropy Theory in Hydraulic Engineering: An Introduction*; ASCE Press: Reston, VA, USA, 2014.
10. Hao, Z.; Singh, V.P. Entropy-based method for bivariate drought analysis. *J. Hydrol. Eng.* **2013**, *18*, 780–786. [CrossRef]

11. Hao, Z.; Singh, V.P. Entropy-copula method for single-site monthly streamflow simulation. *Water Resour. Res.* **2012**, *48*, W06604. [CrossRef]

12. Hao, Z. Application of Entropy Theory in Hydrologic Analysis and Simulation. Ph.D. Thesis, Texas A&M University, College Station, TX, USA, May 2012.

13. Hao, Z.; Singh, V.P. Modeling multi-site streamflow dependence with maximum entropy copula. *Water Resour. Res.* **2013**, *49*, 7139–7143. [CrossRef]

14. Hao, Z.; Singh, V.P. Single-site monthly streamflow simulation using entropy theory. *Water Resour. Res.* **2011**, *47*. [CrossRef]

15. Hao, Z.; Singh, V.P. Entropy-based method for extreme rainfall analysis in Texas. *J. Geophys. Res.* **2013**, *118*, 263–273. [CrossRef]

16. Hao, Z.; Singh, V.P. Entropy-based parameter estimation for extended Burr XII distribution. *Stoch. Environ. Res. Risk Assess.* **2009**, *23*, 1113–1122. [CrossRef]

17. Mishra, A.K.; Özger, M.; Singh, V.P. An entropy-based investigation into the variability of precipitation. *J. Hydrol.* **2009**, *370*, 139–154. [CrossRef]

18. Cui, H.; Singh, V.P. On the Cumulative Distribution Function for Entropy-Based Hydrologic Modeling. *Trans. ASABE* **2012**, *55*, 429–438. [CrossRef]

19. Cui, H.; Singh, V.P. Two-Dimensional Velocity Distribution in Open Channels Using the Tsallis Entropy. *J. Hydrol. Eng.* **2013**, *18*, 331–339.

20. Cui, H.; Singh, V.P. Computation of Suspended Sediment Discharge in Open Channels Using Tsallis Entropy. *J. Hydrol. Eng.* **2013**, *19*, 18–25. [CrossRef]

21. Cui, H.; Singh, V.P. One Dimensional Velocity Distribution in Open Channels Using Tsallis Entropy. *J. Hydrol. Eng.* **2013**, *19*, 290–298. [CrossRef]

22. Cui, H.; Singh, V.P. Configurational Entropy Theory for Streamflow Forecasting. *J. Hydrol.* **2015**, *521*, 1–7. [CrossRef]

23. Cui, H.; Singh, V.P. Minimum Relative Entropy Theory for Streamflow Forecasting with Frequency as a Random Variable. *Stoch. Environ. Res. Risk Assess.* **2016**, *30*, 1545–1563. [CrossRef]

24. Papalexiou, S.M.; Koutsoyiannis, D. Entropy based derivation of probability distributions: A case study to daily rainfall. *Adv. Water Resour.* **2012**, *45*, 51–57. [CrossRef]

25. Brouers, F. The Burr XII Distribution Family and the Maximum Entropy Principle: Power-Law Phenomena Are Not Necessarily Nonextensive. *Open J. Stat.* **2015**, *5*, 730–741. [CrossRef]

26. Shao, Q.; Wong, H.; Xia, J.; Ip, W.C. Models for extremes using the extended three-parameter Burr XII system with application to flood frequency analysis. *Hydrol. Sci. J.* **2004**, *49*, 685–702. [CrossRef]

12

Information Entropy Suggests Stronger Nonlinear Associations between Hydro-Meteorological Variables and ENSO

Tue M. Vu, Ashok K. Mishra * and Goutam Konapala

Glenn Department of Civil Engineering, Clemson University, Clemson, SC 29634, USA;
tuev@g.clemson.edu (T.M.V.); gkonapa@clemson.edu (G.K.)
* Correspondence: ashokm@g.clemson.edu

Abstract: Understanding the teleconnections between hydro-meteorological data and the El Niño–Southern Oscillation cycle (ENSO) is an important step towards developing flood early warning systems. In this study, the concept of mutual information (MI) was applied using marginal and joint information entropy to quantify the linear and non-linear relationship between annual streamflow, extreme precipitation indices over Mekong river basin, and ENSO. We primarily used Pearson correlation as a linear association metric for comparison with mutual information. The analysis was performed at four hydro-meteorological stations located on the mainstream Mekong river basin. It was observed that the nonlinear correlation information is comparatively higher between the large-scale climate index and local hydro-meteorology data in comparison to the traditional linear correlation information. The spatial analysis was carried out using all the grid points in the river basin, which suggests a spatial dependence structure between precipitation extremes and ENSO. Overall, this study suggests that mutual information approach can further detect more meaningful connections between large-scale climate indices and hydro-meteorological variables at different spatio-temporal scales. Application of nonlinear mutual information metric can be an efficient tool to better understand hydro-climatic variables dynamics resulting in improved climate-informed adaptation strategies.

Keywords: information entropy; mutual information; kernel density estimation; ENSO; nonlinear relation

1. Introduction

For many water resources planning and management studies, reliable preliminary estimates of dependence between two hydroclimatic variables are extremely important. For example, knowledge of dependence between large-scale climate patterns such as El Niño–Southern Oscillation (ENSO) [1], the Pacific Decadal Oscillation (PDO) [2], and the Atlantic Multi-decadal Oscillation (AMO) [3] with local precipitation, temperature, or streamflow has resulted in improved longer lead-time forecasting models [4–6]. Large-scale climate patterns can also predict ecological processes better than local weather [7]. In addition to that, several studies indicated the presence of a significant relationship between large-scale climate phenomena and hydrologic extremes, such as extreme precipitation events [8–10], droughts [11–13], and floods [14,15]. Therefore, the presence of any kind of significant dependence forms the preliminary metric to identify appropriate predictors for forecasting streamflow and other hydroclimatic variables in ungauged river basins [16–18]. As a result, the predictability of these large-scale climate patterns much in advance is extremely important to improve the design of early warning systems of extreme events [19].

A wide range of methods is available for detecting the presence of an association between bivariate data set. Among them, Pearson's correlation coefficient is the most widely used metric to quantify the linear dependence between any two variables [20]. Pearson correlation coefficient is based on the assumption that the considered variables follow a Gaussian distribution. Therefore, using Pearson correlation in case of variables that follow non-Gaussian distributions may be suboptimal.

However, several insights were derived from these linear associations. For example, Zhang et al. [21] quantified the Pearson linear correlation between different sea surface temperature (SST) anomalies and seasonal precipitation for Huai River basin in China. The authors identified some positive/negative correlation with the coefficients ranges from absolute 0.2 to 0.3. Whereas, Wrzesinski [22] characterized and confirmed the strong influences of large-scale climate index NAO to seasonal river flow in Poland by comparing the difference in average runoff during the positive and negative NAO phases. On the same lines, Wrzesinski [23] also found the linkage between NAO and European river streamflow at the 140 gauging station by analyzing the high and low water flows according to the positive and negative NAO phases.

In addition, most of the hydro-climatic series do not follow a Gaussian distribution leading to a possible misinterpretation of the dependency between the variables when using linear measures [24–26]. As a result, the non-parametric rank-based correlation metrics of Kendall's tau and Spearman's rho are applied to quantify the relationship between two given variables [27–29]. However, even though the associations evaluated by these two metrics are independent of any probability distribution assumptions, they are more successful in detecting the monotonous relationships between any two variables. In addition, assuming a monotonous relationship among two hydroclimatic variables might be too restrictive to characterize the existing complex dependence structures between the hydroclimatic variables [30–32].

To overcome the limitations of linear dependence, several studies in hydro-climatology adopted the concepts based on nonlinear statistics to evaluate the strength of association among distinct hydroclimatic variables. For instance, Fleming et al. [33] observed a nonlinear association between northern hemisphere river basin streamflow and teleconnection patterns. In regard to climate extremes, Cannon [34] utilized generalized extreme value distributions to investigate the relationship between ENSO and winter extreme station precipitation in North America. Whereas Lin-Ye et al. [35] investigated the relationship between extreme events of wave storms and large-scale climate covariates using generalized additive and linear models. While Kusumastuti [36] studied the nonlinearity between threshold and rainfall-runoff transformation. More recently, Konapala et al. [37] have investigated the nonlinear relationship between low flows in Texas river basins with large scale climatic patterns.

Recently entropy-based approaches are gaining popularity for different applications, such as, climate variability [38], uncertainty analysis [39], hydrometric network analysis [40–42], selection of predictors [43,44], and drought regionalization [45]. Similarly, the entropy theory based mutual information (MI) is widely used for quantifying the non-linear association between multiple variables [40,41]. The MI concepts have been utilized to quantify nonlinear dependence structure between several climate variables across the globe [46,47]. Also, several MI metrics were developed to assist in predictor selection for hydroclimatic forecasting [48,49]. More importantly, MI was able to detect the presence of strong nonlinear associations of streamflow, rainfall, and global mean temperature with several large-scale climatic variables [31'50–54].

Unlike the signals in other research fields, hydroclimatic data being a subset of geophysical research field are non-repeatable, shorter in length and contaminated with significant noise levels [31]. In addition to that, studies have indicated a presence of chaotic component [55]. Therefore, it becomes evident that the selected measure should be relatively robust to noise and chaos and, more importantly, detects signals possibly shorter in length. Thus, the goal of this study is to investigate and compare the performance of nonlinear estimation method based on mutual information (MI) with the traditional linear association metric of Pearson correlation. For this purpose, the hydro-meteorological data are considered such as total annual precipitation, annual average streamflow, and extreme precipitation indices specific to Mekong River Basin in Southeast Asia and estimated its association with the large-scale climate index of ENSO.

2. Materials and Methods

2.1. Hydro-Meteorological Data and ENSO

The Mekong River Basin (MRB), originates from Tibetan Plateau in China and flows through the territory of five other countries in the Southeast Asia (Myanmar, Laos PDR, Thailand, Cambodia, and drains to the sea in Vietnam) (Figure 1a). This is currently home to more than 70 million people and it is expected to increase by 100 million in 2050 [56]. The detailed topography map of study region is displayed in Figure 1a with relative high terrain in the upstream of the basin (upper Mekong). In this study, the gridded precipitation data is provided from Asian Precipitation Highly Resolved Observational Data Integration Towards the Evaluation of Water Resources (APHRODITE) with a spatial resolution of 0.25° and daily temporal resolution with the sufficiently long record period from 1951–2007 covering the Asian monsoon domain. The dataset was created primarily with data obtained independently from rain gauge observation network across all the regions of Asia. The daily precipitation values from rain gauges were interpolated using the sphere map technique [57] and the first six components of the fast Fourier transforms were taken to obtain daily data for all land areas in the Asian monsoon domain [58]. This dataset is referred to as "APH" in this paper, for the purpose of brevity. APH has been previously utilized as a reliable gridded rain data set over this area for various hydrological applications [59–63]. The spatial distribution of total precipitation from APH data over MRB is displayed in Figure 1b with high rainfall volumes (around 2000 to 2500 mm) gathering around the eastern part of the river basin near to the Annamite mountain range.

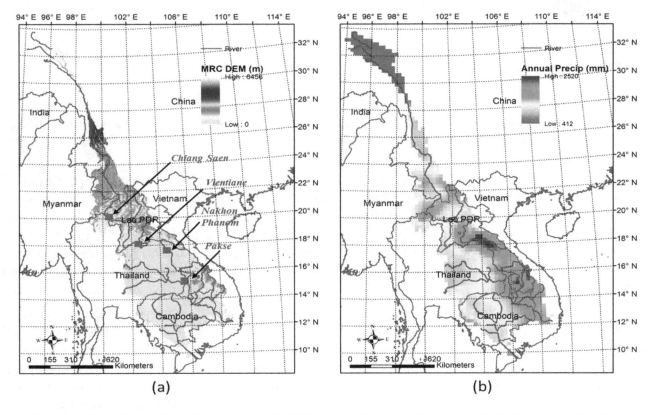

Figure 1. (a) Mekong River Basin map with digital elevation model terrain and digitized river, location of four hydro-meteorology stations are displayed; (b) Annual rainfall distribution over MRB using APHRODITE gridded rainfall 1951–2007.

The southwest monsoon wind originated from the Gulf of Thailand blocked by the Annamite mountain range is the main factor that causes the high seasonal rainfall in summer months (June, July August, September) [61] (Figure 2a). The APH daily gridded rainfall values were extracted to station locations using bilinear interpolation [59,63,64] to four meteorological stations located in MRB—Chiang Saen, Vientiane, Nakhon Phanom, and Pakse (Figure 1a)—for the time period of 1951 to 2007 over the Mekong River Basin. In addition to rainfall data, the corresponding annual streamflow data were also collected at these four hydrological stations from the MRC website for the same period. The monthly mean climatology over 57 years of hydro-meteorological stations is displayed in Figure 2a. Even though different patterns are found over four stations, the hydrographs are quite similar from the upstream station at Chiang Saen to downstream station Pakse, with flooding season from May to November, while the peak discharge months are around August and September. This study considers the seasonal cycle which includes all 12 months of the hydrologic year (starting from the month with the lowest rainfall—i.e., May—and ending in April of the next year; the streamflow month is considered to lag one month which starts from June and ends in May next year).

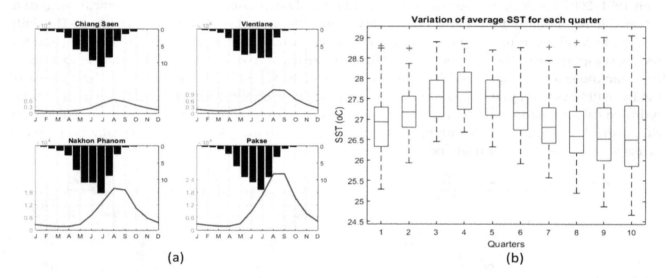

(a)   (b)

Figure 2. (**a**) Seasonal cycle of monthly climatology rainfall and streamflow measured at four stations on the mainstream Mekong river; (**b**) Boxplots of monthly SST Nino 3.4 for different quarters

The study focuses on the total annual rainfall, annual streamflow, as well as extreme precipitation indices [65] for hydrologic years derived from daily precipitation datasets for both wet and dry spell statistics, which are (1) R5d: Max consecutive five-day rainfall, (2) P90p: 90th percentile of the daily precipitation time series for the year, (3) SDII: average daily precipitation on a wet day, and (4) dry spell length computed from daily precipitation for a year. Among four stations, Pakse has the highest streamflow data due to its furthest downstream location. Nakhon Phanom and Pakse have the highest rainfall as seen in Figures 1 and 2.

The assumption has been that the long-term hydro-meteorological data variability can be captured by the fluctuation of the SST over seasonal scales. In this study, the ENSO index has been considered by using the SST over NINO 3.4 region [60,66], which is the area averaged monthly SST over the region bounded by the coordinates 5° N–5° S, 170° W–120° W. This time series could effectively indicate the occurrence of ENSO events [66]. The dataset can be downloaded from the Climate Data Guide website by NCAR [67]. Therefore in this study, the ENSO indices are computed at multiple three-month time windows in terms of quarters [31] consisting of the average SST anomalies over months of JFM, FMA, MAM, etc., for all the years (J: Jan, F: Feb, M: Mar). The SST over Nino 3.4 regions boxplots bound for 57 years for 10 different quarters are displayed in Figure 2b. The red crosses in Figure 2b denote outliers. There are four quarters just before the seasonal cycle, four quarters corresponding to the seasonal cycle, and two quarters after seasonal cycle, computed with the mean monthly SST over Nino 3.4 region. Among all 10 quarters, the fourth quarter (AMJ) has the highest median and whisker values compared to other quarters whilst the last two quarters (9 and 10) have the lowest median values.

2.2. Mutual Information Estimation

The mutual information (*MI*) has been utilized to capture the nonlinear dependence structure between two random variables. When analyzing experimental times series from the non-linear system, the *MI* is especially an important statistics [68]. According to [69], there are three theorems for *MI* between two random variables *X*, *Y*: (1) *MI* is non-negative and is zero if *X* and *Y* are strictly independent; (2) *MI* is infinity if there exists a function "g" such that $X = g(Y)$; (3) *MI* is invariant to separate one to one transformations.

The *MI* can be computed using the relative entropy suggested by Joe [70]. Assuming that a pair of continuous random variables (*X*, *Y*) exist which have a joint probability density function (pdf) p_{XY} and with its marginal pdf accordingly p_X and p_Y. The mutual information or relative entropy can be defined as

$$MI(X,Y) = \int \int p_{XY} \log \frac{p_{XY}}{p_X p_Y} dx dy \tag{1}$$

It is noted that the Equation (1) measures the distance between a joint distribution and the distribution when there is independence [69]. In case of continuous variables, there is no direct way to accurately determine continuous probability distributions. Therefore, several methods have been introduced to approximate the continuous probability distribution functions as discrete distributions. Among them, Khan et al. [71] compared four different methods to estimate the probability distribution function to calculate *MI* using: kernel density estimator (KDE) [68], K-nearest neighbors (KNN) [72], Edgeworth approximation of multivariate differential entropy [73], and adaptive partitioning of the *XY* plane [74]. The authors found that KDE and KNN outperform the other two methods in term of their ability to capture the dependence structure. Khan et al. [31] indicated that KDE is able to capture the underlying nonlinear dependence more consistently compared to KNN and Edgeworth when they are short and noisy assuming such dependence exists. Therefore, this article utilizes the KDE to estimate probability density function using the equation

$$p(x) = \frac{1}{n} \sum_i^n K(u) \tag{2}$$

in which, $K(u)$ is a multivariate kernel function

$$K(u) = \frac{1}{h^d \sqrt{(2\pi)^d \det(S)}} \exp\left(-\frac{u}{2}\right) \tag{3}$$

$$u = \frac{(x - x_i)^T S^{-1}(x - x_i)}{h^2} \tag{4}$$

From (4), x_i a *d*-dimensional random vector for the multivariate data set $(x_1, \dots, x_n)$. *S* is the covariance matrix on the x_i, det(*S*) is a determinant of *S* and *h* is the kernel bandwidth or smoothing variable. The optimal Gaussian bandwidth "*h*" can be computed using Equation (5) [75]

$$h = \left(\frac{4}{d+2}\right)^{\frac{1}{d+4}} n^{-\frac{1}{d+4}} \tag{5}$$

The "*d*" is taken from [76] for the value of *d* = 2, and is similar to [68].

Substituting Equations (3) and (4) to Equation (2) to obtain the approximate probability density function as

$$p(x) = \frac{1}{n} \sum_i^n \frac{1}{h^d \sqrt{(2\pi)^d \det(S)}} \exp\left[-\frac{(x - x_i)^T S^{-1}(x - x_i)}{2h^2}\right] \tag{6}$$

The detailed procedure using KDE to estimate pdf can be found in [31,68,71]. Here, the discrete formulation of MI is shown in Equation (7)

$$MI(X,Y) = \frac{1}{n}\sum_{i=1}^{n} \ln \frac{p_{XY}(x_i, y_i)}{p_X(x_i)p_Y(y_i)} \tag{7}$$

In which $p_{XY}(x_i, y_i)$ is the joint pdf and $p_X(x_i), p_Y(y_i)$ are marginal pdfs at (x_i, y_i). The MI values range between independent ($MI = 0$) to completely dependent ($MI = \infty$). In order to make the generalization of the correlation with a range from 0 (independent) to 1 (dependent), Joe [70] proposed a formula to transform the MI to nonlinear correlation coefficient ($NLCC$)

$$NLCC = \sqrt{[1 - \exp(-2MI)]} \tag{8}$$

The $NLCC$ range is similar to linear correlation coefficients (LCC) and has been used in most of the studies by [31,69,70].

Figure 3 modified [77] using their python code to demonstrate the LCC and $NLCC$ using the scatter plots of two random variables obtained from a different sample of data. The advantage of $NLCC$ is based on the fact that MI makes no assumption on the distribution of the variables or the nature of the relationship between them and is sensitive to nonlinear and non-monotonic effects [77]. It can be observed that (top row of Figure 3) both the LCC and $NLCC$ can able to capture the linear relationship between the two random variables with a very close range of values. However, the advantage of $NLCC$ compared to LCC is due to its ability to recognize the different distributions of the two random variables (bottom row of Figure 3).

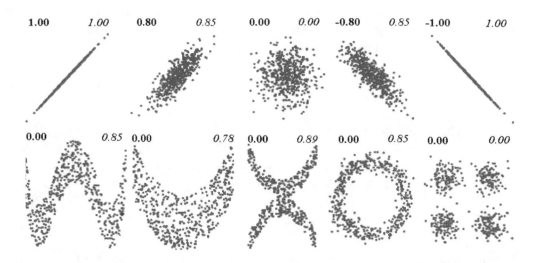

Figure 3. The comparison between linear correlation coefficient (LCC) and nonlinear correlation coefficient ($NLCC$) obtained from mutual information (MI) computed based on two random variables. For each panel, top left value (bold) shows LCC and top right (italic) represents $NLCC$. The top row illustrates the similarity between LCC and $NLCC$, whereas, the bottom row exhibits advantages of non-linear relationship based on different distribution. (Figures partially adopted from [77] using their python code. The $NLCC$ was computed based on MI values).

Subsequently, the $LCCs$ are computed between each of the precipitation indices and compared with $NLCCs$ at four hydro-meteorology stations located in MRB (Figure 1a). The four stations are chosen based on their differences in terms of seasonal cycles, total annual rainfall amount and geographical height. Finally, the gridded linear/non-linear CC is constructed for the MRB to showcase the spatial variability of the correlation coefficient.

3. Results

3.1. Linear and Nonlinear Correlation between Annual Precipitation/Streamflow and ENSO Index

We first illustrate the linear dependence as a bivariate normal distribution and nonlinear dependence as a kernel bivariate distribution following the work of Khan et al. [71]. The bivariate normal and kernel density between the annual average streamflow at different hydrologic gauging stations and different quarters of ENSO indices are computed and plotted in Figure 4 for the highest and lowest linear and nonlinear CCs. For kernel density, a Gaussian kernel with an optimal Gaussian bandwidth computed by $h = N^{-1/6}$ with N is the total number of observed points (57 in this case).

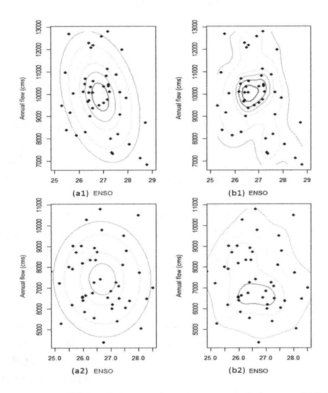

Figure 4. The bivariate normal and kernel density between annual flow and ENSO index for (**a**) normal density; (**b**) kernel density; (1) Quarter 1 at Pakse station (highest linear and nonlinear CCs) (2) Quarter 9 at Nakhon Phanom station (lowest, respectively).

Subsequently, the linear and nonlinear CCs between total annual precipitation and average streamflow were computed at different window lengths of SST Nino 3.4 (quarters) for four different hydro-meteorology stations and display in Figure 5. Based on the formulation Equation (8) *NLCCs* are positive, therefore the absolute value of *LCCs* are computed for comparison plotting in Figure 5 and all other figures henceforth. In order to compute the 90% confidence intervals for the absolute correlation coefficients, the bootstrapping approach [71] was applied using 100 simulations and plotted in Figure 5. It can be seen in Figure 5 that the *NLCCs* have higher values than the *LCCs* for all quarters. It indicates that the KDE captures the more extrabasinal connection between ENSO and precipitation as well as river flows compared to linear modeling [31].

In particular at station scale, Nakhon Phanom has the highest precipitation amount, its *LCCs* and *NLCCs* are also among the highest for the first four quarters Figure 5(c1,c2) which subsequently decrease. The *NLCC* values also have the same patterns as *LCCs* for this station as well as Pakse in Figure 5(d1). Although both stations have highest CCs for the first quarter, however, the lowest CCs values are not found at the same: Quarter 7 (Nakhon Phanom) and Quarter 4 (Pakse). Chiang Saen and Vientiane have variation trends for *LCCs* Figure 5(a1,b1) among all quarters, this is perhaps due to the less rainfall amount observed at these two rain gauges.

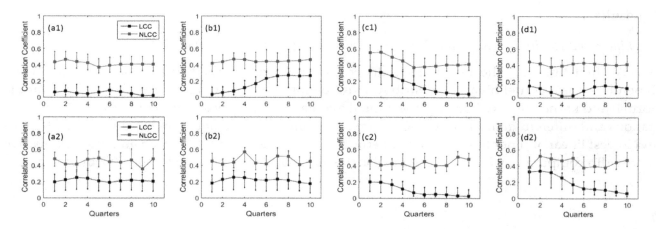

Figure 5. Linear and nonlinear CCs with their 90% confidence intervals between SST Nino 3.4 and (1) total annual precipitation (2) annual average streamflow for (**a**) Chiang Saen; (**b**) Vientiane; (**c**) Nakhon Phanom; and (**d**) Pakse. Note: The correlation coefficients (*CCs*) are calculated between precipitation and ENSO indices in-terms of quarterly temporal windows, such as, JFM, FMA, and MAM and so on. The 90% confidence intervals are computed using 100 simulations by bootstrapping approach.

Among four hydrology gauging stations, Pakse also has highest *LCCs/NLCCs* with ENSO at the first four quarters Figure 5(d2). This is because Pakse is the furthest station downstream and it has the highest streamflow measured. Therefore, it shows more dependences by the fluctuations of ENSO indices. The second highest streamflow station is Nakhon Phanom (see Figure 5(c2)) that also indicates similar patterns of *LCCs* and *NLCCs* with higher dependences with the first four quarters of ENSO indices compared with the rest. When the *LCC* values are close to zero (as in the last four quarters of Figure 5(c2)), the differences between *LCCs* and *NLCCs* should be interpreted with caution because of the exponential scaling of *MI* in *NLCC* as shown in Equation (8). The two upstream gauging stations Chiang Saen and Vientiane (Figure 5(a2,b2)) also have similar patterns for *LCCs* and *NLCCs* at different quarters. Compared to total annual precipitation, the annual average streamflow has slightly higher *LCCs* and *NLCCs*, except for Nakhon Phanom station. Overlapping the 90% confidence intervals of *LCCs* and *NLCCs* for streamflow at Chiang Saen, Vientiane, and Pakse (Figure 5(a2,b2,d2)) for the first three quarters indicates that both KDE and linear regressions effectively capture the strong dependence structure.

3.2. Linear and Nonlinear Correlation between Precipitation Extreme Indices and ENSO

In addition to total annual precipitation and streamflow dependences, further investigations on *LCCs* and *NLCCs* are carried out between precipitation indices for extreme values indices (R5d, SDII, P90p, dry spell) with different quarters of ENSO events. This analysis aims to quantify the dependence structure between annual extreme precipitation events and ENSO indices. The wet indices (R5d, P90p, SDII), as well as dry indices (dry spell), are also taken into account for the analyses. The detail dependences based on *LCCs* and *NLCCs* can be found in Figure 6 for four indices and four stations over 10 quarters. Similar to Figure 5, the *LCCs* are displayed as absolute values comparable to *NLCCs*. The 90% confidence intervals for the absolute correlation are computed based on 100 simulations using bootstrapping approach. The detail analyses on Figure 6 reveals several significant dependencies between annual extreme precipitation events and ENSO indices. Chiang Saen exhibits high *LCCs/NLCCs* for a maximum five consecutive days of rainfall with the last three quarters of ENSO indices and low *CCs* values for the first three quarters Figure 6(a1). Similar patterns are found for Pakse station in Figure 6(d1) but with the highest *CCs* among Quarters 6, 7, 8 and lowest at the first three quarters. This variation is different from the total annual precipitation in Figure 5(a1,d1). In contrast to R5d, the other two indices: P90p and SDII have similar patterns of *LCCs/NLCCs* compared to the total annual precipitation in Figure 5. The last three-quarters of ENSO indices show that it has higher *LCCs/NLCCs* for Vientiane and Pakse Figure 6(b2,b3,d2,d3) compared to other quarters whilst the first three quarters of Nakhon Phanom has the highest values than the last three quarters. The dry spell indices of all four stations show nearly opposite patterns compared to R5d. The overall pattern illustrates that nonlinear *CCs* have higher values (more dependencies) than linear *CCs* for all stations/indices, similar to Figure 5 and [31].

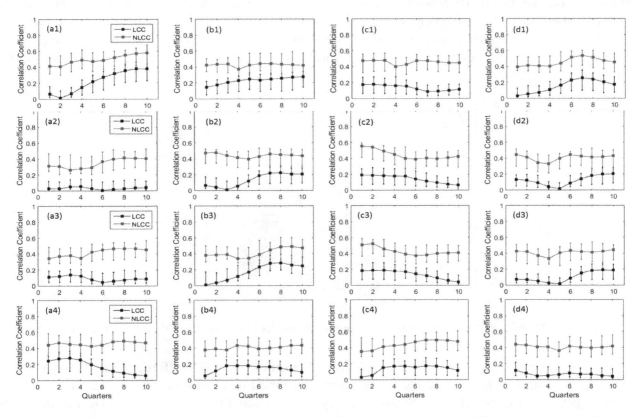

Figure 6. Comparison between linear and nonlinear *CCs* derived between SST Nino 3.4 and (1) R5d; (2) P90p, (3) SDII; (4) dry spell for selected stations located at (**a**) Chiang Saen; (**b**) Vientiane; (**c**) Nakhon Phanom; and (**d**) Pakse. Note: The correlation coefficients (*CCs*) are calculated between precipitation and ENSO indices in-terms of quarterly temporal windows, such as JFM, FMA, MAM, and so on. The 90% confidence intervals are computed using 100 simulations by bootstrapping approach.

The analyses with total annual precipitation, annual average streamflow, and extreme precipitation indices on wet and dry conditions reveal that there exists a nonlinear extrabasinal connection between ENSO and the hydro-meteorological data over Mekong river basin. The analyses based on *LCCs/NLCCs* exhibit the increasing trend in the variation of annual statistics on hydro-meteorology in connection with ENSO by computing nonlinear relationship as compared to linear measures. Therefore, it is expected to give additional support for early prediction (compared to traditional linear measurement) based on the ENSO forecast with the hydro-meteorology connection when *MI*-based approaches are utilized. This approach, somehow, would be helpful in water resources management for drought mitigation, flood control, as well as an irrigation system for agricultural areas.

3.3. Spatial Linear and Nonlinear Correlation Maps

Based on the correlation analysis at four rainfall stations, it was observed that the *NLCCs* have higher values in comparison to *LCCs* at all the selected stations located in MRB. Further analyses of the spatial pattern using the gridded data from APH based on the selected statistics of annual precipitation and extreme indices have been carried out. The linear and nonlinear *CCs* between ENSO Quarter 1 and annual rainfall are displayed in Figure 7 (using the same color bar), and the correlation values range from 0 to 0.6. Quarter 1 was selected arbitrarily as the first quarter, even though, it shows the highest CCs over Nakhon Phanom and Pakse for total annual precipitation but it also displays the lowest CCs for other analyses such as for Vientiane extreme indices (Figure 6b) or Pakse in Figure 6(d1,d3). Figure 7a exhibits the higher CC values between total annual precipitation and ENSO Quarter 1 (about 0.4) from Nakhon Phanom to Pakse station, along with Laos, eastern Thailand, and northern Cambodia. This can be clearly observed during the first quarter time frame Figure 5(c1,d1). On the other hand, the NLCCs have higher CC values (about 0.5 to 0.6) for the same locations compared to other grids. The use of spatial distribution map can be extended to generate the teleconnection patterns in ungauged regions. That, in turn, helps to better inform local stakeholders in building better tools for water resource management.

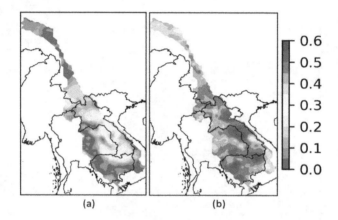

Figure 7. (**a**) Linear and (**b**) nonlinear CCs between Nino 3.4 at Quarter 1 and annual precipitation.

Similarly, the dependency between selected precipitation indices (R5d, P90p, SDII, and dry spell) and Quarter 1 of ENSO are computed and displayed in Figure 8. The linear CCs obtained from precipitation indices (Figure 8) are slightly lower than that of annual precipitation in the previous analysis (Figure 7) for Laos, eastern Thailand, and northern Cambodia. There is slightly higher LCCs/NLCCs magnitude observed at southern Cambodia during a dry spell Figure 8(a4,b4). The magnitudes of NLCCs (Figure 8) are also slightly lower than NLCCs based on the annual precipitation (Figure 7). However, the values obtained from NLCCs are still significantly higher than LCCs for all extreme indices. The lower Mekong basin (the southern part of river basin spread over Laos, Thailand, and Cambodia to Vietnam) has higher correlations to ENSO indices compared to the northern/upper basin, similar to [78].

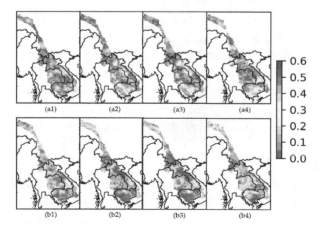

Figure 8. (**a**) Linear and (**b**) nonlinear CCs between Nino 3.4 at quarter 1, and (1) R5d, (2) P90p, (3) SDII, (4) dry spell.

4. Discussion

Although ENSO has a direct influence on rainfall anomalies over the tropical and subtropical regions, only a portion of the variation in the annual flow of rivers located in these regions is associated with ENSO events [31]. This study discusses in detail the possible dependences between different quarters of ENSO indices and hydro-meteorological dataset over Mekong river basin. Other existing studies on MRB highlighted that the linear correlations between ENSO and streamflow over the selected locations in Chiang Saen, Vientiane, and Pakse are around −0.4 to 0.2 [78] and maximum linear correlations between ENSO and precipitation over several stations in Thailand are around −0.18 to 0.22 [79]. The above correlation figures are in line with the study that the maximum LCCs obtained are around 0.35 for both annual precipitation and streamflow as displayed in Figure 5. The results from mutual information derived *NLCCs* therefore suggests an additional approach to look into the dependence structure for multivariate hydro-meteorological data with the large-scale climate patterns. ENSO is a periodic climatic phenomenon with 3–7 years of the cycle and can be predicted several seasons in advances [80]. For instance, Ludescher et al. [81] were able to predict the likelihood of El Niño conditions in 2014 almost a year in advance indicating an improvement in our prediction capacity. In addition, other researchers [82,83] also revealed the intensification of ENSO related precipitation and El Niño frequency in the future due to the warming associated with an increase in greenhouse gas emission. Moreover, Ward et al. [15] indicated that the global flood risk exists during El Niño or La Niña years, or both, in basins spanning 44% of the land surface of the Earth. Therefore, the predictability of ENSO has significant implications and the ability to predict in advance might lead to better local water resources management towards developing more efficient flood and drought early warning systems.

This study limits to using only the kernel density estimation approach to estimate the mutual information values. However, there are several methods that have been used to compute the mutual information, such as K-nearest neighbors (KNN) [72], Edgeworth approximation of multivariate differential entropy [73] and adaptive partitioning of the XY plane [74]. The detailed comparisons among these estimator approaches were carried out in [71] for a different type of data such as linear, quadratic, periodic function, chaotic system. Khan et al. [71] concluded that KDE is the best choice for very short data (50–100 data points). Therefore, KDE approach was utilized to estimate the bivariate mutual information in this study. In addition to mutual information method, there are several approaches used to build the nonlinear relationship between multivariate. Lin-Ye et al. [35] applied the VGAM/VGLM to quantify the nonlinear relationship between storm components and large-scale climate indices (NAO and others) using Global Climate Model data via the regression coefficients that were used to build the location and scale parameters of their statistical model. Zhang et al. [84] investigated the nonlinear relationship by employing the concept of mutual information to evaluate the dependency between the normalized difference vegetation index (NDVI) and meteorological variables for the middle reach of the Hei river basin. Higher dependency between NDVI and coupled precipitation/temperature was observed for the desert area whilst for oasis region, groundwater is an important factor for driving vegetation growth. The authors utilized the mutual information as a method to classify the study region into a smaller area (desert, oasis, artificial oasis).

The *NLCCs* and *LCCs* analysis between ENSO and hydro-meteorological data suggest that there exists a nonlinear correlation for the Mekong River Basin. This finding agrees well with a previous study [28] with a focus on river flow analysis over tropical and sub-tropical river basins. It can be observed that although the linear CCs can be 0, the smallest *NLCCs* can be around 0.3 to 0.4. This needs caution from an artifact of Equation (8) which scales nonlinear CCs exponentially with *MI* [31] when linear CCs are close to 0. It, however, does not affect the other values of CCs higher than 0.

5. Conclusions

This study analyses the possible influence of the large-scale climate index ENSO and hydro-meteorological data in the form of total annual precipitation, annual average streamflow, and extreme precipitation indices using linear and nonlinear correlation over the Mekong River Basin. The nonlinear correlation structure was computed based on mutual information approach using marginal and joint entropy. The kernel density estimation approach was selected among several other techniques to estimate the mutual information values as this approach works best for very short data length of 50 to 100 points. Nonlinear correlations were obtained by transforming the mutual information using Joe formula [70] and thus are comparable with absolute linear correlation coefficients. Bootstrapping approach based on 100 simulations was applied to find the confidence intervals for these absolute correlation coefficients. In-depth analysis was carried out at four hydro-meteorological stations located in the Mekong River Basin. The major conclusions that can be drawn from this study are listed below:

1. Nonlinear correlation is able to reveal the additional dependence structures between hydro-meteorological data over Mekong River Basin and ENSO indices.

2. Both linear and nonlinear correlation exhibits similar varying patterns among different ENSO quarters for most of the stations/indices.

3. The results reveal that higher correlation coefficients can be found using the nonlinear correlation coefficients in comparison to the traditional linear correlation analysis.

4. Spatial correlation structures for *LCCs* and *NLCCs* are also constructed based on extreme precipitation indices and ENSO. The use of spatial maps further complements our analyses based on a single station to other ungagged regions to better inform local stakeholders in building better tools for water resource management.

5. Further analyses are required to reveal the non-linear association between other large-scale climate phenomena (SOI, PDO, NAO, etc.) with local meteorological variables. The mutual information between these indices and local meteorological variables can further help policymakers to improve climate-informed adaptation studies.

Acknowledgments: We appreciate the suggestions provided by the associate editor and three reviewers that helped us to improve quality of our manuscript. Authors would also like to thank the American International Group and the Risk Engineering and Systems Analytics Center, Clemson University for providing financial support.

Author Contributions: Tue Vu conducted experiments analyzed the data/results, Goutam Konapala involved in data collection and preparing Sections 1 and 2. Ashok Mishra supervised the overall procedure and revised the paper. All authors have read and approved the final manuscript.

References

1. Gu, D.; Philander, S.G.H. Interdecadal climate fluctuations that depend on exchanges between the tropics and extratropics. *Science* **1997**, *275*, 805–807. [CrossRef] [PubMed]

2. Mantua, N.J.; Hare, S.R.; Zhang, Y.; Wallace, J.M.; Francis, R.C. A Pacific interdecadal climate oscillation with impacts on salmon production. *Bull. Am. Meteorol. Soc.* **1997**, *78*, 1069–1079. [CrossRef]

3. Enfield, D.B.; Mestas-Nuñez, A.M.; Trimble, P.J. The Atlantic multidecadal oscillation and its relation to rainfall and river flows in the continental US. *Geophys. Res. Lett.* **2001**, *28*, 2077–2080. [CrossRef]

4. Wood, A.W.; Maurer, E.P.; Kumar, A.; Lettenmaier, D.P. Long-range experimental hydrologic forecasting for the eastern United States. *J. Geophys. Res. Atmos.* **2002**, *107*. [CrossRef]

5. Tootle, G.A.; Piechota, T.C.; Singh, A. Coupled oceanic-atmospheric variability and US streamflow. *Water Resour. Res.* **2005**, *41*. [CrossRef]

6. Kalra, A.; Ahmad, S. Using oceanic-atmospheric oscillations for long lead time streamflow forecasting. *Water Resour. Res.* **2009**, *45*. [CrossRef]

7. Hallett, T.B.; Coulson, T.; Pilkington, J.G.; Clutton-Brock, T.H.; Pemberton, J.M.; Grenfell, B.T. Why large-scale climate indices seem to predict ecological processes better than local weather. *Nature* **2004**, *430*, 71–75. [CrossRef] [PubMed]

8.	Cayan, D.R.; Redmond, K.T.; Riddle, L.G. ENSO and hydrologic extremes in the western United States. *J. Clim.* **1999**, *12*, 2881–2893. [CrossRef]

9.	Jones, C. Occurrence of extreme precipitation events in California and relationships with the Madden-Julian oscillation. *J. Clim.* **2000**, *13*, 3576–3587. [CrossRef]

10.	DeFlorio, M.J.; Pierce, D.W.; Cayan, D.R.; Miller, A.J. Western U.S. Extreme Precipitation Events and Their Relation to ENSO and PDO in CCSM4. *J. Clim.* **2013**, *26*, 4231–4243. [CrossRef]

11.	Barlow, M.; Nigam, S.; Berbery, E.H. ENSO, Pacific Decadal Variability, and U.S. Summertime Precipitation, Drought, and Stream Flow. *J. Clim.* **2001**, *14*, 2105–2128. [CrossRef]

12.	Mo, K.C. Drought onset and recovery over the United States. *J. Geophys. Res.* **2011**, *116*. [CrossRef]

13.	Özger, M.; Mishra, A.K.; Singh, V.P. Low frequency drought variability associated with climate indices. *J. Hydrol.* **2009**, *364*, 152–162. [CrossRef]

14.	Andrews, E.D.; Antweiler, R.C.; Neiman, P.J.; Ralph, F.M. Influence of ENSO on flood frequency along the California coast. *J. Clim.* **2004**, *17*, 337–348. [CrossRef]

15.	Ward, P.J.; Jongman, B.; Kummu, M.; Dettinger, M.D.; Weiland, F.C.S.; Winsemius, H.C. Strong influence of El Niño Southern Oscillation on flood risk around the world. *Proc. Natl. Acad. Sci. USA* **2014**, *111*, 15659–15664. [CrossRef] [PubMed]

16.	Sivapalan, M. Prediction in ungauged basins: A grand challenge for theoretical hydrology. *Hydrol. Process.* **2003**, *17*, 3163–3170. [CrossRef]

17.	Hrachowitz, M.; Savenije, H.H.G.; Blöschl, G.; McDonnell, J.J.; Sivapalan, M.; Pomeroy, J.W.; Arheimer, B.; Blume, T.; Clark, M.P.; Ehret, U.; et al. A decade of Predictions in Ungauged Basins (PUB)—A review. *Hydrol. Sci. J.* **2013**, *58*, 1198–1255. [CrossRef]

18.	Samaniego, L.; Bárdossy, A.; Kumar, R. Streamflow prediction in ungauged catchments using copula-based dissimilarity measures. *Water Resour. Res.* **2010**, *46*. [CrossRef]

19.	Chikamoto, Y.; Timmermann, A.; Luo, J.J.; Mochizuki, T.; Kimoto, M.; Watanabe, M.; Ishii, M.; Xie, S.P.; Jin, F.F. Skilful multi-year predictions of tropical trans-basin climate variability. *Nat. Commun.* **2015**, *6*. [CrossRef] [PubMed]

20.	Hlinka, J.; Hartman, D.; Vejmelka, M.; Novotná, D.; Paluš, M. Non-linear dependence and teleconnections in climate data: Sources, relevance, nonstationarity. *Clim. Dyn.* **2014**, *42*, 1873–1886. [CrossRef]

21.	Zhang, Q.; Wang, Y.; Sing, V.P.; Gua, X.; Kong, D.D.; Xiao, M.Z. Impacts of ENSO and ENSO Modoki+A regimes on seasonal precipitation variations and possible underlying causes in the Huai River basin, China. *J. Hydrol.* **2016**, *533*, 308–319. [CrossRef]

22.	Wrzesinski, D. Regional differences in the influence of the North Atlantic Oscillation on seasonal river runoff in Poland. *Quaest. Geogr.* **2011**, *30*, 127–136. [CrossRef]

23.	Wrzesinski, D. Typology of spatial patterns seasonality in European rivers flow regime. *Quaest. Geogr.* **2008**, *27A*, 87–98.

24.	Lanzante, J.R. Resistant, robust and non-parametric techniques for the analysis of climate data: Theory and examples, including applications to historical radiosonde station data. *Int. J. Clim.* **1996**, *16*, 1197–1226. [CrossRef]

25.	Yue, S.; Pilon, P.; Cavadias, G. Power of the Mann–Kendall and Spearman's rho tests for detecting monotonic trends in hydrological series. *J. Hydrol.* **2002**, *259*, 254–271. [CrossRef]

26.	Konapala, G.; Mishra, A.K. Three-parameter-based streamflow elasticity model: Application to MOPEX basins in the USA at annual and seasonal scales. *Hydrol. Earth Syst. Sci.* **2016**, *20*, 2545–2556. [CrossRef]

27.	Belle, G.; Hughes, J.P. Nonparametric tests for trend in water quality. *Water Resour. Res.* **1984**, *20*, 127–136. [CrossRef]

28.	Li, J.; Tan, S. Nonstationary flood frequency analysis for annual flood peak series, adopting climate indices and check dam index as covariates. *Water Resour. Manag.* **2015**, *29*, 5533–5550. [CrossRef]

29.	Zhang, Y.; Cabilio, P.; Nadeem, K. Improved Seasonal Mann–Kendall Tests for Trend Analysis in Water Resources Time Series. In *Advances in Time Series Methods and Applications*; Li, W.K., Stanford, D.A., Yu, H., Eds.; Springer: New York, NY, USA, 2016; pp. 215–229, ISBN 978-1-4939-6568-7.

30.	Coulibaly, P.; Baldwin, C.K. Nonstationary hydrological time series forecasting using nonlinear dynamic methods. *J. Hydrol.* **2005**, *307*, 164–174. [CrossRef]

31. Khan, S.; Ganguly, A.R.; Bandyopadhyay, S.; Saigal, S.; Erickson, D.J.; Protopopescu, V.; Ostrouchov, G. Nonlinear statistics reveals stronger ties between ENSO and the tropical hydrological cycle. *Geophys. Res. Lett.* **2006**, *33*. [CrossRef]

32. Hao, Z.; Singh, V.P. Review of dependence modeling in hydrology and water resources. *Prog. Phys. Geogr.* **2016**, *40*, 549–578. [CrossRef]

33. Fleming, S.W.; Dahlke, H.E. Parabolic northern-hemisphere river flow teleconnections to El Niño-Southern Oscillation and the Arctic Oscillation. *Environ. Res. Lett.* **2014**, *9*. [CrossRef]

34. Cannon, A.J. Revisiting the nonlinear relationship between ENSO and winter extreme station precipitation in North America. *Int. J. Climatol.* **2015**, *35*, 4001–4014. [CrossRef]

35. Lin-Ye, J.; Garcia-Leon, M.; Gracia, V.; Ortego, M.I.; Lionello, P.; Sanchez-Arcilla, A. Multivariate statistical modeling of future marine storms. *Appl. Ocean Res.* **2017**, *65*, 192–205. [CrossRef]

36. Kusumastuti, D.I.; Struthers, I.; Sivapalan, M.; Reynolds, D.A. Threshold effects in catchment storm response and the occurrence and magnitude of flood events: Implications for flood frequency. *Hydrol. Earth Syst. Sci.* **2007**, *11*, 1515–1528. [CrossRef]

37. Konapala, G.; Veettil, A.V.; Mishra, A.K. Teleconnection between low flows and large-scale climate indices in Texas River basins. *Stoch. Environ. Res. Risk Assess.* **2017**, 1–14. [CrossRef]

38. Mishra, A.K.; Özger, M.; Singh, V.P. An entropy based investigation into the variability of precipitation. *J. Hydrol.* **2009**, *370*, 139–154. [CrossRef]

39. Mishra, A.K.; Özger, M.; Singh, V.P. Association between uncertainty in meteorological variables and water resources planning for Texas. *J. Hydrol. Eng.* **2011**, *16*, 984–999. [CrossRef]

40. Mishra, A.K.; Coulibaly, P. Hydrometric network evaluation for Canadian watersheds. *J. Hydrol.* **2010**, *380*, 420–437. [CrossRef]

41. Mishra, A.K.; Coulibaly, P. Variability in Canadian Seasonal Streamflow Information and its Implication for Hydrometric Network Design. *J. Hydrol. Eng.* **2014**, *19*. [CrossRef]

42. Li, C.; Singh, V.P.; Mishra, A.K. Entropy theory-based criterion for hydrometric network evaluation and design: Maximum information minimum redundancy. *Water Resour. Res.* **2012**, *48*. [CrossRef]

43. Mishra, A.K.; Singh, V.P. Analysis of drought severity-area-frequency curves using a general circulation model and scenario uncertainty. *J. Geophys. Res. Atmos.* **2009**, *114*. [CrossRef]

44. Mishra, A.K.; Özger, M.; Singh, V.P. Trend and persistence of precipitation under climate change scenarios. *Hydrol. Proc.* **2009**, *23*, 2345–2357. [CrossRef]

45. Rajsekhar, D.; Mishra, A.K.; Singh, V.P. Regionalization of drought characteristics using an entropy approach. *J. Hydrol. Eng.* **2013**, *18*, 870–887. [CrossRef]

46. Sharma, A. Seasonal to interannual rainfall probabilistic forecasts for improved water supply management: Part 1—A strategy for system predictor identification. *J. Hydrol.* **2000**, *239*, 232–239. [CrossRef]

47. Harrold, T.I.; Sharma, A.; Sheather, S. Selection of a kernel bandwidth for measuring dependence in hydrologic time series using the mutual information criterion. *Stoch. Environ. Res. Risk Assess.* **2001**, *15*, 310–324. [CrossRef]

48. Song, X.; Zhang, J.; Zhan, C.; Xuan, Y.; Ye, M.; Xu, C. Global sensitivity analysis in hydrological modeling: Review of concepts, methods, theoretical framework, and applications. *J. Hydrol.* **2015**, *523*, 739–757. [CrossRef]

49. Han, M.; Ren, W.; Liu, X. Joint mutual information-based input variable selection for multivariate time series modeling. *Eng. Appl. Artif. Intell.* **2015**, *37*, 250–257. [CrossRef]

50. Knuth, K.H.; Gotera, A.; Curry, C.T.; Huyser, K.A.; Wheeler, K.R.; Rossow, W.B. Revealing relationships among relevant climate variables with information theory. *arXiv* **2013**, arXiv:1311.4632.

51. Khokhlov, V.N.; Glushkov, A.V.; Loboda, N.S. On the nonlinear interaction between global teleconnection patterns. *Q. J. R. Meteorol. Soc.* **2006**, *132*, 447–465. [CrossRef]

52. Hurtado, A.F.; Poveda, G. Linear and global space-time dependence and Taylor hypotheses for rainfall in the tropical Andes. *J. Geophys. Res.* **2009**, *114*. [CrossRef]

53. Naumann, G.; Vargas, W.M. Joint diagnostic of the surface air temperature in southern South America and the Madden–Julian oscillation. *Weather Forecast* **2010**, *25*, 1275–1280. [CrossRef]

54. Yoon, S.; Lee, T. Investigation of hydrological variability in the Korean Peninsula with the ENSO teleconnections. *Proc. IAHS* **2016**, *374*. [CrossRef]

55.	Sivakumar, B. *Chaos in Hydrology: Bridging Determinism and Stochasticity*; Springer: Berlin/Heidelberg, Germany, 2016; ISBN 978-90-481-2552-4. [CrossRef]

56.	Varis, O.; Kummu, M.; Salmivaara, A. Ten major rivers in monsoon Asia-Pacific: An assessment of vulnerability. *Appl. Geogr.* **2012**, *32*, 441–454. [CrossRef]

57.	Wilmott, C.J.; Rowe, C.M.; Philpot, W.D. Small-scale climate maps: A sensitivity analysis of some common assumptions associated with grid-point interpolation and contouring. *Am. Cartogr.* **1985**, *12*, 5–16. [CrossRef]

58.	Yatagai, A.; Kamiguchi, K.; Arakawa, O.; Hamada, A.; Yasutomi, N.; Kitoh, A. APHRODITE: Constructing a long-term daily gridded precipitation dataset for Asia based on a dense network of rain gauges. *Bull. Am. Meteorol. Soc.* **2012**, *93*, 1401–1415. [CrossRef]

59.	Vu, M.T.; Raghavan, V.S.; Liong, S.Y. SWAT use of gridded observations for simulating runoff—A Vietnam river basin study. *Hydrol. Earth Syst. Sci.* **2012**, *16*, 2801–2811. [CrossRef]

60.	Vu, M.T.; Mishra, A.K. Spatial and Temporal Variability of Standardized Precipitation Index over Indochina Peninsula. *Cuad. Investig. Geogr.* **2016**, *42*, 221–232. [CrossRef]

61.	Vu, M.T.; Raghavan, S.V.; Liong, S.-Y.; Mishra, A.K. Uncertainties in gridded precipitation observations in characterizing spatio-temporal drought and wetness over Vietnam. *Int. J. Climatol.* **2017**. [CrossRef]

62.	Raghavan, V.S.; Vu, M.T.; Liong, S.Y. Ensemble climate projections of mean and extreme rainfall over Vietnam. *Glob. Planet. Chang.* **2017**, *148*, 96–104. [CrossRef]

63.	Raghavan, V.S.; Vu, M.T.; Liong, S.Y. Impact of climate change on future stream flow in the Dakbla river. *J. Hydroinform.* **2014**, *16*, 231–244. [CrossRef]

64.	Raghavan, V.S.; Liong, S.Y.; Vu, M.T. Assessment of future stream flow over the Sesan catchment of the Lower Mekong Basin in Vietnam. *Hydrol. Proc.* **2012**, *26*, 3661–3668. [CrossRef]

65.	Vu, M.T.; Aribarg, T.; Supratid, S.; Raghavan, V.S.; Liong, S.Y. Statistical downscaling rainfall over Bangkok using Artificial Neural Network. *Theor. Appl. Climatol.* **2016**, *126*, 453–467. [CrossRef]

66.	Trenberth, K.E. The Definition of El Niño. *Bull. Am. Meteorol. Soc.* **1997**, *78*, 2771–2777. [CrossRef]

67.	Trenberth, K.E. The Climate Data Guide: Nino SST Indices (Nino 1+2, 3, 3.4, 4; ONI and TNI). Available online: https://climatedataguide.ucar.edu/climate-data/nino-sst-indices-nino-12-3-34-4-oni-and-tni (accessed on 6 January 2018).

68.	Moon, Y.I.; Rajagopalan, B.; Lall, U. Estimation of mutual information using kernel density estimators. *Phys. Rev. E* **1995**, *52*, 2318–2321. [CrossRef]

69.	Granger, C.; Lin, J.L. Using the mutual information coefficient to identify lags in nonlinear models. *J. Time Ser. Anal.* **1994**, *15*, 371–384. [CrossRef]

70.	Joe, H. Relative entropy measures of multivariate dependence. *J. Am. Stat. Assoc.* **1989**, *84*, 157–164. [CrossRef]

71.	Khan, S.; Bandyopadhyay, S.; Ganguly, A.R.; Saigal, S.; Erickson, D.J., III; Protopopescu, V.; Ostrouchov, G. Relative performance of mutual information estimation methods for quantifying the dependence among short and noisy data. *Phys. Rev. E* **2007**, *76*. [CrossRef] [PubMed]

72.	Kraskov, A.; Stögbauer, H.; Grassberger, P. Estimating mutual information. *Phys. Rev. E* **2004**, *69*. [CrossRef] [PubMed]

73.	Hull, M.M.V. Edgeworth approximation of multivariate differential entropy. *Neural Comput.* **2005**, *17*, 1903–1910. [CrossRef]

74.	Cellucci, C.J.; Albano, A.M.; Rapp, P.E. Statistical validation of mutual information calculations: Comparison of alternative numerical algorithms. *Phys. Rev. E* **2005**, *71*. [CrossRef] [PubMed]

75.	Silverman, B.W. *Density Estimation for Statistics and Data Analysis*; Chapman and Hall/CRC: London, UK, 1986.

76.	Wand, M.P.; Jones, M.C. Multivariate plug-in bandwidth selection. *Comput. Stat.* **1994**, *9*, 97–117.

77.	Ince, R.A.A.; Giordano, B.L.; Kayser, C.; Rousselet, G.A.; Gross, J.; Schyns, P.G. A statistical framework for neuroimaging data analysis based on mutual information estimated via a Gaussian Copula. *Hum. Brain Mapp.* **2017**, *38*, 1541–1573. [CrossRef] [PubMed]

78.	Räsänen, T.A.; Kummu, M. Spatiotemporal influences of ENSO on precipitation and flood pulse in the Mekong River Basin. *J. Hydrol.* **2013**, *476*, 154–168. [CrossRef]

79.	Xu, Z.X.; Takeuchi, K.; Ishidaira, H. Correlation between El Niño–Southern Oscillation (ENSO) and precipitation in Southeast Asia and the Pacific region. *Hydrol. Proc.* **2003**, *18*, 107–123. [CrossRef]

80. Chen, D.; Cane, M.A.; Kaplan, A.; Zebiak, S.E.; Huang, D. Predictability of El Niño over the past 148 years. *Nature* **2004**, *428*, 733–736. [CrossRef] [PubMed]

81. Ludescher, J.; Gozolchiani, A.; Bogacheva, M.I.; Bunde, A.; Havlin, S.; Schellnhuber, H.J. Very early warning of next El Niño. *Proc. Natl. Acad. Sci. USA* **2014**, *111*, 2064–2066. [CrossRef] [PubMed]

82. Power, S.; Delage, F.; Chung, C.; Kociuba, G.; Keay, K. Robust twenty-first-century projections of El Niño and related precipitation variability. *Nature* **2013**, *502*, 541–545. [CrossRef] [PubMed]

83. Cai, W.; Borlace, S.; Lengaigne, M.; Van Rensch, P.; Collins, M.; Vecchi, G.; Timmermann, A.; Santoso, A.; McPhaden, M.J.; Wu, L.; et al. Increasing frequency of extreme El Niño events due to greenhouse warming. *Nat. Clim. Chang.* **2014**, *4*, 111–116. [CrossRef]

84. Zhang, G.; Su, X.; Singh, V.P.; Ayantobo, O.O. Modeling NDVI using Joint Entropy method considering hydro-meteorological driving factors in the middle reaches of Hei river basin. *Entropy* **2017**, *19*, 502. [CrossRef]

13

Modeling Multi-Event Non-Point Source Pollution in a Data-Scarce Catchment using ANN and Entropy Analysis

Lei Chen, Cheng Sun, Guobo Wang, Hui Xie and Zhenyao Shen *

State Key Laboratory of Water Environment Simulation, School of Environment, Beijing Normal University, Beijing 100875, China; chenlei1982bnu@bnu.edu.cn (L.C.); 13756892980@163.com (C.S.); wangguobo@yeah.net (G.W.); imbahui@163.com (H.X.)
* Correspondence: zyshen@bnu.edu.cn

Abstract: Event-based runoff–pollutant relationships have been the key for water quality management, but the scarcity of measured data results in poor model performance, especially for multiple rainfall events. In this study, a new framework was proposed for event-based non-point source (NPS) prediction and evaluation. The artificial neural network (ANN) was used to extend the runoff–pollutant relationship from complete data events to other data-scarce events. The interpolation method was then used to solve the problem of tail deviation in the simulated pollutographs. In addition, the entropy method was utilized to train the ANN for comprehensive evaluations. A case study was performed in the Three Gorges Reservoir Region, China. Results showed that the ANN performed well in the NPS simulation, especially for light rainfall events, and the phosphorus predictions were always more accurate than the nitrogen predictions under scarce data conditions. In addition, peak pollutant data scarcity had a significant impact on the model performance. Furthermore, these traditional indicators would lead to certain information loss during the model evaluation, but the entropy weighting method could provide a more accurate model evaluation. These results would be valuable for monitoring schemes and the quantitation of event-based NPS pollution, especially in data-poor catchments.

Keywords: non-point source pollution; ANN; entropy weighting method; data-scarce; multi-events

1. Introduction

Non-point source (NPS) pollution has resulted in the deterioration of water bodies and has become a major environmental threat among most counties [1,2]. The quantification of the rainfall–runoff process and the resulting NPS pollutants is essential for developing mitigation strategies, which are the basis for watershed management [3]. The rainfall process is the major driving force for NPS, thus rainfall-runoff-pollutant (R-R-P) relationships have become the focus of watershed research [4,5]. Many studies have been conducted in the fields of rainfall–runoff relationships but have rarely involved the runoff–pollutant relationship, especially for the event-based estimation of NPS loads [6–8].

The NPS processes can be expressed from the event-step to long-term steps. Event-based NPS exports and the resulting change in water quality can provide detailed features of the NPS, which is more appropriate for the design of storm-based management practices [9]. Models are developed to construct the runoff–pollutant relationship, and the discrepancies of the collected measured data in different rainfall patterns would have a considerable influence on the model construction. Identifying the correlation among the series of rainfall, runoff and pollutant loads for multiple rainfall events is inevitable for NPS model construction. Although many models are well suited for offline water quality analyses, Soil and Water Assessment Tool (SWAT) is more representative than any other models [10].

However, owing to limited human resources, data scarcity has become one of the key barriers to establish the R-R-P relationship, especially for event-based process [11,12]. Thus, the application of watershed models such as SWAT for assessing NPS pollution is also limited by temporal resolution which ranges from annual to sub-hourly averages. The SWAT model usually operates continuously at a daily time step, which ensures that the long-term impacts of NPS can be quantified. Sub-daily calculations of runoff, erosion, and sediment transport are also available in new version of SWAT by sub-daily rainfall input and Green and Ampt method, though few attempts have reached to that higher temporal resolution. In the future, we would develop other more appropriate models to solve this problem. Currently, acceptable rainfall and streamflow data sets are more readily available, especially because of the recent development of data centers and satellite data observations. However, hourly or sub-hourly flow data for high-frequency time series are still limited, especially with respect to event-based hydrological studies for data-poor regions [7]. Water quality records, which are based on periodic monitoring by human resources, are thus even scarcer. Therefore, data scarcity for NPS predictions is unavoidable for multiple-rainfall event simulations. Typically, we collected samples during multiple rainfall events in the monitoring process but discarded some of events from further analysis, especially for light rainfall, for which only a few data points exist. This treatment of incomplete data would result in the loss of information, especially for multiple rainfall events among data-scarce regions.

Currently, statistical models have been widely used to estimate rainfall–runoff relationships for its ease of application without considering a large amount of delicate formulas and parameters [13]. For example, unit hydrographs (UH), as one of the most famous methods, is used to estimate a direct runoff hydrograph of a given rainfall duration. Meanwhile, statistical models are used to simulate pollutant loads based on the established runoff–pollutant relationship. For example, Park and Engel [14] developed Load Estimator (LOADEST) to predict pollutant concentration (or load) on days when flow data were measured, and the results showed that absolute values of errors in the annual sediment load estimation decreased from 39.7% to 10.8%. Meanwhile, most of the findings demonstrate that the LOADEST model could provide more accurate results and may be useful for simulating runoff–pollutant processes [15–17]. However, the LOADEST model has strict requirements on the number of data points, which should include continuous flow data and dispersed water quality data, and its calibration process is relatively complicated.

Owing to the limited measured data, the black-box model might be a substitution to construct the logical relationships between runoff and pollutant loads for multiple-events processes. The artificial neural network (ANN) with the characteristics of self-learning and adaptability has become the most commonly used tool in environmental prediction, and it is also available for poor-data regions. This method is applicable to simulate the imaginal thinking of the human brain, for which the most prominent characteristic is the parallel processing of information and distributed storage. As an example, Melesse et al. [18] used the ANN to estimate suspended sediment loads for three major rivers. The results showed that daily predictions were better than weekly predictions. Therefore, it can be seen that ANN models have flexible structures that allow multi-input and multi-output modeling. This is particularly important in streamflow forecasting where inflows at multiple locations are considered within a given catchment [19]. Though the application of ANN in the field of load production has proliferated in recent years, the impact of data scarcity on its prediction capabilities during different rainfall patterns still creates limitations [20].

Simulation evaluation is the most important step for the setup of statistical models [21]. In traditional applications, the model evaluation is usually performed using a single regression goodness-of-fit indicator, the most common of which is the point-to-point pairs (a series of single data pairs) of the predicted and measured data. However, this might lead to the loss of specific information, resulting in dubitable simulation results. In this case, a joint evaluation should be a substitute for the traditional single indicator. With the high precision and objectivity, the entropy regulates the uncertainty of different criteria from different perspectives [22]. Compared with the traditional single indicators, it can combine different indicators to evaluate the discrepancy between causes comprehensively. For instance, Khosravi et al. [23] sought to map the flooding susceptibility using different bivariate methods, including Shannon's entropy, the statistical index and the weighting factor. Yuan et al. [24] developed an entropy method to find the weight sum of the information entropy maximum to allocate the reduction of pollutants for the main seven valleys in China. The entropy weighting method may be an efficient way to evaluate the regulation of the simulation results and to balance the strengths and weaknesses of the results. However, these studies do not provide much attention to event-based NPS predictions, especially for data-scarce catchments.

This study surveys the motivation for a methodology of action, looks at the difficulties posed by data scarcity and outlines the need for the development of possibility methods to cope with data scarcity in multiple rainfall events. The objectives of this work are: (1) to identify the impacts of different rainfall patterns on the model construction using a complete data series; (2) to simulate the scarce pollutant data in other data-scarce rainfall events; and (3) to test the application of the entropy weighting method for the evaluation of ANN.

2. Materials and Methods

A prediction-evaluation framework is proposed for the NPS prediction for data-scarce catchments, the flow chart of methods is shown in Figure 1. The ANN is proposed to simulate the missing data points during multiple-events, and the entropy weighting method is used as a comprehensive indicator to construct the model. As a necessary supplement, the interpolation method is used for tail correction during multiple rainfall events. Data-scarce rainfall events denote the absence of data, especially for measured flow and water quality, in a given period of time due to human mistakes during high-resolution monitoring process. Instead, complete rainfall events are defined if there are no measured flow and water quality data scarcity. The demonstration of traditional indicators is shown in Section 2.2.

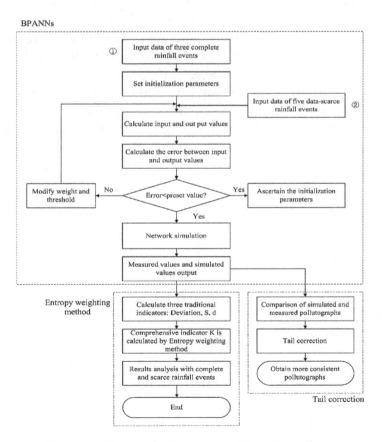

Figure 1. The methods presented in a flow chart.

2.1. The Description of the ANN

The back propagation algorithm is a supervised learning method based on the commonly used steepest descent method to minimize global errors [25], while it is also the multilayer feedforward network based on the error back propagation algorithm [2,26]. It accumulates an abundant mapping relation of the input-output pattern and does not need to reveal mathematical equations to describe the mapping relation before calculation. The ANN may be an efficient method to adjust the weights and thresholds through back propagation to minimize the sum of the squared errors. As shown in Figure 2, the topological structure of the ANN consists of an input layer, a hidden layer and an output layer [27].

The learning mechanism of the ANN is shown in Figure 2, where x_i is the input signal and w_i is the weight coefficient. The outside input samples $x_1, x_2, \ldots, x_n$ are accepted into the input layer, and the network weight coefficients are adjusted during training. The discrete values, 0 and 1, are selected as the input sampling signals. By comparing the network output signals and the expected output signals to generate the error signals, the weight coefficients of the learning system can be rectified based through iterative adjustments to minimize the errors until reaching an acceptable range [28]. In this process, the expected output signals are regarded as the teacher signals, which are compared with the actual output, and the errors produced are applied to rectify the weight coefficients. At the point when the actual output values and expected values are nearly the same, the process is concluded [26]. Finally, the results are produced through and equation of U based on the weight coefficients and are exported by the output layers. In the ANN training process, three prime criteria can be summarized: the error surface gradient can converge rapidly, the mean squared error is below the error of the preset level, and the correlation coefficient of the training results is more than 0.9, indicating that training results are an improvement [29]. This section briefly surveys the measurement for methodology, while the ANN should be judged for whether each indicator can or cannot reach the given standards.

In this study, multiple rainfall events are used as the input conditions. Multiple rainfall events are divided into either the training process or the simulation process based on the data conditions. To establish the black-box model, data of three complete rainfall events are first input into the layer, including light, moderate, and heavy rainfall patterns. The training results also indicate that the ANN is applicable for various rainfall patterns. In the simulation process, the flow data for all the rainfall and water quality information for the data-complete rainfall events for the same rainfall pattern are regarded as the input layer. The hidden layer contains the water quality data for the data-scarce rainfall events which correspond to all the flow and water quality data in the input layer. To obtain the output layer, the training layer feedbacks the results into the prediction interval. Finally, the output layer is simulated using the input data of the input layer.

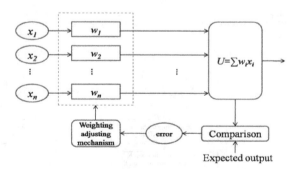

Figure 2. The learning mechanism of the artificial neural network (ANN).

2.2. The Description of the Entropy Weighting Method

Three commonly used indicators, the mean relative error $(\bar{d})$, the standard deviation of the relative error (S), and the load deviation percentage $(deviation)$, are selected to evaluate the simulation results [30,31]. The formulas are shown as followed:

$$deviation = \frac{O_i^{origin} - O_i}{O_i^{origin}} \times 100\% \tag{1}$$

$$\bar{d} = \sum_{i=1}^{n}(O_i - P_i)/n \tag{2}$$

$$S = \sqrt{\frac{\sum_{i=1}^{n}\left(d_i - \bar{d}\right)^2}{n-1}} \tag{3}$$

where O_i is the set of measured data, P_i is the set of predicted data, and O_i^{origin} denotes the total loads of the original conditions, and is the mean value of the measured data.

Each of the three indicators represents the credibility of the measurements based on the discrepancy between the measured and simulated values. Lower indicator values indicate that the fitting between the simulated and measured data is improved, and the model is considered to have a satisfactory performance. However, single indicators have limitation on amount of information loss. Therefore, these indicators are handled with the entropy weighting method for a more comprehensive assessment of the ANN. Based on the fundamental principles of information theory, information is a measurement of the degree of order for a given system, and the entropy is a measurement of the degree of disorder [32]. The entropy weighting method serves as a mathematic method and considers the information provided by each factor [33]. Information entropy is negatively associated with the increase in information provided by different indicators, and a smaller information entropy result in higher weights for each single indicator. As an objective and comprehensive method, the entropy weighting method considers the advantages of every indicator and makes a synthetic evaluation. This principle is as follows:

Firstly, an $n \times m$ origin data matrix is established according to the selected evaluation indicators:

$$X = \begin{bmatrix} x_{11} & \cdots & x_{1n} \\ \vdots & \ddots & \vdots \\ x_{n1} & \cdots & x_{nm} \end{bmatrix}_{n \times m} \tag{4}$$

where m denotes the evaluation indicator, and individual rows represent different evaluation objects. Therefore, matrix X is known.

A second, positive matrix should be established with a transformation following same trend. The transformed matrix is

$$Y = \begin{bmatrix} y_{11} & \cdots & y_{1n} \\ \vdots & \ddots & \vdots \\ y_{n1} & \cdots & y_{nm} \end{bmatrix}_{n \times m} \tag{5}$$

Matrix Y is normalized, and the ratio of each column vector y_{ij} and the sum of all elements in this matrix should be normalized. The formulas for these calculations are:

$$Z_{ij} = \frac{y_{ij}}{\sum_{i=1}^{n} Y_{ij}} (j = 1, 2, \ldots, m) \tag{6}$$

where Z_{ij} are the elements of the normalized matrix.

The operational formula in the process of generating the entropy weights of the evaluation indicators is

$$H(x_j) = -k \sum_{i=1}^{n} z_{ij} \ln z_{ij} (j = 1, 2, \ldots, m) \tag{7}$$

where k is a normalizing constant, $k = 1/\ln n$, and Z_{ij} is the j-th the probability of the element of the i-th evaluation unit. Entropy values of the evaluation indicators should be transformed into the weighted values:

$$w_j = \frac{1 - H(x_j)}{m - \sum_{j=1}^{m} H(x_j)} \quad j = 1, 2, \ldots, m, \tag{8}$$

where $0 \leq w_i \leq 1$ and $\sum_{j=1}^{m} w_j = 1$ are the acquired weighted values. Finally, the comprehensive weighting values for each evaluation indicator should be ensured. The weighted values of each indicator are multiplied with the corresponding indicators and summed. The evaluation model is

$$U = \sum_{j=1}^{m} w_i z_{ij} \ (j = 1, 2, \ldots, n) \tag{9}$$

where U represents the comprehensive evaluation function of the entropy weights for each evaluation indicator. This function reflects the comprehensive characteristics of the evaluation objective, which avoids limiting these indicators [34].

The principle of the entropy weighting method is that information for each evaluation unit will be qualified and synthesized, while every factor is weighted to simplify the evaluation process [35]. Therefore, the weight values can be ascertained with the entropy weighting method, and we choose the *deviation*, $\bar{d}$, and S as the evaluation indicators.

2.3. Method for Tail Correction

Statistical models would result in tail deviation problems if data scarcity exists in this study. This problem addressed through data interpolation for the tail deviation. Therefore, linear interpolation, as a common-used method, is used to obtain the missing values of the other data points. Two values of the function $f(x)$ are used to reduce the errors in the tail of the pollutographs. This approach is relatively straightforward and is used widely in the field of mathematics or computer graphics. The error of the approximate method can be defined as follows:

$$R_T = f(x) - \rho(x) \tag{10}$$

where ρ represents the linear interpolated polynomial:

$$\rho(x) = f(x_0) + \frac{f(x_1) - f(x_0)}{x_1 - x_0}(x - x_0) \tag{11}$$

As a result of Rolle's theorem, if $f(x)$ has two continuous derivatives, the error range is

$$|R_T| \leq \frac{(x_1 - x_0)^2}{8} \max_{x_0 \leq x \leq x_1} |f'(x)| \tag{12}$$

As shown in Formula (12), the approximate error of the linear interpolation increases with the function curvature.

3. Case Study

3.1. Study Areas

As shown in Figure 3, the Zhangjiachong catchment, which is a representative area in the Three Gorges Reservoir Region (TGRR), is selected as a case study [36]. It covers a drainage area of 1.62 km^2, and the landscape is primarily mountainous, with an elevation between 148 m and 530 m above the Yellow Sea level. Agriculture and forests cover the majority of the total area. The main local crops are tea, corn, oil seed rape, and chestnuts [37]. The background values of nitrogen and phosphorus are higher because the fertilizer usage is relatively high, resulting in a high risk of nutrient loss into nearby streams [38].

The average annual temperature is approximately 18 °C, and the average annual precipitation is approximately 1439 m, 80% of which occurs from May to August. Thus, soil erosion frequently occurs during wet seasons, and results in an increase in the pollutant loads with increased runoff. We consider that the variation of rainfall might impact the model accuracy. Therefore, identifying the classification of rainfall patterns should be determined before any simulations. According to the investigation results of existing rainfall data, rainfall patterns are divided into light, moderate, and heavy events. Meanwhile, based on our monitoring data, a majority of rainfall events in the Zhangjiachong catchment are considered moderate events, while heavy events are rare.

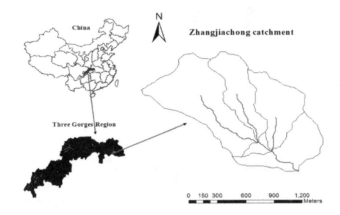

Figure 3. The location of the Zhangjiachong catchment.

3.2. Field Monitoring and Data Record

In this study, field monitoring data were collected from 1 January 2013 to 31 December 2014 and the rainfall, streamflow and pollutant data during eight rainfall events were recorded. The data used in this study represent three complete rainfall events, which include light, middle, and heavy rainfall (21 April 2014, 24 July 2014, and 5 August 2014), and five other data-scarce events (15 April 2014, 23 August 2014, 20 July 2014, 5 July 2013, and 28 August 2013). Data-scarce rainfall events denote the absence of data, especially for measured flow and water quality, in a given period of time due to human mistakes during high-resolution monitoring process. Instead, complete rainfall events are defined if there are no measured flow and water quality data scarcity. The equations with explicit parameters are constructed through a training process with complete data of the three complete rainfall events, and the constructed ANN is used to predict the missing NPS data in the other five data-scarce rainfall events. The output layer includes pollutant load data for five data-scarce rainfall events.

The weather station (Skye Lynx Standard) provided continuous records for climate data and a float-operator sensor (WGZ-1) was located at the catchment outlet, where high-frequency sampling was recorded in approximately 15 min steps. Base flows were measured before the runoff started, and water samples were collected every 15 min in the first hour after runoff began and every 30 min over the following two hours. After water levels had stabilized, water samples were collected once every hour until the end of the event. All water samples were placed in pre-cleaned glass jars with aluminumfoil liners along the lids and stored at −20 °C during transportation to the laboratory for processing and analysis. Specifically, the total nitrogen of NPS (NPS-TN) levels were measured via Alkaline persulfate oxidation-UV spectrophoto metric with the detection limitation from 0.05 mg/L to 4.0 mg/L, while the total phosphorus of NPS (NPS-TP) levels in the samples were measured via Potassium persulfate oxidation-molybdenum blue colorimetric methods. The main instrument is ultraviolet spectrophotometer. Finally, the recorded rainfall, flow and pollutant levels were used for the following analysis.

Table 1. Complete data for the three rainfall events.

21 April 2014				24 July 2014				5 August 2014		
Time	Flow (m³/s)	NPS-TN (mg/L)	NPS-TP (mg/L)	Time	Flow (m³/s)	NPS-TN (mg/L)	NPS-TP (mg/L)	Time	Flow (m³/s)	NPS-TN (mg/L)	NPS-TP (mg/L)
2:45	0.002	4.75	0.063	22:45	0.003	0.84	0.11	21:30	0.008	2.66	0.30
3:00	0.009	8.89	0.193	23:00	0.380	6.26	0.76	21:45	0.678	7.29	0.84
3:15	0.015	15.29	0.300	23:15	0.647	6.96	1.04	22:00	1.107	11.30	1.12
3:30	0.016	13.31	0.301	23:30	0.726	7.06	0.87	22:15	1.400	11.90	1.26
3:45	0.031	25.28	0.369	23:45	0.336	6.77	0.92	22:30	2.227	9.85	0.94
4:00	0.037	14.14	0.297	0:00	0.971	8.77	1.12	23:00	1.647	15.00	1.20
4:30	0.065	22.64	0.652	0:30	0.570	8.08	0.70	23:30	0.585	9.67	0.62
5:00	0.071	24.48	0.441	1:00	0.294	5.53	0.48	0:00	0.945	9.56	0.68
6:00	0.086	26.49	0.469	2:00	0.266	5.05	0.57	1:00	0.410	7.01	0.35
7:00	0.126	23.89	0.443	3:00	0.172	5.11	0.36	2:00	0.237	2.03	0.28
8:00	0.146	16.97	0.286	4:00	0.191	5.79	0.34	3:00	0.183	7.60	0.37
9:00	0.264	11.60	0.171	5:00	0.041	5.85	0.20	4:00	0.166	6.81	0.27
10:00	0.278	11.00	0.117	6:00	0.090	5.59	0.18	5:00	0.115	7.01	0.20
11:00	0.288	10.73	0.104	7:00	0.064	7.78	0.15	6:00	0.126	6.71	0.19
12:00	0.296	10.94	0.109	8:00	0.048	5.29	0.14	7:00	0.102	6.08	0.12
13:00	0.411	9.64	0.113					8:00	0.073	7.03	0.24
14:00	0.593	9.48	0.089					9:00	0.063	5.68	0.11
15:00	0.593	9.65	0.087					10:00	0.086	5.75	0.09
16:00	0.602	9.26	0.072					11:00	0.075	6.44	0.18
17:00	0.770	9.93	0.068					12:00	0.045	6.33	0.21
								13:00	0.029	5.88	0.14

Table 2. Five rainfall events with data scarcity.

15 April 2014 (1.213 mm/h)				28 August 2013 (2.027 mm/h)				20 July 2014 (2.013 mm/h)				5 July 2013 (2.380 mm/h)				28 August 2013 (2.647 mm/h)			
Time	Flow	TN	TP	Time	Flow	TN	TP	Time	Flow	TN	TP	Time	Flow	TN	TP	Time	Flow	TN	TP
6:00	0.0127	7.65	0.050	15:00	0.0375	1.13	0.13	19:00	0.0127	5.51	0.27	14:30	0.0127	5.03	0.06	19:00	0.03	5.59	0.05
6:30	0.024	12.27	0.412	15:30	-	-	-	19:30	0.0163	7.02	0.56	15:00	-	-	-	19:30	0.06	12.65	0.33
6:45	0.0375	11.49	0.366	16:00	0.0964	6.29	0.71	20:00	0.0485	6.92	0.86	15:30	-	-	-	20:00	1.78	15.24	1.39
7:00	0.0427	23.53	0.579	16:30	0.1020	5.57	0.3	20:30	0.1190	5.56	1.15	16:00	0.0127	8.81	1.50	20:30	2.22	16.38	0.84
7:15	0.0401	16.13	0.363	17:00	0.0964	5.81	0.32	21:00	0.0866	5.14	0.53	16:30	0.0127	6.37	2.56	21:00	1.94	13.88	0.69
7:30	0.0406	15.19	0.350	17:30	0.0547	5.2	0.2	21:30	0.0327	7.37	0.4	17:00	0.0375	1.06	2.11	21:30	0.80	14.95	0.44
8:00	0.0327	16.92	0.373	18:00	0.0427	4.95	0.14	22:00	-	-	-	17:30	-	-	-	22:00	0.39	13.43	0.55
8:30	0.0351	14.29	0.363	18:30	-	-	-	22:30	0.0375	5.64	0.24	18:00	0.4140	-	-	22:30	0.26	14.81	0.56
9:00	0.0327	11.26	0.291	19:00	0.0375	5.14	0.13	23:00	-	-	-	18:30	0.4440	-	-	23:00	-	-	-
10:00	-	-	-	19:30	-	-	-	23:30	0.0182	6.49	0.19	19:00	0.2220	3.20	0.41	23:30	-	-	-
10:30	0.0351	8.67	0.228	20:00	0.0427	5.21	0.13					19:30	-	-	-	0:00	-	-	-
11:00	0.0127	7.65	0.050									20:00	0.1590	4.08	0.28	0:30	-	-	-
												20:30	-	-	-	1:00	0.21	10.17	0.14
												21:00	0.0964	4.26	0.11	1:30	-	-	-
												21:30	-	-	-	2:00	0.15	9.83	0.19
												22:00	0.0775	3.78	0.22	2:30	-	-	-
												22:30	-	-	-	3:00	0.09	6.54	0.08
												23:00	0.0616	3.80	0.13				
												23:30	-	-	-				
												0:30	0.0547	-	-				
												1:30	0.0427	-	-				

Note: the units of flow are in m^3/s; the units of NPS-TN and NPS-TP are mg/L; - denotes that data are missing at this time.

However, flow and water quality data were limited because of the use of flow instruments via manual collection. Rainfall levels were recorded to divide the rainfall into light, moderate, and heavy events. The rainfall levels for 21 April 2014, 24 July 2014, and 5 August 2014 are 1.308 mm/h, 3.000 mm/h, and 6.054 mm/h, respectively. The flow data were replenished with unit hydrographs as the basis for the ANN. In addition, this catchment is dominated by agriculture, so fertilizer use results in deteriorated water quality. Therefore, the NPS-TN and NPS-TP are selected as the evaluation indicators. All the data for the three complete rainfall events and five typical data-scarce rainfall events are shown in Tables 1 and 2, respectively, including the rainfall intensity, flow data, and the pollutant concentration of the NPS-TN and NPS-TP. As shown in Table 1, complete data are used as the input of the ANN and represent the impacts of the rainfall patterns on the model applicability. As shown in Table 2, data scarcity of the five random rainfall events is simulated, and the impacts of the data scarcity are quantified.

4. Results and Discussion

4.1. Training Results of the ANNUsing the Complete Data

This section demonstrates the training process with data for the three complete rainfall events, illustrating that the applicability of ANN in different rainfall patterns. The training results for the ANN areas followed (the figure is shown in the Supplementary Materials): the error surface gradient rapidly converges to a flat surface for both the NPS-TN and NPS-TP. The mean squared error of the training results for the NPS-TP prediction reaches the 10^{-3}, 10^{-2}, and 10^{-1} orders of magnitude for the light, moderate, and heavy rainfall events, respectively. However, the mean squared error for the NPS-TN prediction reaches the 1.0, 10^{-3}, and 1.0 orders of magnitude during the light, moderate, and heavy rainfall events, respectively, indicating that all the results fall within the range of permissible errors or rapidly reach a flat surface. The correlation coefficients are more than 0.9, indicating that all the training results are good. In this respect, it can be said that the ANN is applicability to simulate the NPS for different rainfall patterns, and we extrapolated ANN for pollutant load simulations in the data-scarce rainfall events.

To better understand the simulation results, the entropy weighting method was used in the evaluation process. As shown in Table 3, K results are all higher than 0.9, indicating that there is no obvious deviation between the simulated and measured values. Meanwhile, the K values for the NPS-TP are higher than the NPS-TN for different rainfall patterns, and the K value for the light rainfall is higher than the other rainfall patterns. Therefore, it is apparent that the NPS-TP simulation is an improvement over the NPS-TN, and the simulation is better suited for the light rainfall events for both the NPS-TP and NPS-TN. It is obvious that the flow have different shear force in different rainfall patterns. The soil particles and pollutants act differently with different rainfall levels and intensities. It is possible that our monitoring scheme is more appropriate in light rainfall patterns in this experiment, and the peak data cannot be monitored during heavy rainfall patterns [39]. The NPS-TN concentration peak and flow peak appear to be consistent. When one of the flow or load peaks is missing, it is the same as both of them missing simultaneously, resulting in a poor simulation effect. However, the apparent time of the NPS-TP concentration peak and flow peak is inconsistent in different rainfall patterns. Xu et al. [40] introduced the support vector regression (SVR) model to develop a quantitative relationship between the environmental factors and the eutrophic indices compared with the ANN. The results show that the correlation coefficients of the NPS-TP are greater than those for the NPS-TN, indicating that the model effect of the NPS-TP is improved over the NPS-TN. This study verifies this conclusion with the ANN model.

Table 3. Evaluation of the simulation results of the pollutant loads for different rainfall patterns.

Rainfall Events	Comprehensive Indicators K	
	NPS-TP	NPS-TN
21 April 2014	0.986	0.953
24 July 2014	0.973	0.938
5 August 2014	0.958	0.921

4.2. Simulated Results of the ANN for Data-Scarce Events

Five typical data-scarce rainfall events were used to discuss the impact of different data-scarce patterns on the NPS predictions. As shown in Figure 4, the NPS-TP training results for the NPS-TN have a faster convergence rate for the grads and lower mean squared errors. The training values for the NPS-TN are represented by an R^2 value that is more than 0.9, and the mean squared errors are under the permissible values or reach the flat surface rapidly. However, only one event (5 July 2013) was observed to have lower grads beyond the preset value, and its training effect was the worst because this rainfall event has peak scarcity.

The entropy values in the five data-scarce rainfall patterns are shown in Table 4. Combined with the complete data events, it is apparent that the simulated effect for 5 July 2013 has a worse fit compared with the other rainfall events, which reflects the poor training effect when there is a scarcity of peak concentration data. The peak data are the key information, and reflect the overall process of the rainfall events. However, the peak scarcity is unintentional and due to system errors. In addition, the training effect of the NPS-TP is improved over the NPS-TN.

Table 4. Evaluation effect of the data scarcity on the models for the five data-scarce rainfall events.

Rainfall Events	Comprehensive Indicators K		Traditional Indicators (NPS-TP)		
	NPS-TN	NPS-TP	Deviation	$\bar{d}$	S
15 April 2014	0.546	0.971	0.989	0.956	0.966
23 August 2014	0.937	0.924	0.934	0.892	0.943
20 July 2014	0.959	0.930	0.964	0.882	0.941
5 July 2013	0.340	0.948	0.997	0.908	0.938
28 August 2013	0.948	0.982	0.996	0.980	0.970

We further compared the evaluation results between the traditional methods and the entropy weighting method, which are shown in Table 4. As shown in the results, the rank order of the effects of the simulation results with the traditional indicators (high to low) as the following: deviations: 5 July 2013, 23 August 2013, 15 April 2014, 20 July 2014, and 23 August 2014; $\bar{d}$: 23 August 2014, 5 July 2013, 20 July 2014, 23 August 2013, and 15 April 2014; and S: 23 August 2014, 5 July 2013, 23 August 2013, 15 April 2014, and 20 July 2014. The application of a single indicator is limited by the indicator selection so that we cannot sum them up simply or select one of them. For instance, the effect of 5 July 2013 showed the best *deviation* but the worst S, which represents the rainfall amount and the average rainfall, respectively. Therefore, choosing these traditional indicators would lead to information loss during the model evaluation. Conversely, the entropy weighting method considers the advantages and characteristics of each traditional indicators and assesses the simulation results comprehensively form different perspectives [34,35]. Thus, the K values are more accurate and easier to compare.

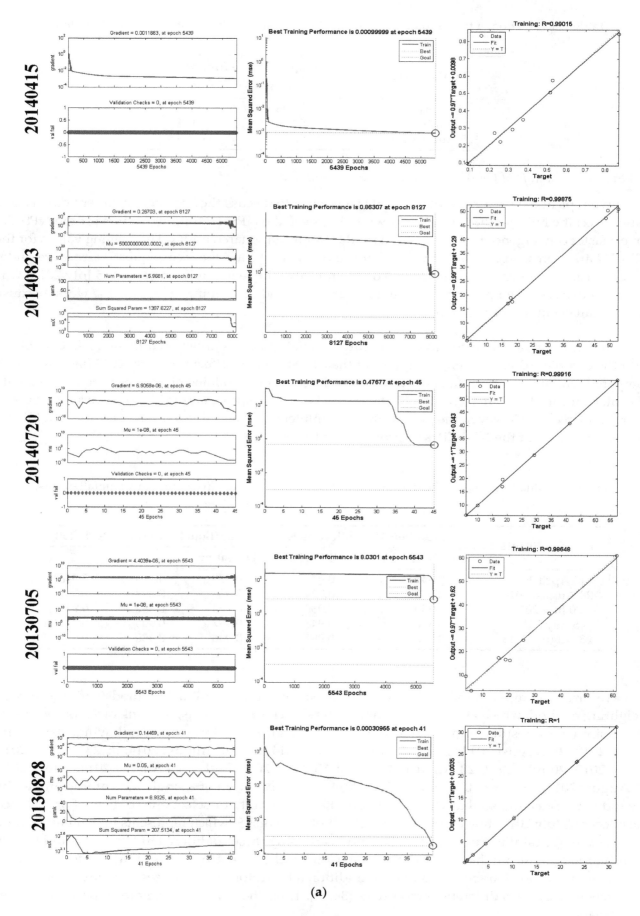

(a)

Figure 4. *Cont.*

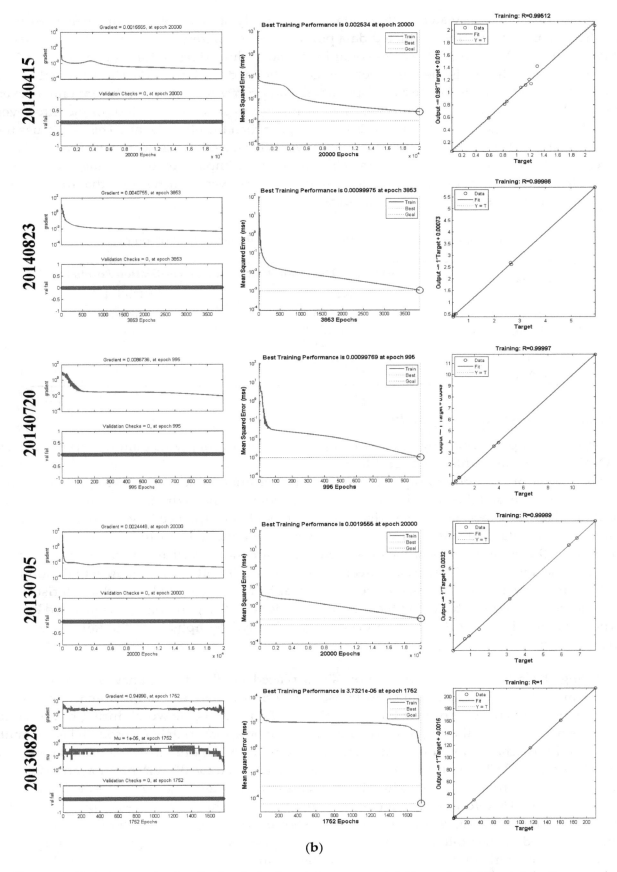

(b)

Figure 4. Training results for the loads in five data-scarce rainfall events: (**a**) the total nitrogen of non-point source (NPS-TN); and (**b**) the total phosphorus of non-point source (NPS-TP). Note: the pink line with dots represent the epochs are smaller.

Owing to the limited water quality data, we randomly selected 30% of the measured values as verification points, and the simulated data points were compared to the selected data to test the accuracy of the ANN during data-scarce conditions. The evaluated results are shown in Table 5, and the intuitionistic indicator is the mean percentage of the load deviation. As shown in Table 5, the effects of the training results for the runoff–pollutant load process are better in different rainfall events, and each of the load deviations is smaller. The mean percentage load deviations of the three events (15 April 2014, 23 August 2014, 28 August 2013) are higher than the other events. This is because these pollutant load data for the three individual rainfall events have peak loads nearby the flow peaks. It is apparent that the flow peak and these high values have major impacts on the training and predictive values of the ANN. In general, the simulation effects are improved, which shows that this method is feasible for estimating scarce pollutant load data.

Table 5. Evaluation of the predicted effects of the verification points.

Rainfall Events	Mean Load Deviation Percentage of Verification Points (%)	
	NPS-TP	NPS-TN
15 April 2014	0.150	0.0770
23 August 2014	0.153	0.029
20 July 2014	0.041	0.094
5 July 2013	0.002	0.038
28 August 2013	0.120	0.022

4.3. Implication for NPS Studies of Multi-Events

Figure 5 compares pollutographs of complete rainfall events with pollutographs simulated by ANN in the same rainfall pattern (data-scarce rainfall events in light and moderate rainfall patterns). Most of the pollutographs conform to the ordinary rules (pollutographs in complete data rainfall events), and the overall tendency is consistent with the hydrographs with complete data, indicating that the method is reliable. Moreover, the model performance is worse under conditions of missing peak data, which is consistent with the abovementioned conclusions. According to the comparisons, the pollutographs of the measured points are more consistent with the ordinary rules than when the tails have missing data. Meanwhile, tail scarcity often appears in actual monitoring to reduce manpower [41]. The tails of the pollutographs have stronger linear characteristics, so a linear interpolation is used to amend the incomplete tails [42]. The pollutographs amended by linear interpolation are shown in Figure 6, indicating that the hydrographs with the tail correction are more coincident.

During the monitoring process, emphasis is placed on the discrepancy in the monitoring mechanism under different rainfall conditions. Based on the abovementioned analysis, the NPS prediction performs the best during the light rainfall events and is the worst during heavy rainfall events. Therefore, the monitoring process for the NPS can be appropriately focused on heavy rainfall conditions. Researchers should pay more attention to monitoring time to avoid peak data scarcity, especially for the NPS-TN monitoring [43]. As already suggested, peak concentration appeared after nearly five hours of runoff during light rainfall events and after nearly three hours of runoff during moderate events. Therefore, we promote peak monitoring techniques, for example, anautomatic sampler with programming, we can appropriately shorten sampling intervals for the peak lag times. Meanwhile, based on the pollutographs improved by this study, we can design the sampling scheme and avoid the risk time in order to require complete water quality data. In addition, the entropy weighting method can be effectively used to evaluate the measured and simulated data [44], showing that it can be used to comprehensively assess the discrepancies more accurately and to easily compare the results, which can be generalized to other catchments.

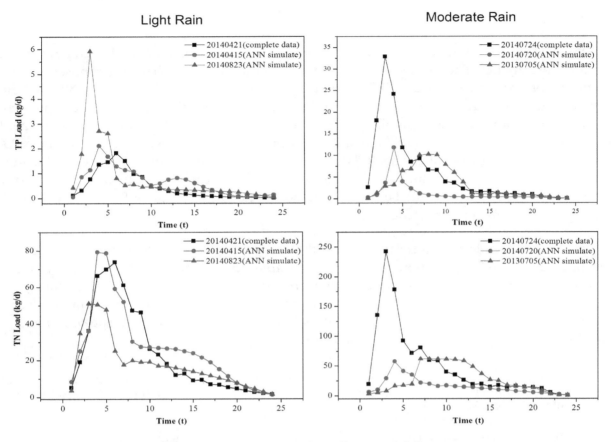

Figure 5. The pollutographs for different rainfall patterns.

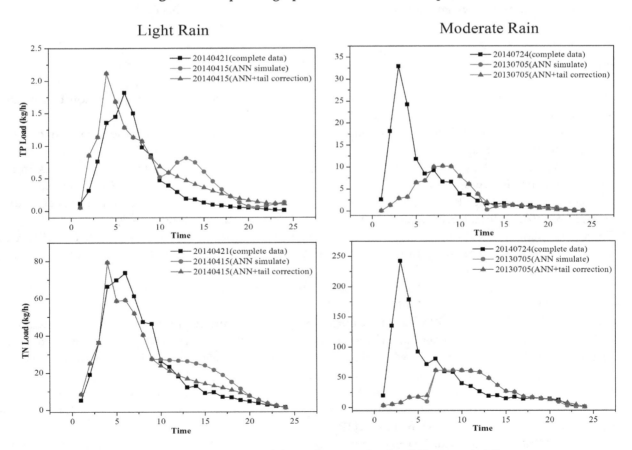

Figure 6. The tail amendment of the pollutographs for different rainfall patterns.

5. Conclusions

In this study, a new framework is proposed for the event-based NPS prediction and evaluation in data-scarce catchments. The results obtained from this study indicate that the proposed ANN had an improved performance over the NPS simulation of light rainfall events, and the NPS-TP model was always more accurate than the NPS-TN under scarce data conditions. In addition, the scarcity of the peak pollutant data has a significant impact on the model performance, so more attention should be given to the monitoring scheme of the event-based NPS studies, especially for the NPS-TN monitoring and the lag time of the peak data. Compared to the traditional indicators, the entropy weighting method can provide a more accurate ANN by considering all of the information during model evaluation. These tools could be extended to other catchments to quantify the event-based NPS pollution, especially data-poor catchments.

However, we should pay more attention to the mechanism of the NPS during multiple rainfall events because the NPS pollution was not the simple consequence of current rainfall events. Additionally, because of the computational burden, the errors and the related uncertainty of the model results were not explored, so more studies are suggested to test this new framework among more diverse regions. Meanwhile, data-driven black-box models are not good at long-term forecasting, nor are they good for examining the effect of BMPs.

Acknowledgments: This research was funded by the National Natural Science Foundation of China (Nos. 51579011 and 51409003) and the Fund for the Innovative Research Group of the National Natural Science Foundation of China (No. 51421065).

Author Contributions: Lei Chen constructed the research framework and designed the study. Cheng Sun was responsible for the data analysis and code programming. Guobo Wang performed the experiments and conducted data analysis. Hui Xie and Zhenyao Shen reviewed and edited the manuscript. All of the authors have read and approved the final manuscript.

References

1.   Ongley, E.D.; Zhang, X.L.; Yu, T. Current status of agricultural and rural non-point source Pollution assessment in China. *Environ. Pollut.* **2010**, *158*, 1159–1168. [CrossRef] [PubMed]

2.   Li, X.F.; Xiang, S.Y.; Zhu, P.F.; Wu, M. Establishing a dynamic self-adaptation learning algorithm of the BP neural network and its applications. *Int. J. Bifurc. Chaos* **2015**, *25*, 1540030. [CrossRef]

3.   Gong, Y.W.; Liang, X.Y.; Li, X.N.; Li, J.Q.; Fang, X.; Song, R.N. Influence of rainfall characteristics on total suspended solids in urban runoff: A case study in Beijing, China. *Water* **2016**, *8*, 278. [CrossRef]

4.   Coulliette, A.D.; Noble, R.T. Impacts of rainfall on the water quality of the Newport River Estuary (Eastern North Carolina, USA). *J. Water Health* **2008**, *6*, 473–482. [CrossRef] [PubMed]

5.   Chen, C.L.; Gao, M.; Xie, D.T.; Ni, J.P. Spatial and temporal variations in non-point source losses of nitrogen and phosphorus in a small agricultural catchment in the Three Gorges Region. *Environ. Monit. Assess.* **2016**, *188*, 257. [CrossRef] [PubMed]

6.   Sajikumar, N.; Thandaveswara, B.S. A non-linear rainfall-runoff model using an artificial neural network. *J. Hydrol.* **1999**, *216*, 32–55. [CrossRef]

7.   Bulygina, N.; McIntyre, N.; Wheater, H. Conditioning rainfall-runoff model parameters for ungauged catchments and land management impacts analysis. *Hydrol. Earth Syst. Sci.* **2009**, *13*, 893–904. [CrossRef]

8.   Maniquiz, M.C.; Lee, S.; Kim, L.H. Multiple linear regression models of urban runoff pollutant load and event mean concentration considering rainfall variables. *J. Environ. Sci.* **2010**, *22*, 946–952. [CrossRef]

9.   Chen, N.W.; Hong, H.S.; Cao, W.Z.; Zhang, Y.Z.; Zeng, Y.; Wang, W.P. Assessment of management practices in a small agricultural watershed in Southeast China. *J. Environ. Sci. Health Part A Toxic Hazard. Subst. Environ. Eng.* **2006**, *41*, 1257–1269. [CrossRef] [PubMed]

10.   Sun, A.Y.; Miranda, R.M.; Xu, X. Development of multi-meta models to support surface water quality management and decision making. *Environ. Earth Sci.* **2015**, *73*, 423–434. [CrossRef]

11. Yuceil, K.; Baloch, M.A.; Gonenc, E.; Tanik, A. Development of a model support system for watershed modeling: A case study from Turkey. *CLEANSoil Air Water* **2007**, *35*, 638–644. [CrossRef]

12. Shope, C.L.; Maharjan, G.R.; Tenhunen, J.; Seo, B.; Kim, K.; Riley, J.; Arnhold, S.; Koellner, T.; Ok, Y.S.; Peiffer, S.; et al. Using the SWAT model to improve process descriptions and define hydrologic partitioning in South Korea. *Hydrol. Earth Syst. Sci.* **2014**, *18*, 539–557. [CrossRef]

13. Jeong, J.; Kannan, N.; Arnold, J.; Glick, R.; Gosselink, L.; Srinivasan, R. Development and integration of sub-hourly rainfall-runoff modeling capability within a watershed model. *Water Resour. Manag.* **2010**, *24*, 4505–4527. [CrossRef]

14. Park, Y.S.; Engel, B.A. Identifying the correlation between water quality data and LOADEST model behavior in annual sediment load estimations. *Water* **2016**, *8*, 368. [CrossRef]

15. Park, Y.S.; Engel, B.A.; Frankenberger, J.; Hwang, H. A web-based tool to estimate pollutant loading using LOADEST. *Water* **2015**, *7*, 4858–4868. [CrossRef]

16. Das, S.K.; Ng, A.W.M.; Perera, B.J.C.; Adhikary, S.K. Effects of climate and landuse activities on water quality in the Yarra River catchment. In Proceedings of the 20th International Congress on Modelling and Simulation (Modsim2013), Adelaide, Australia, 1–6 December 2013; pp. 2618–2624.

17. Chen, D.J.; Hu, M.P.; Guo, Y.; Dahlgren, R.A. Reconstructing historical changes in phosphorus inputs to rivers from point and nonpoint sources in a rapidly developing watershed in eastern China, 1980–2010. *Sci. Total Environ.* **2015**, *533*, 196–204. [CrossRef] [PubMed]

18. Melesse, A.M.; Ahmad, S.; Mcclain, M.E.; Wang, X.; Lim, Y.H. Suspended sediment load prediction of river systems: An artificial neural network approach. *Agric. Water Manag.* **2011**, *98*, 855–866. [CrossRef]

19. Tran, H.D.; Muttil, N.; Perera, B.J.C. Investigation of artificial neural network models for streamflow forecasting. In Proceedings of the 19th International Congress on Modelling and Simulation (Modsim 2011), Perth, Australia, 12–16 December 2011; pp. 1099–1105.

20. Hassan, M.; Shamim, M.A.; Sikandar, A.; Mehmood, I.; Ahmed, I.; Ashiq, S.Z.; Khitab, A. Development of sediment load estimation models by using artificial neural networking techniques. *Environ. Monit. Assess.* **2015**, *187*, 686. [CrossRef] [PubMed]

21. Dhiman, N.; Markandeya; Singh, A.; Verma, N.K.; Ajaria, N.; Patnaik, S. Statistical optimization and artificial neural network modeling for acridine orange dye degradation using in-situ synthesized polymer capped ZnO nanoparticles. *J. Colloid Interface Sci.* **2017**, *493*, 295–306. [CrossRef] [PubMed]

22. Wang, H.W.; Ai, Z.W.; Cao, Y. Information-entropy based load balancing in parallel adaptive volume rendring. In Proceedings of the International Conferences on Interfaces and Human Computer Interaction 2015, Game and Entertainment Technologies 2015, and Computer Graphics, Visualization, Computer Vision and Image Processing 2015, Las Palmas de Gran Canaria, Spain, 22–24 July 2015; pp. 163–169.

23. Khosravi, K.; Pourghasemi, H.R.; Chapi, K.; Bahri, M. Flash flood susceptibility analysis and its mapping using different bivariate models in Iran: A comparison between Shannon's entropy, statistical index, and weighting factor models. *Environ. Monit. Assess.* **2016**, *188*, 656. [CrossRef] [PubMed]

24. Yuan, Y.M.; Wei, G.A. Empirical studies of unblocked index for urban freeway traffic flow states. In Proceedings of the 2009 12th International IEEE Conference on Intelligent Transportation Systems, St. Louis, MO, USA, 4–7 Octorber 2009; pp. 1–6.

25. Kan, J.M.; Liu, J.H. Self-Tuning PID controller based on improved BP neural network. In Proceedings of the 2009 Second International Conference on Intelligent Computation Technology and Automation, Changsha, China, 10–11 Octorber 2009; pp. 95–98.

26. Chen, X.Y.; Chau, K.W. A hybrid double feedforward neural network for suspended sediment load estimation. *Water Resour. Manag.* **2016**, *30*, 2179–2194. [CrossRef]

27. Guo, Z.H.; Wu, J.; Lu, H.Y.; Wang, J.Z. A case study on a hybrid wind speed forecasting method using BP neural network. *Knowl. Based Syst.* **2011**, *24*, 1048–1056. [CrossRef]

28. Ju, Q.; Yu, Z.B.; Hao, Z.C.; Ou, G.X.; Zhao, J.; Liu, D.D. Division-based rainfall-runoff simulations with BP neural networks and Xinanjiang model. *Neurocomputing* **2009**, *72*, 2873–2883. [CrossRef]

29. Jing, J.T.; Feng, P.F.; Wei, S.L.; Zhao, H. Investigation on surface morphology model of Si3N4 ceramics for rotary ultrasonic grinding machining based on the neural network. *Appl. Surf. Sci.* **2017**, *396*, 85–94. [CrossRef]

30. Ullrich, A.; Volk, M. Influence of different nitrate-N monitoring strategies on load estimation as a base for model calibration and evaluation. *Environ. Monit. Assess.* **2010**, *171*, 513–527. [CrossRef] [PubMed]

31. Wilson, D.R.; Apreleva, M.V.; Eichler, M.J.; Harrold, F.R. Accuracy and repeatability of a pressure measurement system in the patellofemoral joint. *J. Biomech.* **2003**, *36*, 1909–1915. [CrossRef]

32. Ai, Y.T.; Guan, J.Y.; Fei, C.W.; Tian, J.; Zhang, F.L. Fusion information entropy method of rolling bearing fault diagnosis based on n-dimensional characteristic parameter distance. *Mech. Syst. Signal Process.* **2017**, *88*, 123–136. [CrossRef]

33. Liu, F.; Zhao, S.; Weng, M.; Liu, Y. Fire risk assessment for large-scale commercial buildings based on structure entropy weight method. *Saf. Sci.* **2017**, *94*, 26–40. [CrossRef]

34. Sun, L.Y.; Miao, C.L.; Yang, L. Ecological-economic efficiency evaluation of green technology innovation in strategic emerging industries based on entropy weighted TOPSIS method. *Ecol. Indic.* **2017**, *73*, 554–558. [CrossRef]

35. Huang, Z.Y. Evaluating intelligent residential communities using multi-strategic weighting method in China. *Energy Build.* **2014**, *69*, 144–153. [CrossRef]

36. Shen, Z.Y.; Gong, Y.W.; Li, Y.H.; Liu, R.M. Analysis and modeling of soil conservation measures in the Three Gorges Reservoir Area in China. *Catena* **2010**, *81*, 104–112. [CrossRef]

37. Shen, Z.Y.; Gong, Y.W.; Li, Y.H.; Hong, Q.; Xu, L.; Liu, R.M. A comparison of WEPP and SWAT for modeling soil erosion of the Zhangjiachong Watershed in the Three Gorges Reservoir Area. *Agric. Water Manag.* **2009**, *96*, 1435–1442. [CrossRef]

38. Shen, Z.; Qiu, J.; Hong, Q.; Chen, L. Simulation of spatial and temporal distributions of non-point source pollution load in the Three Gorges Reservoir Region. *Sci. Total Environ.* **2014**, *493*, 138–146. [CrossRef] [PubMed]

39. Gottschalk, L.; Weingartner, R. Distribution of peak flow derived from a distribution of rainfall volume and runoff coefficient, and a unit hydrograph. *J. Hydrol.* **1998**, *208*, 148–162. [CrossRef]

40. Xu, Y.F.; Ma, C.Z.; Liu, Q.; Xi, B.D.; Qian, G.R.; Zhang, D.Y.; Huo, S.L. Method to predict key factors affecting lake eutrophication—A new approach based on Support Vector Regression model. *Int. Biodeterior. Biodegrad.* **2015**, *102*, 308–315. [CrossRef]

41. Wagner, P.D.; Fiener, P.; Wilken, F.; Kumar, S.; Schneider, K. Comparison and evaluation of spatial interpolation schemes for daily rainfall in data scarce regions. *J. Hydrol.* **2012**, *464–465*, 388–400. [CrossRef]

42. Croke, B.; Islam, A.; Ghosh, J.; Khan, M.A. Evaluation of approaches for estimation of rainfall and the unit hydrograph. *Hydrol. Res.* **2011**, *42*, 372–385. [CrossRef]

43. Ryu, J.; Jang, W.S.; Kim, J.; Jung, Y.; Engel, B.A.; Lim, K.J. Development of field pollutant load estimation module and linkage of QUAL2E with watershed-scale L-THIA ACN model. *Water* **2016**, *8*, 292. [CrossRef]

44. Li, P.Y.; Qian, H.; Wu, J.H. Groundwater quality assessment based on improved water quality index in Pengyang County, Ningxia, Northwest China. *J. Chem.* **2010**, *7*, S209–S216.

14

Entropy Applications to Water Monitoring Network Design

Jongho Keum [1],*, Kurt C. Kornelsen [2], James M. Leach [1] and Paulin Coulibaly [1,2]

[1]   Department of Civil Engineering, McMaster University, Hamilton, ON L8S4L8, Canada;
     leachjm@mcmaster.ca (J.M.L.); couliba@mcmaster.ca (P.C.)

[2]   School of Geography and Earth Sciences, NSERC Canadian FloodNet, McMaster University, Hamilton,
     ON L8S4L8, Canada; kornelkc@mcmaster.ca

*   Correspondence: jkeum@mcmaster.ca

Abstract: Having reliable water monitoring networks is an essential component of water resources and environmental management. A standardized process for the design of water monitoring networks does not exist with the exception of the World Meteorological Organization (WMO) general guidelines about the minimum network density. While one of the major challenges in the design of optimal hydrometric networks has been establishing design objectives, information theory has been successfully adopted to network design problems by providing measures of the information content that can be deliverable from a station or a network. This review firstly summarizes the common entropy terms that have been used in water monitoring network designs. Then, this paper deals with the recent applications of the entropy concept for water monitoring network designs, which are categorized into (1) precipitation; (2) streamflow and water level; (3) water quality; and (4) soil moisture and groundwater networks. The integrated design method for multivariate monitoring networks is also covered. Despite several issues, entropy theory has been well suited to water monitoring network design. However, further work is still required to provide design standards and guidelines for operational use.

Keywords: entropy; water monitoring; network design; hydrometric network; information theory; entropy applications

---

1. Introduction

Water monitoring networks account for all aspects of the water-related measurement system including precipitation, streamflow, water quality, groundwater, soil moisture, etc. [1–3]. Adequate water monitoring networks and quality data from them comprise one of the first and primary steps towards efficient water resource management. The basic principles of water monitoring network design have simply been a number of monitoring stations, locations of the stations and data period or sampling frequency [4,5]. Recent technological advances have allowed gradual transitions from manual sampling to the automated observations, while some water quality parameters still require field and/or lab analyses of water or other environmental samples. One may expect that the more data we collect, the more water resource problems are solved efficiently. However, this is not always true because irrelevant, inadequate or inefficient data in the wrong location or at the wrong time can inhibit the quality of a dataset [1,6,7]. More seriously, the decline of water monitoring networks has been a general trend due to financial limitations and changes of monitoring priority [8–10]. Therefore, determining the adequate number of monitoring stations and their locations has become critical to network design. However, a standardized methodology for a proper water monitoring network design process has not been drawn yet due to the practical and socioeconomic complexity in diverse design cases [1,11].

The existing reviews have investigated the broad range of the water monitoring network design methodologies, such as statistical analysis, spatial interpolation, application of information theory, optimization techniques, physiographic analysis, user survey or expert recommendations and combinations of multiple methods [4,6,10–15]. A prior comprehensive review by Mishra and Coulibaly [10] reviewed evidence of declining hydrometric network density, highlighted the importance of quality data from well-designed networks and considered a range of approaches by which networks were designed. They also compared statistical, spatial interpolation, physiographic, sampling-driven and entropy-based approaches to hydrometric network design. Mishra and Coulibaly [10] were able to draw several conclusions about the importance of high quality hydrometric data for water resource management that remain valid. They also concluded that one of the most promising approaches for network design was the application of entropy methods highlighting early studies using the principle of maximum entropy and information transfer. Therefore, this review focuses on the recent studies that have applied information theory. Information theory was initially developed by Shannon in 1948 [16] to measure the information content in a dataset and has been applied to solve water resource problems. Recently, its applications extended to water monitoring network design by adopting the concept that the entropy would be able to explain the inherent information content in a monitoring station or a monitoring network. The basic objective has naturally been to have the maximum amount of information. In other entropy approaches, stations in a monitoring network would have the least sharable or common information, which is called transinformation. To achieve this, the stations should be as independent from each other as possible.

The scope of this paper includes water monitoring network design and evaluation studies that (1) applied entropy theory in the design process and (2) were published after the existing comprehensive review by Mishra and Coulibaly in 2009 [10]. To the best of our knowledge, no review exists that has focused on entropy applications to water monitoring network design previous to the 2009 study by Mishra and Coulibaly [10]. However, there has been considerable progress in the application of entropy theory to monitoring network design following the previous review, including new entropy-based measures, optimization techniques and approaches to estimating information content at ungauged stations. Therefore, a need was identified to consolidate knowledge and recent advances on the subject. For publications prior to 2010, the reader is referred to Mishra and Coulibaly [10]. This review firstly describes entropy concepts and various terms that are typically used in network design. The previous studies are then summarized by categorizing the type of networks; i.e., precipitation, streamflow and water level, soil moisture and groundwater and water quality monitoring networks. The integrated design method for multiple types of networks is also reviewed. We define some terminology hereafter to ensure a common understanding for the readers.

The term network evaluation is used when the network quality is assessed without changing any station, while network design is a general term that suggests some changes in stations. Specifically, network design includes network reduction, network expansion and network redesign. Network reduction is applied where some monitoring stations should be removed from the network. On the other hand, if financial flexibility meets the monitoring needs, further stations can be added to the existing network, called network expansion. Network redesign refers to rearranging stations without changing the number of stations. The term optimal network is to be used only if the network consists of optimal locations of stations that are identified by the actual use of an optimization technique.

2. Definitions of Entropy Terms as Applied to Water Monitoring Networks

2.1. Entropy Concept

In thermodynamics, entropy has been understood as a measure of disorder or randomness of a system. Shannon [16] extended the entropy concept to information theory by recognizing that uncertainty in a system will be decreased when information is added to the system. Therefore, the term entropy in information theory introduced by Shannon [16] in 1948 describes the amount of information content in a random variable. The likelihood of an event is typically given by probability p. If a probability of an event is very high, such as 0.9999 or one, one will not be surprised, but can certainly anticipate the outcome. On the other hand, any low probability event is highly uncertain, so that a considerable amount of information can be given if this happens. Hence, the information from an event that occurred is inversely related to its probability, $1/p$ [17]. Suppose that there are two independent events A and B with their probabilities p_A and p_B, respectively. The probability of the joint occurrence of the events A and B can be $p_A p_B$, and the information gained by the joint event is then $1/(p_A p_B)$. However, the sum of information from each individual event is not equal to the information from the joint event, that is:

$$\frac{1}{p_A p_B} \neq \frac{1}{p_A} + \frac{1}{p_B} \tag{1}$$

The only transition that will make both sides of Equation (1) be equal is the logarithm [17–19], which can be written as:

$$\log \frac{1}{p_A p_B} = \log \frac{1}{p_A} + \log \frac{1}{p_B} = -\log p_A - \log p_B \tag{2}$$

Likewise, Tribus [20] showed that the uncertainty of an event with probability p is $-\log p$, which became a basis of the Shannon entropy, which is further described hereafter.

2.2. Marginal Entropy

When information is provided in a system, one can expect that the uncertainty of the system would be reduced; therefore, the amount of information that was given to a system by knowing a variable is called marginal entropy. If a random variable X is expected to have N outcomes with a probability distribution $P = \{p_1, p_2, \cdots, p_N\}$, the (weighted) average information provided by the N joint events is given by:

$$H(X) = -p_1 \log p_1 - p_2 \log p_2 - \cdots - p_N \log p_N = -\sum_{i=1}^{N} p_i \log p_i, \quad \sum_{i=1}^{N} p_i = 1, \ p_i \geq 0 \tag{3}$$

where $H(X)$ is the marginal entropy of a random variable X. Any base of the logarithm can be used in Equation (3), the choice depending on the problem given. In binary questions (i.e., yes or no questions), the base of two should be used, and the corresponding unit of entropy is bit. Similarly, unit trit for base 3, unit nat for base e and unit decibels or decit for base 10 are some example units of information. Recall that this review covers entropy applications for hydrometric network design, and the expected answer of the design process can be either "use/include/install the station" and "do not use/include/install the station" for the network to be optimal. Therefore, the logarithm in Shannon entropy calculation for hydrometric network design is most appropriate with a base of two. Then, the $H(X)$ value from Equation (3) will be understood as the information contents of a station X that can be delivered if installed.

If a variable K has a known value, the probability of an event will be one, while all the other alternative probabilities are zero. The information content in the variable K, $H(K)$, will be zero from Equation (3) representing that there is no uncertainty or a certain outcome. On the other hand, if a variable U has a uniform distribution (i.e., probability of each event is equal, $1/N$), the entropy of the variable U will be:

$$H(U) = \log N \tag{4}$$

The value of Equation (4) is often called as maximum entropy or saturated entropy. These two entropies, $H(K)$ and $H(U)$, define the minimum and maximum boundaries of entropy values, that is:

$$0 \leq H(X) \leq \log N \tag{5}$$

2.3. Multivariate Joint Entropy

While the marginal entropy described in Section 2.2 explains a univariate entropy, one can imagine how to calculate entropy values in a bivariate or a multivariate case. The total information contents from N variables can be calculated by using joint probability instead of univariate probability in Equation (3), given by:

$$H(X_1, X_2, \cdots, X_N) = -\sum_{i_1=1}^{n_1} \sum_{i_2=1}^{n_2} \cdots \sum_{i_N=1}^{n_N} p(x_{1,i_1}, x_{2,i_2}, \cdots, x_{N,i_N}) \log_2 p(x_{1,i_1}, x_{2,i_2}, \cdots, x_{N,i_N}) \quad (6)$$

where $H(X_1, X_2, \cdots, X_N)$ is the joint entropy of N variables, $p(x_{1,i_1}, x_{2,i_2}, \cdots, x_{N,i_N})$ is the joint probability of N variables and $n_1, n_2, \cdots, n_N$ are the numbers of class intervals of corresponding variable distributions [21]. If all variables are stochastically independent, the joint entropy from Equation (6) will be equal to the sum of marginal entropies, which becomes the maximum value of joint entropy. Therefore, the joint entropy is bounded by [21]:

$$0 \leq H(X_1, X_2, \cdots, X_N) \leq \sum_{i=1}^{N} H(X_i) \leq N \log_2 N \quad (7)$$

2.4. Conditional Entropy

Conditional entropy explains a measure of information content of one variable that is not deliverable by other variables. If two random variables, A and B, are correlated, providing information from one variable may clear some uncertainty that the other variable has. In the case of no correlation between variables, the conditional entropy is equal to marginal entropy. That is:

$$H(A|B) = H(A, B) - H(B) \leq H(A) \quad (8)$$

where $H(A|B)$ is conditional entropy of the variable A when the information contents of the variable B is given. One can rewrite Equation (8) as:

$$H(A, B) = H(A|B) + H(B) = H(B|A) + H(A) \quad (9)$$

Furthermore, conditional entropy can be also presented mathematically using joint and conditional probabilities and Bayes theorem as:

$$H(A|B) = -\sum_{i=1}^{N_A} \sum_{j=1}^{N_B} p(a_i, b_j) \log p(a_i|b_j) = -\sum_{i=1}^{N_A} \sum_{j=1}^{N_B} p(a_i, b_j) \log \frac{p(a_i, b_j)}{p(b_j)} \quad (10)$$

2.5. Transinformation

The two variables, A and B, described in Section 2.4 will have some common or shared information, which is called transinformation or mutual information, because they are correlated.

$$T(A, B) = H(A) - H(A|B) = H(B) - H(B|A) = T(B, A) \quad (11)$$

where $T(A, B)$ is transinformation between the variables A and B. The larger the transinformation is, the higher those variables depend on each other. In other words, the transinformation indicates how much information content is transferrable from other variables. Similar to Equation (10), transinformation shall be written as [19]:

$$T(A, B) = \sum_{i=1}^{N_A} \sum_{j=1}^{N_B} p(a_i, b_j) \log \frac{p(a_i|b_j)}{p(a_i)} = \sum_{i=1}^{N_A} \sum_{j=1}^{N_B} p(a_i, b_j) \log \frac{p(a_i, b_j)}{p(a_i)p(b_j)} \quad (12)$$

Transinformation is typically used for measuring mutual information between two variables or two groups of variables as the generalized form for multivariate transinformation is given as:

$$T[(X_1, X_2, \cdots, X_k); \quad (X_{k+1}, X_{k+2}, \cdots, X_N)] \\ = H(X_1, X_2, \cdots, X_k) - H[(X_1, X_2, \cdots, X_k)|(X_{k+1}, X_{k+2}, \cdots, X_N)] \tag{13}$$

2.6. Total Correlation

While transinformation and mutual information have the same definition, total correlation is not equivalent to them as the total correlation is a simple estimate that defines the amount of shared information typically of multiple variables. Simply, the total correlation is defined by the difference between the sum of marginal entropy of N variables and their joint entropy [22,23], which is given as:

$$C(X_1, X_2, \cdots, X_N) = \sum_{i=1}^{N} H(X_i) - H(X_1, X_2, \cdots, X_N) \tag{14}$$

If $N = 2$ in Equation (14), the total correlation will be equal to the transinformation or mutual information. However, the transinformation is only meaningful to two random variables as shown in Equations (11) to (13); therefore, the total correlation and the transinformation values would be different if $N > 2$.

2.7. Other Entropy Terms

The entropy terms described above (i.e., marginal entropy, joint entropy, conditional entropy, transinformation and total correlation) are the basic measures that have been typically used in entropy applications to water monitoring network design. While many studies developed specific approaches and applied for case studies using the basic entropy terms, some have extended the terms beyond them by deriving from or combining the basic measures. The detailed descriptions of the extended entropy terms are not included in this review, but briefly explained when needed in Section 3. Interested readers may refer to the original references.

3. Applications of Entropy to Water Monitoring Network Design

This section summarizes the recent applications of entropy theory to design water monitoring networks. The review was categorized by the types of networks, such as precipitation, streamflow or water level, soil moisture or groundwater and water quality networks. Then later, a hybrid design method for multivariate water monitoring networks was discussed. Table 1 presents brief summaries including network types, methods and key findings of the selected research articles that applied entropy theory for designing the water monitoring network and were published in 2010 or after to cover the most recent contributions since the existing review [10].

Table 1. Summary of significant contributions to water monitoring network design using entropy (author alphabetical order).

Authors/Year	Network Types	Study Areas	Methods/Entropy Measures	Key Findings
Alameddine et al., 2013 [24]	Water quality	Neuse River Estuary, NC, USA	-Total system entropy -Standard violation entropy -Multiple attribute decision making process -Analytical hierarchical process	-Networks designed using total system entropy and violation entropy of dissolved oxygen were similar -When measured water quality parameters have a low probability of violating water quality standards, their violation entropy is less informative
Alfonso et al., 2010 [25]	Water level	Pijnacker Region, The Netherlands	-Directional information transfer (DIT)	-Introduced total correlation for determining multivariate dependence in water monitoring network design -Information content and redundancy is dependent on the DIT between monitoring stations (DIT$_{XY}$ or DIT$_{YX}$)
Alfonso et al., 2010 [26]	Water level	Pijnacker Region, The Netherlands	-Max(Joint Entropy) min(Total Correlation) -Non-dominated Sorting Genetic Algorithm II (NSGA-II)	-Total correlation should be combined with joint entropy to get most information out of monitoring network
Alfonso et al., 2013 [27]	Streamflow	Magdalena River, Colombia	-Max(Joint Entropy) min(Total Correlation) -Rank-based iterative approach	-Rank method is useful in finding extremes on Pareto front -When iteratively selecting stations, the information content of the network is not guaranteed to be maximum if the network contains the station with the most information
Alfonso et al., 2014 [28]	Water level	North Sea, The Netherlands	-Max(Joint Entropy) min(Total Correlation) -Ensemble entropy -NSGA-II	-By creating an ensemble of solutions through varying the bin size of the initial Pareto optimal solution set, the authors highlight the uncertainty related to choosing bin size
Boroumand and Rajaee, 2017 [29]	Water quality	San Francisco Bay, CA, USA	-Transinformation-distance (T-D) curve	-Using T-D curve they were able to reduce the network from 37 to 21 monitoring stations. -New network covered entire study area without having redundant data
Brunsell, 2010 [30]	Precipitation	Continental United States	-Relative entropy -Wavelet multi-resolution analysis	-The temporal scaling regions identified (1) synoptic, (2) monthly to annual, (3) interannual patterns -Little correlation between relative entropy and annual precipitation except for breakpoint at 95° W Lat
Fahle et al., 2015 [31]	Water level/Groundwater level	Spreewald region, Germany	-MIMR, max(Joint Entropy + Transinformation − Total Correlation) -Subsets of time series data	-Found using subsets of the available time series data could better identify important stations -Showed water levels across network react similarly during high precipitation and are more unique during dry periods -Consequently method can allow for design of network which focuses on floods or droughts

Table 1. *Cont.*

Authors/Year	Network Types	Study Areas	Methods/Entropy Measures	Key Findings
Hosseini and Kerachian, 2017 [32]	Groundwater level	Dehgolan plain, Iran	-Marginal entropy -Data fusion of spatiotemporal kriging and ANN model -Value of information (VOI)	-Network reduction from 52 to 42 (35 high priority and 7 low priority) stations while standard deviation of average estimation error variance stayed the same -Found sampling frequency of high priority stations should be every 20 days and low priority should be every 32, based on analysis of stations selected using VOI
Hosseini and Kerachian, 2017 [33]	Groundwater level	Dehgolan plain, Iran	-Bayesian maximum entropy (BME) -Multi-criteria decision making based on ordered weighted averaging	-Network reduction from 52 to 33 stations while standard deviation of average estimation error variance stayed the same -Sampling frequency increased from 4 weeks to 5 weeks
Keum and Coulibaly, 2017 [34]	Precipitation/Streamflow	Columbia River basin, BC, Canada. Southern Ontario, Canada	-Dual Entropy and Multiobjective Optimization (DEMO) to max(Joint Entropy) and min(Total Correlation)	-Found that networks obtain significant amount of information from 5 to 10 years of data periods, and total correlation tends to be stabilized within 5 years by applying daily time series -Recommended minimum 10 years data periods for designing precipitation or streamflow networks using daily time series
Keum and Coulibaly, 2017 [35]	Integrated	Southern Ontario, Canada	-DEMO to max(Joint Entropy), min(Total Correlation), and max(Conditional Entropy) -Sturge, Scott and rounding binning methods	-Precipitation and streamflow networks were designed simultaneously. -Binning methods were compared and concluded that the optimal networks can be altered due to the binning methods
Kornelsen and Coulibaly, 2015 [36]	Soil Moisture	Great Lakes Basin, Canada-USA	-DEMO to Max(Joint Entropy) min(Total Correlation) -SMOS satellite data	-Optimum networks were different for ascending and descending overpasses -Combining overpass data resulted in complimentary spatial distribution of stations
Leach et al., 2015 [37]	Streamflow	Columbia River basin, BC, Canada. Southern Ontario, Canada	-DEMO to Max(Joint Entropy) min(Total Correlation) -Streamflow signatures -Indicators of hydrologic alteration (IHA)	-Found that including streamflow signatures as design objective increases network coverage in headwater areas. -Found including IHAs increases network coverage in downstream and urban areas.
Leach et al., 2016 [38]	Groundwater level	Southern Ontario, Canada	-DEMO to Max(Joint Entropy) min(Total Correlation) -Annual recharge	-Found that considering spatial distribution of annual recharge can improve network coverage

Table 1. *Cont.*

Authors/Year	Network Types	Study Areas	Methods/Entropy Measures	Key Findings
Lee, 2013 [39]	Water quality	Hagye Basin, South Korea	-Marginal entropy analogous cost function -Genetic algorithm	-Developed computationally efficient way to design a monitoring network in an ungauged basin
Lee et al., 2014 [40]	Water quality	Sanganmi Basin, South Korea	-Multivariate transinformation -Genetic algorithm	-Developed method based on maximizing information content to design a water quality monitoring network in a sewer system
Li et al., 2012 [41]	Streamflow / Water level	Brazos River basin, USA. Pijnacker, The Netherlands	-MIMR, max[Joint Entropy + Transinformation − Total Correlation)	-Developed maximum information minimum redundancy method (MIMR) -Found it to better at locating high information content stations for a monitoring network
Mahjouri and Kerachian, 2011 [42]	Water quality	Jajrood River, Iran	-Information transfer index (ITI) distance and time curves -Micro genetic algorithm (MGA)	-The MGA was used to find the optimal combination of monitoring stations which minimize the temporal and spatial ITI -Found that the sampling frequency and number of stations could be increased in the monitoring network
Mahmoudi-Meimand et al., 2016 [43]	Precipitation	Karkheh, Iran	-Transinformation entropy -Kriging error variance -Weighted cost function to select from Monte Carlo generated networks	-Consideration of spatial analysis error and transinformation entropy improved network design
Masoumi and Kerachian, 2010 [44]	Groundwater quality	Tehran, Iran	-Transinformation-distance (T-D) curve -Transinformation-time (T-T) curve -C-mean clustering -Hybrid genetic algorithm (HGA)	-Developed different T-D curves based on homogeneous clusters of existing monitoring stations -Used HGA to find optimal network with maximum spatial coverage and minimum transinformation -Showed that sampling frequency could be optimized in the same way
Memarzadeh et al., 2013 [45]	Water quality	Karoon River, Iran	-Information transfer index (ITI) distance curve -Homogenous zone clustering -Dynamic factor analysis (DFA)	-Increased monitoring network without increasing redundant information

Table 1. *Cont.*

Authors/Year	Network Types	Study Areas	Methods/Entropy Measures	Key Findings
Mishra and Coulibaly, 2010 [46]	Streamflow	Selected basins across Canada	-Transinformation index -Marginal, joint, and transinformation	-Used information theory to highlight critical areas across Canada in need of monitoring -Found that several watersheds are information deficient and would benefit from increased monitoring
Mishra and Coulibaly, 2014 [47]	Streamflow	Selected basins across Canada	-Transinformation index -Seasonal streamflow information (SSI)	-Evaluated and highlighted the effects of seasonal climate on streamflow network design
Mondal and Singh, 2012 [48]	Groundwater level	Kodaganar River basin, India	-Marginal entropy, joint entropy, transinformation -Information transfer index (ITI)	-Identified high priority monitoring stations using marginal entropy -ITI was used to evaluate monitoring network, showed that it could be reduced
Samuel et al., 2013 [49]	Streamflow	St. John and St. Lawrence River basins, Canada	-Combined Regionalization-DEMO -Max(Joint Entropy) min(Total Correlation)	-Proposed combined regionalization dual entropy multi-objective optimization approach to design of minimum optimal network that meets World Meteorological Organization (WMO) guidelines -Found that the location of new monitoring stations added to a network depends on the current network density
Santos et al., 2013 [50]	Precipitation	Portugal	-ANN sensitivity analysis -Mutual Information criteria -K-means with Euclidean distance	-Compared three clustering methods to reduce station density -Best method was case dependent -All subset networks reproduced spatial precipitation pattern
Stosic et al., 2017 [51]	Streamflow	Brazos River, TX, USA	-Joint permutation entropy	-Used joint permutation entropy to account for ordering of time series data to better account for station information -Found that the most efficient measurement window was seven days when compared to daily and monthly
Su and You, 2014 [52]	Precipitation	Shihmen Reservoir Taiwan	-Developed 2D transinformation-distance (T-D) model -T-D model used to interpolate network information	-Network designed by maximizing additional information provided by station given regionalized transinformation -Temporal scale has significant influence on information delivery

Table 1. *Cont.*

Authors/Year	Network Types	Study Areas	Methods/Entropy Measures	Key Findings
Uddameri and Andruss, 2014 [53]	Groundwater level	Victoria County Groundwater Conservation District, TX, USA	-Marginal entropy -Monitoring priority index (MPI)	-Compared MPI found using kriging to MPI found using marginal entropy -Showed entropy derived MPI to be more conservative measure
Wei et al., 2014 [54]	Precipitation	Taiwan University Experimental Forest, Taiwan	-Joint Entropy of hourly, monthly, dry/wet months and annual rainfall at 1, 3, 5 km grids	-Station priority changes at different spatiotemporal scales -Temporal scales have more significant changes on joint entropy values than spatial scales -Long time and short spatial scales require fewer stations for stable joint entropy
Werstuck and Coulibaly, 2016 [55]	Streamflow	Ottawa River Basin, Canada	-Transinformation index -DEMO to Max(Joint Entropy) min(Total Correlation)-Streamflow signatures -Indicators of hydrologic alteration (IHA)	-Compared regionalized data from McMaster University-Hydrologiska Byråns Vattenbalansavdelnin (MAC-HBV) and Inverse Distance Weighting—Drainage Area Ratio (IDW-DAR) and found IDW-DAR to be more adequate for generating synthetic time series for potential monitoring stations -Critical areas highlighted by TI index method were the same areas where additional stations were added using DEMO method
Werstuck and Coulibaly, 2017 [56]	Streamflow	Ottawa River Basin, Canada	-Transinformation index -DEMO to max(Joint Entropy) min(Total Correlation)	-Transinformation index analysis is not significantly affected by scaling -Scaling effects are noticeable when DEMO method was applied
Xu et al., 2015 [57]	Precipitation	Xiangjiang River Basin, China	-Mutual Information (MI) of rain gauges -Designed network by min(Σ[MI]), min(bias), max(NSE) -Resampled rainfall used in Xinanjiang and SWAT models	-Lumped model performance was stable with different Pareto optimal networks -Distributed model performance improves with number of stations
Yakirevich et al., 2013 [58]	Groundwater quality	OPE3 research site, Maryland, USA	-Principle of minimum cross entropy (POMCE) -Hydrus-3D	-Using POMCE with two variants of Hydrus-3D, additional monitoring stations were added where the difference between the models was greatest

3.1. Precipitation Networks

The design of a representative precipitation monitoring network is an important and still challenging task for which an entropy approach is well suited. High quality precipitation information is necessary for streamflow and flood forecasting, surface water management, agricultural management, climate process understanding and many other applications. However, precipitation is well known to be highly variable in both space and time [59] and often statistically represented by highly skewed distributions [60] making the application of parametric analysis methods difficult. These challenges also extend to entropy-based approaches for precipitation monitoring. For example, the marginal entropy has been found to be well correlated with total precipitation in northern Brazil because the probability distribution in regions with higher rainfall tended to be more uniform and less skewed [61]. In contrast, Mishra et al. [59] found that the marginal disorder index (MDI), which is the ratio of observed entropy to the maximum possible entropy at a given site, was inversely related to mean annual rainfall in the U.S. state of Texas, where MDI was found to vary seasonally. Brunsell [30] studied the entropy from monitoring stations across the United States where little correlation was found between precipitation and marginal entropy with the exception of a breakpoint in entropy at $-95°$ longitude corresponding to high temporal variability precipitation patterns. It has also been noted by several studies that the temporal sampling of precipitation is an important consideration for calculating entropy and for designing precipitation networks [54,59]. At finer timescales (hourly to daily), precipitation is highly variable resulting in higher overall entropy, whereas longer time periods (monthly to annual) have less variability resulting in lower marginal entropy [30,52,59]. The dependence on spatial and temporal scales has also been identified in a network design application. Wei et al. [54] prioritized potential stations in Central Taiwan to maximize the joint entropy of the network at hourly, monthly and annual temporal scales, as well as 1-, 3- and 5-km spatial scales. They found that priority stations changed with both spatial and temporal scales, where changes in temporal scales resulted in more significant changes in station priority than spatial-rescaling. The decrease in entropy at longer timescales also had an impact on station density where fewer stations were required to reach a stable joint entropy value for longer time scales [54]. These findings demonstrate the important first consideration of network objectives when determining the spatial and temporal sampling used to calculate entropy. However, the research on this topic is still limited, and more work is needed to provide robust guidance on sampling strategies.

Several approaches have been proposed to design or redesign a precipitation monitoring network using one or more entropic measures. Many of these approaches are initialized by building a network around a central station usually selected as the station with the highest marginal entropy [43,62–64]. In urban Rome, Ridolfi et al. [62] selected stations for the precipitation network by sequentially finding the next station that minimized the conditional entropy of the network and adding that station to the network. A similar approach was taken by Yeh et al. [63] to expand a precipitation network in Taiwan. Hourly rainfall data were normalized with a Box-Cox transform and kriging used to interpolate rainfall to candidate grid cells. The joint entropy of the network was calculated using an analytic equation for joint entropy valid for normal data [65], and stations were added sequentially that had the lowest conditional entropy with the rest of the network. The final number of stations needed by the network was accepted when 95% of the network information was captured [63]. Awadallah [64] applied multiple entropy measures sequentially to add stations to a precipitation network. The first new stations were selected as those with the highest entropy. The second station was chosen to minimize the mutual information and the third as the station that maximized conditional entropy.

The aforementioned approaches all sequentially add single stations to a monitoring network based on a single criterion. Mahmoudi-Meimand et al. [43] presented a methodology to add stations to a network based on a multi-variate cost function. Precipitation data were spatially interpolated from existing stations using the kriging approach where the kriging error associated with the rainfall estimation is calculated as the kriging error variance. Their method selected the station that maximized transinformation entropy and minimized error variance using a weighted average of both measures as an objective [43]. This approach balanced the information content in the network with the errors in the interpolation method. Xu et al. [57] used a multi-objective approach to simultaneously select a subset of stations that minimized the sum of pairwise mutual information, minimized bias and maximized Nash–Sutcliffe efficiency. Solutions were generated via Monte Carlo sampling, and network solutions falling along the Pareto front were found as compromise solutions. Coulibaly and Keum [66] and Samuel and Coulibaly [67] also used a multi-objective approach to add stations to snow monitoring networks in Canada. Their approach used a genetic algorithm to find networks that maximized the joint entropy and minimized the total correlation of the network to form a Pareto front of optimal network designs, some of which also included network cost in the optimization [35,67].

A challenge to an entropy-based approach to adding stations to a precipitation network is the requirement to have data available for candidate points. For precipitation, this can be challenging because data at shorter time scales in particular are well known to be non-normal. Most studies use the kriging approach for interpolation [43,63,64] and address the need for normally distributed data using a Box-Cox transform. Samuel and Coulibaly [67] addressed the interpolation problem by using the external data from the Snow Data Assimilation System (SNODAS) for candidate stations. Su and You [52] presented a unique approach to adding stations that maximized the information content of the network. In most literature cases, entropic measures at ungauged sites are determined by interpolating observations of precipitation across a watershed. Su and You [52] calculated the transinformation between neighbouring stations to develop a 2D transinformation-distance relationship. In contrast to transferring data to ungauged stations, this approach transferred transinformation to ungauged stations and selected a site with the maximum transinformation. This approach should be further tested and contrasted with the data transfer approach.

As previously stated, precipitation data are of critical importance for a variety of applications. Despite this, few studies have explored the impact of precipitation networks designed with an entropy approach for actual water resource applications. Applications found in the literature have taken the reasonable approach of using entropy to reduce network density for comparison to a network that included all stations. In Portugal, Santos et al. [50] compared artificial neural networks, K-means clustering and mutual information (MI) criteria for reducing the density of a precipitation network for drought monitoring at different time scales. They found the best performing reduction method was case dependent depending on the region and time scale applied, but noted that all methods performed well. They also found that all subset networks could reliably reproduce the spatial precipitation pattern. Xu et al. [57] used the multi-objective approach previously described to select a subset of precipitation stations from a dense network in the Xiangjiang River Basin in China. Rainfall from the subset networks was used to force the lumped Xinanjiang hydrological model [68] and the distributed SWAT hydrological model [69]. The author's found that lumped model performance became stable with a subset of 20 to 25 stations, whereas the distributed model's performance continued to increase as more stations were added to the network [57]. These analyses are important to demonstrate the utility of precipitation networks and the advantages of entropy-based approaches in designing precipitation networks.

3.2. Streamflow and Water Level Networks

Water quantity monitoring, such as streamflow rates and water level, is one of the essential tasks for water management to prevent damage to nature and human beings from flooding. A successful floodplain management or flood forecasting and warning system can be feasible through expert forecasters who implement well-calibrated models and reliable tools using quality data [70]. The design of water quantity monitoring network has been well implemented because of not only the good performance of entropy-based methods, but also the unaffectedness by the zero effect, which is caused by discontinuity of probability density function due to zero values in data, except for the ephemeral or intermittent streams. To deal with the zero effect in entropy calculations, Chapman [71] and Gong et al. [60] separated the marginal entropy Equation (3) to nonzero terms and zero values, which are certain. While Gong et al. [60] summarized the possible issues in entropy calculations from hydrologic data as effects due to zero values, histogram binning including skewness consideration and measurement errors, some studies noticed that the length and the location of time window also affect entropy calculations and the corresponding network design. Fahle et al. [31] observed the temporal variability of station rankings by shifting the time window for the design of water level network of a ditch system in Germany. Mishra and Coulibaly [47] also found the dependency of the seasonality on the efficiency of hydrometric networks. Stosic et al. [51] found an inverse relationship between the network density and sampling time interval as the larger number of monitoring stations is required if the time interval is shorter and vice versa. Keum and Coulibaly [34] analyzed the temporal changes of entropy measures and optimal networks by applying daily time series for streamflow network design. They found that the information gain of a monitoring network is not significant when the length of time series is longer than 10 years, and the total correlation tends to stabilize within five years of data. The optimal networks using the data lengths of 5, 10, 15 and 20 years also show that there are no significant differences in the results from 10 years or longer while the optimal network using five years of data was evidently different from others. Werstuck and Coulibaly [56] analyzed scaling effects by considering two study areas. Specifically, one study area is a small watershed, which is a part of another study area. After applying the transinformation analysis and the multi-objective optimization, they concluded that the optimal networks tend to be affected by scaling while transinformation index does not.

Mishra and Coulibaly [46] evaluated the effects of the class intervals and the infilling missing data by applying the linear regression method to daily time series and concluded that the station rankings based on the transinformation values were not significantly changed. Li et al. [41] also investigated the changes of station rankings based on the maximum information minimum redundancy (MIMR) approach and obtained the similar conclusion. However, Fahle et al. [31] and Keum and Coulibaly [35] drew the opposite opinion that station rankings can be affected by the binning method that defines the class intervals. The conflict comes from the selection of the binning methods compared. The former group applied different parameters to a single binning method, the mathematical floor function. However, the latter group compared other binning methods with the floor function. Considering that Alfonso et al. [25] found that the design solutions were not common in some cases from the sensitivity analysis of the parameter of the mathematical floor function, it is not recommended to use a specific binning method without any consideration.

As discussed in the review of precipitation networks in Section 3.1, network redesign and network expansion require data at candidate locations, which are ungauged. Alfonso et al. [27] applied a one-dimensional hydrodynamic model to generate the discharge time series. The model estimated discharge at each segment, which divides rivers with approximately 200 m increments longitudinally. The use of hydrodynamic model enabled to determine the critical monitoring locations in the main stream and its tributaries. On the other hand, Samuel et al. [49] combined regionalization techniques with entropy calculation in order to estimate the discharge at candidate locations. They compared the performance of various regionalization methods including not only a conceptual hydrologic model, but also spatial proximity, physical similarity and their combinations with drainage area ratio. Based on the performance statistics by applying multiple basins, inverse distance weighting coupled with drainage area ratio performed the best, and this conclusion has been adopted in several studies [34,35,37,55].

Some studies have extended the entropy applications for the streamflow monitoring network design. Stosic et al. [51] proposed the concept of permutation entropy, which is able to differentiate based on the order of sequential observations, as well as the histogram frequency in basic Shannon entropy measures. Even though histograms from two different observations are the same, the permutation entropy value tends to be higher if there are more variations between time steps. However, the network design studies using the permutation entropy are still limited. On the other hand, Leach et al. [37] applied additional features to the network design. While the common objectives in water monitoring network design using an optimization technique are to maximize the information and to minimize the redundancy in the network, they additionally considered the physical properties of watersheds, such as the streamflow signatures [72,73] and the indicators of hydrologic alterations (IHAs) [74,75]. After the comparison of the optimal streamflow monitoring networks with and without considerations of the streamflow signatures and IHAs, it was concluded that inclusion of basin physical characteristics yielded a better coverage of the selected locations of the optimal networks.

3.3. Soil Moisture and Groundwater Networks

Soil moisture is a critical water variable as the interface between the atmosphere and subsurface. Unfortunately, the monitoring of soil moisture is very sparse compared to its spatial variability. To design an optimum network for monitoring soil moisture in the Great Lakes Basin, Kornelsen and Coulibaly [36] proposed using data from the Soil Moisture and Ocean Salinity (SMOS) satellite [76] to design a soil moisture monitoring network using the DEMO algorithm of Samuel et al. [49]. Grid cells were selected to add monitoring stations that optimally maximized joint entropy while minimizing total correlation using only the satellite data. The ascending and descending overpasses were found to contain different information, and the spatial distribution of a network designed with both overpasses was found to contain complimentary features from both datasets [36].

Groundwater monitoring allows for a better understanding of the hydrogeology in an area. This is achieved through groundwater quality and quantity monitoring. Groundwater quality monitoring is used to detect contaminant plumes or for long-term monitoring (LTM) of post remediation effects, and groundwater quantity monitoring is used to determine available water for drinking, irrigation and industry. However, monitoring groundwater is inherently difficult due to physical barriers between observers and the water. Through the understanding of subsurface flow physics and with flow and contaminant transport models such as MODFLOW, MODPATH and MT3D [77–79], we can simulate the behaviour of groundwater. Unfortunately, our simulations are not always accurate, and the models require real-world observations to be calibrated and validated. Due to constraints such as accessibility and financial cost, it is not feasible to monitor at every possible location in an area of interest. It is instead ideal for an optimal monitoring network to be designed to allow for the best placement of monitoring stations and to determine the ideal measurement frequencies. The merit of using information theory entropy has been shown in several cases of groundwater network design [31–33,38,44,48,53,58].

Various methods that utilize information theory entropy have been developed for use in designing optimal groundwater monitoring networks. These include the use of entropy measures in both single and multi-objective optimization problems and are used in network reduction [32,33], expansion [38,58] and redesign [44], as well as have been used to highlight vulnerable areas in an area that should be monitored [53]. In identifying vulnerable areas in the Victoria County Groundwater Conservation District (VCGCD) in Texas, USA, Uddameri and Andruss [53] developed a monitoring priority index (MPI) based on a weighted stakeholder preference to highlight the areas of interest. They compared kriging standard deviation and marginal entropy as metrics to characterize groundwater variability and found entropy to be the more conservative metric.

In areas where there is excessive monitoring, Mondal and Singh [48] showed the information transfer index (ITI), the quotient of joint entropy and transinformation, could be used to evaluate the existing monitoring network. Through this evaluation, redundant monitoring stations (wells) could be identified and removed from the groundwater monitoring network. It may also be the case that the existing groundwater monitoring network is not adequate and additional monitoring stations are needed. Yakirevich et al. [58] developed a method that utilizes minimum cross entropy (MCE) to sequentially add monitoring stations to a network. MCE was used as a metric to quantify the difference between two variants of a Hydrus-3D model [80], and the monitoring stations were added to the network where the difference between models was largest. A multi-objective approach for adding monitoring stations to a groundwater monitoring network was applied by Leach et al. [38], which utilized two entropy measures, total correlation and joint entropy and a metric used to quantify the spatial distribution of annual recharge; the results of which were used to develop maps that highlight areas in which additional monitoring stations should be added. The majority of network design experiments look at the entire available time series when calculating entropy measures; however, Fahle et al. [31] showed that using a combination of MIMR and subsets of the data series could be more ideal. The subsets were used to represent the intra-annual variability of groundwater levels. This method identified locations which were consistently important through each subset and found that monitoring stations showed similarities during wet periods and uniqueness during dry periods. Fahle et al. [31] also suggest that a consequence of using subsets of data allows for the design of a network, which can be focused on floods or droughts.

One issue that can arise with entropy-based methods is the need for lengthy data series to produce accurate measures of entropy. Unfortunately, the area of interest for new monitoring stations will not have available data for all possible locations. To work around this limitation transinformation-distance (T-D) curves have been applied in the design of optimal groundwater monitoring networks [44,81]. In these studies, T-D curves were developed for sub areas within the desired study area based in different clustering methods. Additionally, Masoumi and Kerachian [44] showed that this method could be applied temporally as transinformation-time curves which could then be used to optimize the temporal sampling frequency of the stations. It should be noted that both previously mentioned studies were applied in the same study area using slightly different methods for clustering monitoring stations, and both produced different groundwater monitoring networks that could be considered optimal. This highlights an issue with optimal monitoring network design in that it can be subjective and does not have a singular solution. A comparison of Hosseini and Kerachian [32,33] also illustrates this issue, where through the use of different entropy measures, marginal entropy and Bayesian maximum entropy and optimization techniques, one experiment found the optimal monitoring network included 42 monitoring stations while the other only included 33 stations.

3.4. Water Quality Networks

The importance of water quality monitoring networks is their ability to assist in identifying those parameters that exceed water quality standards. Several water quality monitoring strategies, including two methods that utilized entropy measures [42,45], were recently reviewed by Behmel et al. [15]. This review found that identifying a single approach to water quality monitoring network design would be virtually impossible. Despite this, various applications of the transinformation-distance curve methods have shown promise in the optimal redesign and reduction of water quality monitoring networks [29,42,45]. Lee [39] found that by maximizing the multivariate transinformation between chosen and unchosen stations, using the storm water management model to simulate the total suspended solids and a GA for optimization, an optimal water quality network could be designed for a sewer system. Banik et al. [82] compared information theory, detection time and reliability measures for the design of a sewer system monitoring network through both single and multi-objective optimization approaches. It was shown that for a small monitoring network, the methods had similar performances, while the single objective detection time-based method had slightly better performance when the number of monitoring station is larger. Alameddine et al. [24] used exceedance probabilities to determine violation entropy of dissolved oxygen and chlorophyll-a in the Neuse River estuary. Along with violation entropy, the total system entropy was used as a measure to identify areas of importance of monitoring. A multi-objective optimization scheme based on expert assigned weights was used to develop a compromise solution from the three entropy measures. Ultimately, the method allowed for the identification of high uncertainty areas, which would benefit from future water quality monitoring. Data availability is an issue when using entropy methods, particularly when attempting to use them in the design of a monitoring network in an ungauged basin. To address this, Lee et al. [40] developed a method that uses a measure analogous to marginal entropy. This method uses characteristics of the basin such as the length and number of reaches in the river network as part of the cost function, which is then optimized using a combined GA and filtering algorithm. This was shown to be a computationally-efficient method for use in optimal network design of an ungauged river basin.

3.5. Integrated Network Design

To the best of our knowledge, almost all of the previous studies about water monitoring network design have focused on a specific network type (i.e., considering a single hydrologic variable in each study) as reviewed in the previous sections. However, considering that hydrologic processes are interconnected in a water cycle, there are causes and effects between hydrologic variables. For instance, if a noticeable amount of precipitation occurs, streamflow or groundwater level is likely increased; hence, the information content of a variable may affect that of other variables. Keum and Coulibaly [35] developed a multivariate network design method by taking conditional entropy as the measure of information that is independent to a given variable. In their study, the method designed precipitation and streamflow monitoring networks simultaneously. Specifically, the method followed the traditional multi-objective approach that maximizes joint entropy and minimizes total correlation, but added another objective that maximizes conditional entropy of streamflow network given precipitation network to mimic the direction of the water cycle as streamflow may fluctuate due to precipitation. After comparing the integrated design with the single-variable design, their results showed that the effectiveness of network integration mostly came from reducing the number of additional precipitation stations. It was also found that the integrated network design approach allows adding a precipitation station at a location that will benefit the stream gauge network.

4. Conclusions and Recommendations

It is evident that successful water management cannot be achieved without proper water monitoring networks. Although there has been much progress in network design methods and applications, a standardized design methodology has not yet emerged. After the pioneering invention of information theory in the 1940s, entropy concepts have been applied in various applications with recent efforts on network design problems. The unique benefit of this approach is that a water monitoring network can be evaluated or designed based on the information the network monitors, which is in contrast to the set station densities proposed by WMO guidelines; the advantage of the former being that a network could be better tailored to specific applications or optimized to provide the most gain at densities lower than those suggested in WMO guidelines. In addition, when combined with multi-objective optimization techniques, users' specific criteria can be included in the optimal network design process.

This manuscript provides a comprehensive review of the recent research attainments and their applications in entropy-based water monitoring network design. The literature has demonstrated the use of various information theory measures and adaptations thereof for use in network design with an emerging consensus that the goal of these network design methods is to select the stations that provide the most information to the monitoring network while simultaneously being independent of each other. Through rigorous testing, information theory has proven to be a robust tool to use when evaluating and designing an optimal water monitoring network. However, when it comes to evaluating the optimal design, there are still issues that need to be addressed.

The first is that an optimal monitoring network design can be found based on specified design criteria; however, the practical application of the new optimal monitoring network is rarely evaluated in a hydrologic or other model [11,57]. This type of numerical experiment is an important requirement to evaluate the utility of a network rather than just identifying its optimality or information content. Further, it is an important exercise to identify the benefits of entropy-based network designs in order to convince decision makers of the importance of adopting entropy approaches.

Another issue with the optimal network is that it can be subjective, based on choices made in the calculation of entropy and the design method chosen, especially when additional objective functions are considered in the design. This extends to the method selected for finding the optimal monitoring network, whether it is found using an iterative method where one station is added at a time or a collection of stations is added all at once. Research has also shown that data length, catchment scale and ordering can influence the design of an optimal network [31,34,51]. Finally, when using discrete entropy, the binning method has been shown to influence the final network design [35]. The influence of binning on entropy calculation has received greater attention in other geophysical network design applications [83–85], and similar consideration should be given in the field of water resources, particularly owing to the unique and difficult nature of water variables (e.g., streamflow, precipitation) spatial and temporal distribution [30,60]. Thus, explicit consideration is needed when choosing the bins based on the intended application of the monitoring network and further research to provide guidance specific to water monitoring networks. Therefore, despite the possibility of finding an optimal network design in a formal sense, the subjectivity induced by the designer's choices, and the lack of standardized design methods, must be recognized. Future research should focus on comparative studies among multiple entropy design methods, discretization approaches and data characteristics. The current literature provides many novel entropy design approaches and the evolution of concepts, but rigorous comparisons are critical to provide generic guidelines for network design. Despite the potential sources of subjectivity identified, entropy methods remain one of the most objective approaches for network design.

In particular, more work is needed on spatial and temporal scaling of data for entropy calculation to provide robust guidance to decision makers. Many new methods and optimization techniques have been reviewed herein, but few examples were found in the literature that explored the data characteristics used in those techniques. Further research is required to provide guidance on the proper length of data in water monitoring network design [34], the sampling frequency of the data [54] and the spatial scale at which information should be measured for various monitoring network applications.

The aforementioned issues are considered crucial gaps that need to be filled to enable practical recommendations or guidelines for a widespread adoption of entropy approaches for designing optimal water monitoring networks. In addition, the comparative studies of entropy-based methods reviewed herein should be robustly compared to network design methods from other disciplines, such as geostatistics, to identify areas of equivalence and disparity [10]. Considerable advances have occurred over the past decade as reviewed herein, and measures derived from Shannon's base equation [16] have reached a high level of maturity for the task of network design. We challenge the research community to put a similar creativity into the joint consideration of the nexus of data characteristics, network design and applications, all of which are intricately linked.

Acknowledgments: This research was supported by the Natural Science and Engineering Research Council (NSERC) of Canada, Grant NSERC Canadian FloodNet (NETGP-451456).

Author Contributions: The structure of this review paper was built and the contents were written by all authors. Specifically, Jongho Keum contributed Sections 1, 2, 3.2 and 3.5. Kurt C. Kornelsen contributed Sections 3.1 and 3.3. James M. Leach contributed Sections 3.3, 3.4 and 4. Paulin Coulibaly defined the scope of the review and supervised and reviewed the entire manuscript. Although the primary contributing sections are as indicated above, all authors cross-reviewed other co-authors' sections. All authors have read and approved the final manuscript.

References

1. Langbein, W.B. Overview of Conference on Hydrologic Data Networks. *Water Resour. Res.* **1979**, *15*, 1867–1871. [CrossRef]
2. Herschy, R.W. *Hydrometry: Principles and Practice*, 2nd ed.; John Wiley and Sons Ltd.: Chichester, UK, 1999.
3. Boiten, W. *Hydrometry*; A.A. Balkema Publishers: Lisse, The Netherlands, 2003.
4. Nemec, J.; Askew, A.J. Mean and variance in network-design philosophies. In *Integrated Design of Hydrological Networks (Proceedings of the Budapest Symposium)*; Moss, M.E., Ed.; International Association of Hydrological Sciences Publication: Washington, DC, USA, 1986; pp. 123–131.

5. Rodda, J.C.; Langbein, W.B. *Hydrological Network Design—Needs, Problems and Approaches*; World Meteorological Organization: Geneva, Switzerland, 1969.

6. World Meteorological Organization. *Casebook on Hydrological Network Design Practice*; Langbein, W.B., Ed.; World Meteorological Organization: Geneva, Switzerland, 1972.

7. Davis, D.R.; Duckstein, L.; Krzysztofowicz, R. The Worth of Hydrologic Data for Nonoptimal Decision Making. *Water Resour. Res.* **1979**, *15*, 1733–1742. [CrossRef]

8. Pilon, P.J.; Yuzyk, T.R.; Hale, R.A.; Day, T.J. Challenges Facing Surface Water Monitoring in Canada. *Can. Water Resour. J.* **1996**, *21*, 157–164. [CrossRef]

9. U.S. Geological Survey. *Streamflow Information for the Next Century—A Plan for the National Streamflow Information Program of the U.S. Geological Survey*; U.S. Geological Survey: Denver, CO, USA, 1999.

10. Mishra, A.K.; Coulibaly, P. Developments in Hydrometric Network Design: A Review. *Rev. Geophys.* **2009**, *47*. [CrossRef]

11. Chacon-hurtado, J.C.; Alfonso, L.; Solomatine, D.P. Rainfall and streamflow sensor network design: A review of applications, classification, and a proposed framework. *Hydrol. Earth Syst. Sci.* **2017**, *21*, 3071–3091. [CrossRef]

12. Moss, M.E. *Concepts and Techniques in Hydrological Network Design*; World Meteorological Organization: Geneva, Switzerland, 1982.

13. Van der Made, J.W.; Schilperoort, T.; van der Schaaf, S.; Buishand, T.A.; Brouwer, G.K.; van Duyvenbooden, W.; Becinsky, P. *Design Aspects of Hydrological Networks*; Commissie voor Hydrologisch Onderzoek TNO: The Hague, The Netherlands, 1986.

14. Pyrce, R.S. *Review and Analysis of Stream Gauge Networks for the Ontario Stream Gauge Rehabilitation Project*, 2nd ed.; Watershed Science Centre: Peterborough, ON, Canada, 2004.

15. Behmel, S.; Damour, M.; Ludwig, R.; Rodriguez, M.J. Water quality monitoring strategies—A review and future perspectives. *Sci. Total Environ.* **2016**, *571*, 1312–1329. [CrossRef] [PubMed]

16. Shannon, C.E. A Mathematical Theory of Communication. *Bell Syst. Tech. J.* **1948**, *27*, 379–423. [CrossRef]

17. Batty, M. Space, scale, and scaling in entropy maximizing. *Geogr. Anal.* **2010**, *42*, 395–421. [CrossRef]

18. Singh, V.P. *Entropy Theory in Hydrologic Science and Engineering*; McGraw-Hill Education: New York, NY, USA, 2015.

19. Lathi, B.P. *An Introduction to Random Signals and Communication Theory*; International Textbook Company: Scranton, PA, USA, 1968.

20. Tribus, M. *Rational Descriptions, Decisions and Designs*; Irvine, T.F., Hartnett, J.P., Eds.; Pergamon Press: Oxford, UK, 1969.

21. Krstanovic, P.F.; Singh, V.P. Evaluation of rainfall networks using entropy: I. Theoretical development. *Water Resour. Manag.* **1992**, *6*, 279–293. [CrossRef]

22. McGill, W.J. Multivariate information transmission. *Psychometrika* **1954**, *19*, 97–116. [CrossRef]

23. Watanabe, S. Information Theoretical Analysis of Multivariate Correlation. *IBM J. Res. Dev.* **1960**, *4*, 66–82. [CrossRef]

24. Alameddine, I.; Karmakar, S.; Qian, S.S.; Paerl, H.W.; Reckhow, K.H. Optimizing an estuarine water quality monitoring program through an entropy-based hierarchical spatiotemporal Bayesian framework. *Water Resour. Res.* **2013**, *49*, 6933–6945. [CrossRef]

25. Alfonso, L.; Lobbrecht, A.; Price, R. Information theory-based approach for location of monitoring water level gauges in polders. *Water Resour. Res.* **2010**, *46*. [CrossRef]

26. Alfonso, L.; Lobbrecht, A.; Price, R. Optimization of water level monitoring network in polder systems using information theory. *Water Resour. Res.* **2010**, *46*. [CrossRef]

27. Alfonso, L.; He, L.; Lobbrecht, A.; Price, R. Information theory applied to evaluate the discharge monitoring network of the Magdalena River. *J. Hydroinform.* **2013**, *15*, 211–228. [CrossRef]

28. Alfonso, L.; Ridolfi, E.; Gaytan-Aguilar, S.; Napolitano, F.; Russo, F. Ensemble Entropy for Monitoring Network Design. *Entropy* **2014**, *16*, 1365–1375. [CrossRef]

29. Boroumand, A.; Rajaee, T. Discrete entropy theory for optimal redesigning of salinity monitoring network in San Francisco bay. *Water Sci. Technol. Water Supply* **2017**, *17*, 606–612. [CrossRef]

30. Brunsell, N.A. A multiscale information theory approach to assess spatial-temporal variability of daily precipitation. *J. Hydrol.* **2010**, *385*, 165–172. [CrossRef]

31. Fahle, M.; Hohenbrink, T.L.; Dietrich, O.; Lischeid, G. Temporal variability of the optimal monitoring setup assessed using information theory. *Water Resour. Res.* **2015**, *51*, 7723–7743. [CrossRef]

32. Hosseini, M.; Kerachian, R. A data fusion-based methodology for optimal redesign of groundwater monitoring networks. *J. Hydrol.* **2017**, *552*, 267–282. [CrossRef]

33. Hosseini, M.; Kerachian, R. A Bayesian maximum entropy-based methodology for optimal spatiotemporal design of groundwater monitoring networks. *Environ. Monit. Assess.* **2017**, *189*, 433. [CrossRef] [PubMed]

34. Keum, J.; Coulibaly, P. Sensitivity of Entropy Method to Time Series Length in Hydrometric Network Design. *J. Hydrol. Eng.* **2017**, *22*. [CrossRef]

35. Keum, J.; Coulibaly, P. Information theory-based decision support system for integrated design of multi-variable hydrometric networks. *Water Resour. Res.* **2017**, *53*, 6239–6259. [CrossRef]

36. Kornelsen, K.C.; Coulibaly, P. Design of an Optimal Soil Moisture Monitoring Network Using SMOS Retrieved Soil Moisture. *IEEE Trans. Geosci. Remote Sens.* **2015**, *53*, 3950–3959. [CrossRef]

37. Leach, J.M.; Kornelsen, K.C.; Samuel, J.; Coulibaly, P. Hydrometric network design using streamflow signatures and indicators of hydrologic alteration. *J. Hydrol.* **2015**, *529*, 1350–1359. [CrossRef]

38. Leach, J.M.; Coulibaly, P.; Guo, Y. Entropy based groundwater monitoring network design considering spatial distribution of annual recharge. *Adv. Water Resour.* **2016**, *96*, 108–119. [CrossRef]

39. Lee, J.H. Determination of optimal water quality monitoring points in sewer systems using entropy theory. *Entropy* **2013**, *15*, 3419–3434. [CrossRef]

40. Lee, C.; Paik, K.; Yoo, D.G.; Kim, J.H. Efficient method for optimal placing of water quality monitoring stations for an ungauged basin. *J. Environ. Manag.* **2014**, *132*, 24–31. [CrossRef] [PubMed]

41. Li, C.; Singh, V.P.; Mishra, A.K. Entropy theory-based criterion for hydrometric network evaluation and design: Maximum information minimum redundancy. *Water Resour. Res.* **2012**, *48*. [CrossRef]

42. Mahjouri, N.; Kerachian, R. Revising river water quality monitoring networks using discrete entropy theory: The Jajrood River experience. *Environ. Monit. Assess.* **2011**, *175*, 291–302. [CrossRef] [PubMed]

43. Mahmoudi-Meimand, H.; Nazif, S.; Abbaspour, R.A.; Sabokbar, H.F. An algorithm for optimisation of a rain gauge network based on geostatistics and entropy concepts using GIS. *J. Spat. Sci.* **2016**, *61*, 233–252. [CrossRef]

44. Masoumi, F.; Kerachian, R. Optimal redesign of groundwater quality monitoring networks: A case study. *Environ. Monit. Assess.* **2010**, *161*, 247–257. [CrossRef] [PubMed]

45. Memarzadeh, M.; Mahjouri, N.; Kerachian, R. Evaluating sampling locations in river water quality monitoring networks: Application of dynamic factor analysis and discrete entropy theory. *Environ. Earth Sci.* **2013**, *70*, 2577–2585. [CrossRef]

46. Mishra, A.K.; Coulibaly, P. Hydrometric Network Evaluation for Canadian Watersheds. *J. Hydrol.* **2010**, *380*, 420–437. [CrossRef]

47. Mishra, A.K.; Coulibaly, P. Variability in Canadian Seasonal Streamflow Information and Its Implication for Hydrometric Network Design. *J. Hydrol. Eng.* **2014**, *19*. [CrossRef]

48. Mondal, N.C.; Singh, V.P. Evaluation of groundwater monitoring network of Kodaganar River basin from Southern India using entropy. *Environ. Earth Sci.* **2011**, *66*, 1183–1193. [CrossRef]

49. Samuel, J.; Coulibaly, P.; Kollat, J.B. CRDEMO: Combined Regionalization and Dual Entropy-Multiobjective Optimization for Hydrometric Network Design. *Water Resour. Res.* **2013**, *49*, 8070–8089. [CrossRef]

50. Santos, J.F.; Portela, M.M.; Pulido-Calvo, I. Dimensionality reduction in drought modelling. *Hydrol. Process.* **2013**, *27*, 1399–1410. [CrossRef]

51. Stosic, T.; Stosic, B.; Singh, V.P. Optimizing streamflow monitoring networks using joint permutation entropy. *J. Hydrol.* **2017**, *552*, 306–312. [CrossRef]

52. Su, H.T.; You, G.J.Y. Developing an entropy-based model of spatial information estimation and its application in the design of precipitation gauge networks. *J. Hydrol.* **2014**, *519*, 3316–3327. [CrossRef]

53. Uddameri, V.; Andruss, T. A GIS-based multi-criteria decision-making approach for establishing a regional-scale groundwater monitoring. *Environ. Earth Sci.* **2014**, *71*, 2617–2628. [CrossRef]

54. Wei, C.; Yeh, H.C.; Chen, Y.C. Spatiotemporal scaling effect on rainfall network design using entropy. *Entropy* **2014**, *16*, 4626–4647. [CrossRef]

55. Werstuck, C.; Coulibaly, P. Hydrometric network design using dual entropy multi-objective optimization in the Ottawa River Basin. *Hydrol. Res.* **2016**, *48*, 1–13. [CrossRef]

56. Werstuck, C.; Coulibaly, P. Assessing Spatial Scale Effects on Hydrometric Network Design Using Entropy and Multi-Objective Methods. *JAWRA J. Am. Water Resour. Assoc.* **2017**, in press.

57. Xu, H.; Xu, C.-Y.; Sælthun, N.R.; Xu, Y.; Zhou, B.; Chen, H. Entropy theory based multi-criteria resampling of rain gauge networks for hydrological modelling—A case study of humid area in southern China. *J. Hydrol.* **2015**, *525*, 138–151. [CrossRef]

58. Yakirevich, A.; Pachepsky, Y.A.; Gish, T.J.; Guber, A.K.; Kuznetsov, M.Y.; Cady, R.E.; Nicholson, T.J. Augmentation of groundwater monitoring networks using information theory and ensemble modeling with pedotransfer functions. *J. Hydrol.* **2013**, *501*, 13–24. [CrossRef]

59. Mishra, A.K.; Özger, M.; Singh, V.P. An entropy-based investigation into the variability of precipitation. *J. Hydrol.* **2009**, *370*, 139–154. [CrossRef]

60. Gong, W.; Yang, D.; Gupta, H.V.; Nearing, G. Estimating information entropy for hydrological data: One dimensional case. *Water Resour. Res.* **2014**, *50*, 5003–5018. [CrossRef]

61. Silva, V.; da Silva, V.d.P.R.; Belo Filho, A.F.; Singh, V.P.; Almeida, R.S.R.; da Silva, B.B.; de Sousa, I.F.; de Holanda, R.M. Entropy theory for analysing water resources in northeastern region of Brazil. *Hydrol. Sci. J.* **2017**, *62*, 1029–1038.

62. Ridolfi, E.; Montesarchio, V.; Russo, F.; Napolitano, F. An entropy approach for evaluating the maximum information content achievable by an urban rainfall network. *Nat. Hazards Earth Syst. Sci.* **2011**, *11*, 2075–2083. [CrossRef]

63. Yeh, H.C.; Chen, Y.C.; Wei, C.; Chen, R.H. Entropy and kriging approach to rainfall network design. *Paddy Water Environ.* **2011**, *9*, 343–355. [CrossRef]

64. Awadallah, A.G. Selecting optimum locations of rainfall stations using kriging and entropy. *Int. J. Civ. Environ. Eng* **2012**, *12*, 36–41.

65. Ahmed, N.A.; Gokhale, D.V. Entropy expressions and their estimators for multivariate distributions. *IEEE Trans. Inf. Theory* **1989**, *35*, 688–692. [CrossRef]

66. Coulibaly, P.; Keum, J. *Snow Network Design and Evaluation for La Grande River Basin*; Hydro-Quebec: Hamilton, ON, Canada, 2016.

67. Samuel, J.; Coulibaly, P. *Design of Optimum Snow Monitoring Networks Using Dual Entropy Multiobjective Optimization (DEMO) with Remote Sensing Data: Case Study for Columbia River Basin*; BC-Hydro: Hamilton, ON, Canada, 2013.

68. Zhao, R.J. The Xinanjiang model applied in China. *J. Hydrol.* **1992**, *135*, 371–381.

69. Arnold, J.G.; Srinivasan, R.; Muttiah, R.S.; Williams, J.R. Large Area Hydrologic Modeling and Assessment Part I: Model Development. *J. Am. Water Resour. Assoc.* **1998**, *34*, 73–89. [CrossRef]

70. United Nations. *Guidelines for Reducing Flood Losses*; Pilon, P.J., Ed.; United Nations: Geneva, Switzerland, 2004.

71. Chapman, T.G. Entropy as a measure of hydrologic data uncertainty and model performance. *J. Hydrol.* **1986**, *85*, 111–126. [CrossRef]

72. Yadav, M.; Wagener, T.; Gupta, H. Regionalization of constraints on expected watershed response behavior for improved predictions in ungauged basins. *Adv. Water Resour.* **2007**, *30*, 1756–1774. [CrossRef]

73. Sawicz, K.; Wagener, T.; Sivapalan, M.; Troch, P.A.; Carrillo, G. Catchment classification: Empirical analysis of hydrologic similarity based on catchment function in the eastern USA. *Hydrol. Earth Syst. Sci.* **2011**, *15*, 2895–2911. [CrossRef]

74. Richter, B.D.; Baumgartner, J.V.; Powell, J.; Braun, D.P. A Method for Assessing Hydrologic Alteration within Ecosystems. *Conserv. Biol.* **1996**, *10*, 1163–1174. [CrossRef]

75. Monk, W.A.; Peters, D.L.; Allen Curry, R.; Baird, D.J. Quantifying trends in indicator hydroecological variables for regime-based groups of Canadian rivers. *Hydrol. Process.* **2011**, *25*, 3086–3100. [CrossRef]

76. Kerr, Y.H.; Waldteufel, P.; Wigneron, J.P.; Delwart, S.; Cabot, F.; Boutin, J.; Escorihuela, M.J.; Font, J.; Reul, N.; Gruhier, C.; et al. The SMOS Mission: New Tool for Monitoring Key Elements of the Global Water Cycle. *Proc. IEEE* **2010**, *98*, 666–687. [CrossRef]

77. Harbaugh, A.W. *MODFLOW-2005, The U.S. Geological Survey Modular Ground-Water Model—The Ground-Water Flow Process*; U.S. Geological Survey: Reston, VA, USA, 2005.

78. Pollock, D.W. *User Guide for MODPATH Version 7—A Particle-Tracking Model for MODFLOW*; U.S. Geological Survey: Reston, VA, USA, 2016.

79. Bedekar, V.; Morway, E.D.; Langevin, C.D.; Tonkin, M.J. *MT3D-USGS Version 1: A U.S. Geological Survey Release of MT3DMS Updated with New and Expanded Transport Capabilities for Use with MODFLOW*; U.S. Geological Survey: Reston, VA, USA, 2016.

80. Šimůnek, J.; van Genuchten, M.T.; Šejna, M. *The HYDRUS Software Package for Simulating Two- and Three-Dimensional Movement of Water, Heat, and Multiple Solutes in Variably-Saturated Porus Media*; PC Progress: Prague, Czech Republic, 2012.

81. Owlia, R.R.; Abrishamchi, A.; Tajrishy, M. Spatial-temporal assessment and redesign of groundwater quality monitoring network: A case study. *Environ. Monit. Assess.* **2011**, *172*, 263–273. [CrossRef] [PubMed]

82. Banik, B.K.; Alfonso, L.; di Cristo, C.; Leopardi, A.; Mynett, A. Evaluation of Different Formulations to Optimally Locate Sensors in Sewer Systems. *J. Water Resour. Plan. Manag.* **2017**, *143*. [CrossRef]

83. Ruddell, B.L.; Kumar, P. Ecohydrologic process networks: 1. Identification. *Water Resour. Res.* **2009**, *45*. [CrossRef]

84. Ruddell, B.L.; Kumar, P. Ecohydrologic process networks: 2. Analysis and characterization. *Water Resour. Res.* **2009**, *45*, 1–14. [CrossRef]

85. Kang, M.; Ruddell, B.L.; Cho, C.; Chun, J.; Kim, J. Agricultural and Forest Meteorology Identifying CO_2 advection on a hill slope using information flow. *Agric. For. Meteorol.* **2017**, *232*, 265–278. [CrossRef]

PERMISSIONS

LIST OF CONTRIBUTORS

Xun Liu, Lingna Lin and Lianbo Zhu
School of Civil Engineering, Suzhou University of Science and Technology, Suzhou 215000, China

Fei Qian and Kun Zhang
Institute of Engineering Management, Hohai University, Nanjing 211100, China

Kuo-Wei Liao, Jia-Jun Guo, Jen-Chen Fan and Shao-Hua Chang
Department of Bioenvironmental Systems Engineering, National Taiwan University, No. 1, Section 4, Roosevelt Rd., Taipei 10617, Taiwan

Chien Lin Huang
Hetengtech Company Limited, New Taipei City 24250, Taiwan

Lu Chen
College of Hydropower & Information Engineering, Huazhong University of Science & Technology, Wuhan 430074, China

Vijay P. Singh
Department of Biological & Agricultural Engineering and Zachry Department of Civil Engineering, Texas A&M University, 2117 TAMU, College Station, TX 77843, USA

Feng Huang, Xunzhou Chunyu, Dayong Zhao and Ziqiang Xia
State Key Laboratory of Hydrology-Water Resources and Hydraulic Engineering, Hohai University, Nanjing 210098, China
College of Hydrology and Water Resources, Hohai University, Nanjing 210098, China

Yuankun Wang
School of Earth Sciences and Engineering, Nanjing University, Nanjing 210023, China

Yao Wu
College of Hydrology and Water Resources, Hohai University, Nanjing 210098, China
Poyang Lake Hydro Project Construction Office of Jiangxi Province, Nanchang 330046, China

Bao Qian
Bureau of Hydrology, Changjiang River Water Resources Commission, Wuhan 430012, China

Lidan Guo
International River Research Centre, Hohai University, Nanjing 210098, China

Hui-Chung Yeh
Department of Natural Resources, Chinese Culture University, Taipei 11114, Taiwan

Yen-Chang Chen, Che-Hao Chang and Cheng-Hsuan Ho
Department of Civil Engineering, National Taipei University of Technology, Taipei 10608, Taiwan

Chiang Wei
Experimental Forest, National Taiwan University, NanTou 55750, Taiwan

Zhenghong Zhou, Juanli Ju, Xiaoling Su and Gengxi Zhang
College of Water Resources and Architectural Engineering, Northwest A&F University, Yangling 712100, China

Hossein Foroozand and Steven V. Weijs
Department of Civil Engineering, University of British Columbia, Vancouver, BC V6T 1Z4, Canada

Valentina Radić
Department of Earth, Ocean and Atmospheric Sciences, University of British Columbia, Vancouver, BC V6T 1Z4, Canada

Turkay Baran, Nilgun B. Harmancioglu, Cem Polat Cetinkaya and Filiz Barbaros
Faculty of Engineering, Civil Engineering Department, Dokuz Eylul University, Tinaztepe Campus, Buca, 35160 Izmir, Turkey

Carlo Giudicianni
Dipartimento di Ingegneria Civile, Design, Edilizia e Ambiente, Università degli Studi della Campania "Luigi Vanvitelli", via Roma 29, 81031 Aversa, Italy

Giovanni Francesco Santonastaso, Armando Di Nardo, Michele Di Natale and Roberto Greco
Dipartimento di Ingegneria Civile, Design, Edilizia e Ambiente, Università degli Studi della Campania "Luigi Vanvitelli", via Roma 29, 81031 Aversa, Italy
Action Group CTRL+SWAN of the European Innovation Partnership on Water, EU, B-1049 Brussels, Belgium

Hernán D. Salas, Germán Poveda and Oscar J. Mesa
Facultad de Minas, Departamento de Geociencias y Medio Ambiente, Universidad Nacional de Colombia, Sede Medellín, Carrera 80 # 65-223, Medellín 050041, Colombia

Songbai Song, Xiaoyan Song and Yan Kang
College of Water Resources and Architectural Engineering, Northwest A & F University, Yangling 712100, China

Tue M. Vu, Ashok K. Mishra and Goutam Konapala
Glenn Department of Civil Engineering, Clemson University, Clemson, SC 29634, USA

Lei Chen, Cheng Sun, Guobo Wang, Hui Xie and Zhenyao Shen
State Key Laboratory of Water Environment Simulation, School of Environment, Beijing Normal University, Beijing 100875, China

Jongho Keum and James M. Leach
Department of Civil Engineering, McMaster University, Hamilton, ON L8S4L8, Canada

Kurt C. Kornelsen
School of Geography and Earth Sciences, NSERC Canadian FloodNet, McMaster University, Hamilton, ON L8S4L8, Canada

Paulin Coulibaly
Department of Civil Engineering, McMaster University, Hamilton, ON L8S4L8, Canada
School of Geography and Earth Sciences, NSERC Canadian FloodNet, McMaster University, Hamilton, ON L8S4L8, Canada

Index

Printed in the USA
CPSIA information can be obtained
at www.ICGtesting.com
JSHW052313231023
50683JS00006BA/90

9 781647 404291